RAND McNALLY
Deluxe Road Atlas and Travel Guide

Contents

How to Use this Book   2

Photo Credits

3, Alabama Tourism & Travel; 4, Ray Atkeson; 5, Aston; 6, Opryland/Tennessee Tourism; 7, Travel Montana; 8 & 9, Oklahoma Tourism; 10, Travel Montana; 14, North Carolina Travel & Tourism; 15, Texas Tourism Division; 16, Alberta Tourism Parks & Recreation; 17, Texas Tourism Division; 18, Rand McNally photo; 19, Arkansas Department of Parks & Tourism; 20, Arizona Tourism; 21, Erica Lauf; 22, Killington Resorts; 23, Bill Thomas; and 25 (Seattle), Washington State Tourism Division

How to Use This Book

Today's leisure traveler faces a wealth of choices when planning a trip—not only what the destination is to be, but how best to get there, where to stay, which of countless sightseeing opportunities to engage in. Many travelers also want to plan their trips around recreational interests such as a theme park visit, fishing, or camping.

Rand McNally's *Deluxe Road Atlas & Travel Guide* brings together the details, facts, and travel information necessary for the planning and execution of a successful, enjoyable leisure trip. The Travel and Recreation Guide provides pertinent information in an easy-to-use format, valuable both in the initial trip planning stages and once the destination has been reached. The Major Cities Guide relates important facts and describes the amenities of principal cities in the United States and Canada, plus Mexico City. Together with the accurate and detailed city maps contained in this section, and those of the United States, Canada, and Mexico contained in the Atlas section, these features create an all-in-one guide and atlas that takes the leisure traveler from the planning stage to the destination—and back home again.

Travel and Recreation Guide

Arranged in easy-to-use charts, this guide provides ready access to recreation and travel opportunities and resources around the country. The following topics are covered in this section.

United States Mileage Chart. An aid to estimating driving time, this chart provides the mileage between a host of major United States cities.

Resorts. This listing of selected resorts includes address/phone and details regarding locale and range of facilities.

Theme Parks. Hours of operation and other pertinent information on 31 theme parks around the U.S. and Canada.

Fishing/Hunting. License fee information is provided for all 50 states.

Special Events. A selected list of special events taking place throughout the United States and Canada.

National Parks. Outdoor sports enthusiasts, campers, nature lovers, and history buffs can use this chart to determine activities, facilities, and accommodations available at over 200 National Park areas, monuments, and historic sites.

Tourism Information and Auto/RV Laws. Keep abreast of state regulations pertaining to trailer towing, child restraints, seat belts, and more.

Canada Travel Information. Canadian-U.S. border crossing regulations and information about fishing and hunting in both countries is summarized here.

Road Condition "Hot Lines." Current status for road conditions and construction projects can be obtained by calling these "hot line" numbers.

Intercity Toll Road Information. An aid to trip planning, this chart gives the location, mileage, and cost of using toll roads in the United States and Mexico.

Car Rental Toll-free Numbers. Locate your rental car before starting your trip, using these convenient toll-free numbers.

Hotel/Motel Toll-free Numbers. These toll-free telephone numbers for more than 65 hotel and motel chains can save both time and money.

Clear Channel Radio Stations. Specially organized for quick-reference use while driving; dial one of these stations and enjoy clear reception over an extended range.

Climate. In-depth climate information is provided in this chart for all 50 states plus the District of Columbia.

Mileage and Driving Time Map. Mileages and driving times between major places in the United States, Canada, and Mexico are provided in a handy map format.

Major Cities Guide

Information is provided for more than 85 major cities in the United States, Canada, and Mexico, including:

Selected Hotels, Restaurants and Attractions. Listed for each city, these hotels, restaurants, and attractions have been chosen for their broad appeal to the traveling public.

City Maps. Large-scale maps are included for each city to aid in orienting the first-time visitor, and refreshing the memory of the returning traveler.

Airport Maps. Flyers will appreciate the detailed airport maps provided for 41 of the largest American and Canadian airports.

Information Sources. The city listings also include the address/phone for local convention & visitors bureaus, offices of tourism, and chambers of commerce.

Maps

The atlas section includes a United States map and maps of each of the 50 states, Canada, the Canadian provinces, and Mexico.

Valuable as aids to routing your trip, these maps are also useful for pinpointing sites referred to in the Recreation and Travel section charts, and interpreting that section's mileage and climate information.

Advertising representative:
Fox Associates, Inc.
(800) 345-8670

Comments or suggestions should be directed to:
Rand McNally
Consumer Services
P.O. Box 7600
Chicago, IL 60680

Travel and Recreation Guide

The 14 charts and map included in this guide are designed to assist in both the planning and execution of your trip; information provided covers sightseeing, recreation, places to stay, and on-the-road service information.

Look for the following specifics in the Travel and Recreation charts.

United States Mileage Chart. As an aid to estimating driving time, this chart provides the mileage between a host of major United States cities.

Resorts. Details on selected resorts in the U.S. and Canada: address/phone, waterfront locales, and the range of available facilities from pools to exercise rooms to sports to children's programs/playgrounds.

Theme Parks. This chart covers 31 major theme parks—noting when the parks are open; the types of tickets available; special facilities such as handicap access, picnic; and address/phone numbers.

Fishing/Hunting. Fishing and hunting license fees are detailed in this chart. For further information, the phone numbers of state licensing agencies are noted.

Special Events. A selected number of special events ranging from jazz festivals to a regatta, rodeos to winter carnivals. Included is information as to the specific type of celebration involved, the approximate date(s) held, and addresses/phone numbers to contact for further information.

National Parks. Activities, facilities and accommodations at over 200 National Park areas, monuments, and historic sites. Activities cover various winter and water sports, horseback riding, hiking, guided tours. Facility information includes handicap access, food service, museums, bathhouses. Accommodation information covers campgrounds, motels/lodges, and cabins.

Tourism Information & Auto/RV Laws. This chart provides information on state regulations pertaining to trailer towing, seat belts, child restraints, and more. Also, phone numbers for the individual state tourism departments are listed.

Canada Travel Information. Border crossing information for Canadians and Americans is provided as well as information about hunting and fishing regulations in the U.S. and Canada. Sources to contact for information in each Canadian province are listed.

Road Condition "Hot Lines." Use these "hot line" numbers to obtain up-to-date information on road conditions and construction projects.

Intercity Toll Road Information. United States and Mexico; includes road names, location, mileage, and tolls for both autos alone and those pulling a two-axle trailer.

Hotel/Motel and Car Rental Toll-free Numbers. A list of toll-free reservation numbers for lodging and auto rental firms in the continental U.S. and Canada.

Clear Channel Radio Stations. A listing by state and province and call numbers of AM stations designated to render service over an extended area.

Climate. In-depth climate information for all 50 states plus the District of Columbia, including maximum and minimum normal daily temperatures, total precipitation, and total days with precipitation for all four seasons of the year.

Mileage and Driving Time Map. Mileages and driving times between places and major junctions are provided. Driving times take into account topography, congested urban areas and speed limits imposed.

United States Mileage Chart

95-7501 ©Rand McNally & Co.	Atlanta, Ga.	Boston, Mass.	Cheyenne, Wyo.	Chicago, Ill.	Cincinnati, Ohio	Cleveland, Ohio	Dallas, Texas	Denver, Colo.	Des Moines, Iowa	Detroit, Mich.	Houston, Texas	Indianapolis, Ind.	Kansas City, Mo.	Los Angeles, Calif.	Louisville, Ky.	Memphis, Tenn.	Milwaukee, Wis.	Minneapolis, Minn.	New Orleans, La.	New York, N.Y.	Omaha, Nebr.	Philadelphia, Pa.	Pittsburgh, Pa.	Portland, Oreg.	St. Louis, Mo.	Salt Lake City, Utah	San Francisco, Calif.	Seattle, Wash.	Toledo, Ohio	Tulsa, Okla.	Washington, D.C.	Wichita, Kans.
Albuquerque, N. Mex.	1381	2172	517	1281	1372	1560	638	417	977	1525	834	1266	782	807	1301	1010	1319	1190	1134	1979	858	1899	1619	1371	1038	604	1115	1440	1469	645	1824	593
Amarillo, Texas	1097	1897	511	1043	1096	1285	358	423	742	1269	596	991	547	1091	1019	726	1084	975	850	1704	643	1624	1344	1655	756	888	1399	1724	1210	361	1549	350
Atlanta, Ga.		1037	1442	674	440	672	795	1398	870	732	789	493	798	2182	382	371	761	1068	479	841	986	741	687	2601	541	1878	2496	2618	640	772	608	903
Austin, Texas	919	1911	994	1110	1083	1327	195	906	877	1315	186	1037	682	1374	982	615	1184	1129	517	1715	837	1615	1367	2069	823	1302	1748	2138	1256	450	1482	548
Baltimore, Md.	645	392	1608	668	497	343	1356	1621	981	503	1412	563	1048	2636	598	904	755	1073	1115	196	1113	102	218	2751	798	2044	2796	2681	444	1194	45	1244
Birmingham, Ala.	153	1165	1347	642	465	709	645	1286	787	724	639	475	697	2032	364	249	728	1006	342	969	898	869	741	2505	465	1781	2366	2535	665	647	736	778
Bismarck, N. Dak.	1495	1794	572	831	1118	1166	1141	671	670	1097	1384	1012	777	1617	1123	1228	758	427	1583	1633	581	1569	1283	1265	979	916	1604	1195	1063	958	1502	776
Boise, Idaho	2174	2639	732	1683	1906	2011	1582	811	1359	1942	1778	1800	1382	849	1893	1832	1692	1405	2078	2478	1227	2410	2122	432	1665	340	658	501	1908	1486	2343	1312
Boston, Mass.	1037		1907	994	840	628	1748	1949	1280	799	1804	906	1391	2979	941	1296	1050	1368	1507	222	1412	296	574	3046	1141	2343	3095	2976	739	1537	429	1587
Buffalo, N. Y.	859	446	1466	522	431	191	1346	1508	839	253	1460	481	966	2554	532	899	609	927	1217	372	971	353	224	2605	750	1937	2654	2535	298	1112	356	1198
Charleston, S. Car.	289	929	1722	877	603	730	1072	1678	1150	842	1054	696	1078	2459	591	660	964	1282	720	733	1266	633	666	2881	821	2158	2785	2890	783	1061	500	1192
Cheyenne, Wyo.	1442	1907		954	1190	1279	869	100	627	1211	1107	1068	650	1137	1161	1101	987	788	1361	1746	495	1678	1390	1159	901	436	1188	1228	1176	765	1611	583
Chicago, Ill.	674	994	954		287	335	917	996	327	279	1067	185	499	2054	292	530	90	405	912	802	459	738	452	2083	289	1390	2142	2013	244	683	671	696
Cincinnati, Ohio	440	840	1190	287		244	920	1173	589	269	1029	111	591	2179	103	468	374	692	786	675	693	567	295	2333	340	1610	2362	2300	210	736	481	787
Cleveland, Ohio	672	628	1279	335	244		1159	1321	652	170	1273	294	779	2367	345	712	422	740	1030	473	784	413	129	2418	529	1715	2467	2348	116	925	346	975
Columbus, Ohio	533	735	1235	308	111	144	1028	1229	618	192	1137	178	656	2244	209	576	395	713	894	568	750	462	187	2391	406	1671	2423	2321	144	802	427	852
Dallas, Texas	795	1748	869	917	920	1159		781	684	1143	246	865	489	1387	819	452	991	936	496	1552	644	1452	1204	2009	630	1242	1753	2078	1084	257	1319	365
Denver, Colo.	1398	1949	100	996	1173	1321	781		669	1253	1019	1058	606	1059	1120	1040	1029	920	1273	1771	537	1691	1411	1238	857	504	1235	1307	1218	681	1616	509
Des Moines, Iowa	870	1280	627	327	589	652	684	669		584	905	465	194	1727	566	599	361	252	978	1119	135	1051	763	1786	333	1063	1815	1749	549	443	984	392
Detroit, Mich.	732	799	1211	279	269	170	1143	1253	584		1265	278	743	2311	360	719	353	671	1045	649	716	573	287	2349	513	1647	2399	2279	62	909	506	940
Duluth, Minn.	1139	1428	918	465	752	800	1086	994	402	707	1307	646	597	2016	757	943	392	156	1331	1267	510	1203	917	1705	662	1315	2044	1635	697	845	1136	794
El Paso, Texas	1415	2316	754	1430	1515	1704	620	654	1126	1674	748	1410	931	790	1438	1072	1468	1339	1098	2123	1007	2043	1763	1635	1175	868	1164	1704	1618	780	1939	742
Flagstaff, Ariz.	1704	2495	757	1604	1695	1883	961	657	1300	1848	1157	1589	1105	484	1624	1333	1642	1481	1457	2302	1171	2222	1942	1241	1361	511	792	1347	1792	968	2147	916
Fort Wayne, Ind.	593	825	1093	170	153	197	983	1135	466	160	1105	132	586	2175	216	553	243	561	901	662	598	585	362	2239	353	1529	2281	2169	134	749	518	783
Fort Worth, Texas	826	1779	845	941	951	1183	28	757	708	1167	262	889	513	1356	850	483	1015	960	527	1583	655	1483	1235	1978	654	1211	1722	2047	1108	279	1350	359
Harrisburg, Pa.	700	373	1579	639	468	314	1383	1592	952	474	1439	534	1019	2607	569	931	726	1044	1142	198	1084	114	205	2722	769	2015	2767	2652	415	1165	133	1215
Helena, Mont.	2030	2388	685	1425	1712	1760	1554	781	1161	1691	1792	1606	1251	1190	1717	1702	1352	1020	2033	2227	1050	2163	1877	658	1493	477	1098	588	1657	1416	2096	1234
Houston, Texas	789	1804	1107	1067	1029	1273	246	1019	905	1265		987	710	1538	928	561	1142	1157	356	1608	865	1508	1313	2205	839	1438	1912	2274	1206	478	1375	608
Indianapolis, Ind.	493	906	1068	185	110	294	865	1058	465	278	987		485	2073	114	435	268	586	796	713	587	633	353	2227	246	1504	2256	2194	219	631	558	681
Jackson, Miss.	391	1406	1257	742	655	899	404	1169	809	914	406	646	644	1791	554	213	824	1036	180	1210	845	1110	939	2401	495	1646	2157	2470	855	527	977	708
Jacksonville, Fla.	306	1155	1748	980	746	915	990	1704	1176	1003	889	799	1104	2377	688	674	1067	1374	555	959	1292	859	851	2907	847	2184	2743	2924	944	1075	726	1206
Kansas City, Mo.	798	1391	650	499	591	779	489	600	195	743	710	485		1589	520	451	537	447	806	1198	201	1118	838	1809	256	1086	1835	1839	687	248	1043	197
Knoxville, Tenn.	219	911	1372	527	253	485	837	1328	800	512	893	346	728	2202	241	387	614	932	596	715	916	615	511	2531	471	1808	2510	2540	453	786	482	871
Las Vegas, Nev.	1964	2725	855	1772	1941	2097	1221	777	1445	2029	1417	1835	1365	282	1884	1593	1805	1607	1717	2548	1313	2468	2188	981	1621	433	564	1152	1994	1228	2393	1176
Lexington, Ky.	362	896	1233	352	82	317	861	1192	636	337	970	171	592	2180	74	409	439	757	727	703	758	623	343	2392	335	1669	2421	2365	278	731	514	782
Little Rock, Ark.	509	1434	1035	640	606	850	314	947	561	838	427	560	389	1687	505	137	715	813	418	1238	590	1138	890	2179	352	1438	1995	2228	779	271	1005	453
Los Angeles, Calif.	2182	2979	1137	2054	2179	2367	1387	1059	1727	2311	1538	2073	1589		2108	1817	2087	1889	1883	2786	1595	2706	2426	959	1845	715	379	1131	2276	1452	2631	1400
Louisville, Ky.	382	941	1161	292	103	345	819	1120	566	360	928	114	520	2108		367	379	697	685	748	687	668	388	2320	258	1597	2349	2305	301	659	582	710
Mackinaw City, Mich.	935	916	1291	387	495	439	1281	1341	673	284	1427	460	864	2392	562	880	368	508	1247	906	805	842	556	2128	651	1691	2443	2058	328	1047	775	1061
Madison, Wis.	812	1103	912	144	427	475	968	954	286	406	1137	321	483	2012	432	622	77	272	1006	942	418	878	592	1950	358	1348	2100	1880	372	727	811	676
Memphis, Tenn.	371	1296	1101	530	468	712	452	1040	599	719	561	435	451	1817	367		612	826	390	1100	652	1000	752	2259	285	1535	2125	2290	654	401	867	532
Miami, Fla.	655	1504	2097	1329	1095	1264	1300	2037	1525	1352	1190	1148	1448	2687	1037	997	1416	1723	856	1308	1641	1208	1200	3256	1196	2532	3053	3273	1293	1398	1075	1529
Milwaukee, Wis.	761	1050	987	90	374	422	991	1029	361	353	1142	268	537	2087	379	612		332	994	889	493	825	539	2010	363	1423	2175	1940	319	757	758	734
Minneapolis, Minn.	1068	1368	788	405	692	740	936	920	252	671	1157	586	447	1889	697	826	332		1214	1207	380	1143	857	1678	552	1312	1940	1608	637	695	1076	644
Mobile, Ala.	335	1372	1443	851	706	950	590	1355	954	965	478	716	812	1977	605	363	933	1173	143	1176	1013	1076	982	2587	632	1832	2343	2651	906	713	943	893
Montreal, Que.	1181	318	1773	828	805	561	1705	1815	1146	562	1827	840	1305	2873	906	1273	915	1163	1591	378	1278	449	583	2900	1101	2225	2961	2685	621	1471	579	1502
Nashville, Tenn.	242	1088	1200	446	269	513	660	1156	628	528	769	279	556	2025	175	210	532	826	517	892	744	792	553	2359	299	1636	2333	2376	469	609	659	743
New Orleans, La.	479	1507	1361	912	700	1030	496	1273	978	1045	356	796	806	1883	685	390	994	1214		1311	1007	1211	1070	2505	673	1738	2249	2574	986	647	1078	816
New York, N. Y.	841	222	1746	802	675	473	1552	1771	1119	649	1608	713	1108	2786	748	1100	889	1207	1311		1251	101	368	2885	948	2182	2934	2815	578	1344	233	1394
Norfolk, Va.	540	558	1764	831	604	508	1329	1758	1141	666	1328	700	1162	2694	642	877	918	1230	1019	362	1273	272	384	2914	905	2200	2952	2844	607	1278	188	1352
Oklahoma City, Okla.	839	1641	697	787	840	1029	209	609	544	1013	449	735	349	1349	763	468	861	796	668	1448	455	1368	1088	1841	500	1100	1657	1910	954	105	1293	157
Omaha, Nebr.	986	1412	495	459	693	784	644	537	135	716	865	587	201	1595	687	652	493	380	1007	1251		1183	895	1654	449	931	1683	1638	681	435	1116	298
Orlando, Fla.	435	1294	1876	1109	875	1054	1078	1815	1305	1134	968	928	1226	2465	817	775	1196	1503	634	1098	1421	998	990	3034	976	2310	2831	3053	1075	1176	865	1307
Philadelphia, Pa.	741	296	1678	738	567	413	1452	1691	1051	573	1508	633	1118	2706	668	1000	825	1143	1211	101	1183		288	2821	868	2114	2866	2751	514	1264	143	1314
Phoenix, Ariz.	1793	2604	892	1713	1804	1992	998	792	1409	1957	1149	1698	1214	389	1733	1442	1751	1616	1494	2411	1290	2331	2051	1266	1470	648	763	1437	1901	1077	2256	1025
Pierre, S. Dak.	1361	1726	434	763	1050	1098	943	518	492	1029	1186	944	592	1524	1055	1043	690	394	1394	1565	391	1501	1215	1353	824	823	1575	1283	995	760	1434	578
Pittsburgh, Pa.	687	574	1390	452	295	129	1204	1411	763	287	1313	353	838	2426	388	752	539	857	1070	368	895	288		2535	588	1826	2578	2465	228	984	259	1034
Portland, Me.	1139	115	1986	1042	942	707	1850	2028	1359	775	1906	1001	1486	3074	1043	1398	1129	1447	1609	308	1491	398	663	3125	1236	2422	3174	3055	818	1632	531	1682
Portland, Ore.	2601	3046	1159	2083	2333	2418	2009	1238	1786	2349	2205	2227	1809	959	2320	2259	2010	1678	2505	2885	1654	2821	2535		2060	767	636	174	2315	1913	2754	1739
Raleigh, N. Car.	372	685	1695	784	534	561	1166	1661	1092	683	1160	631	1087	2545	541	728	871	1189	851	489	1214	389	445	2854	831	2131	2853	2797	624	1129	256	1215
Rapid City, S. Dak.	1487	1859	313	896	1177	1231	1050	394	618	1162	1288	1071	708	1363	1182	1159	840	565	1507	1698	507	1634	1348	1204	950	662	1414	1134	1128	873	1567	691
Reno, Nev.	2374	2866	959	1913	2133	2238	1668	1011	1586	2170	1864	2027	1609	469	2120	2003	1946	1711	2164	2705	1454	2637	2349	538	1860	523	229	754	2135	1640	2570	1471
Richmond, Va.	510	535	1674	748	514	425	1266	1668	1058	583	1298	610	1072	2631	552	814	835	1153	989	339	1190	239	301	2831	815	2110	2862	2761	524	1211	105	1262
St. Louis, Mo.	541	1141	901	289	340	529	630	857	333	513	839	246	256	1845	258	285	363	552	673	948	449	868	588	2060		1337	2089	2081	454	396	793	447
Salt Lake City, Utah	1878	2343	436	1390	1610	1715	1242	504	1063	1647	1438	1504	1086	715	1597	1535	1423	1312	1738	2182	931	2114	1826	767	1337		752	836	1612	1172	2047	1003
San Antonio, Texas	983	1988	1027	1187	1160	1404	270	939	954	1392	197	1114	759	1363	1059	692	1261	1206	550	1792	914	1692	1444	2086	900	1319	1737	2155	1333	527	1559	625
San Diego, Calif.	2126	2955	1186	2064	2155	2343	1331	1108	1760	2308	1482	2049	1565	124	2084	1783	2102	1938	1827	2762	1641	2682	2402	1084	1821	764	504	1256	2252	1428	2607	1376
San Francisco, Calif.	2496	3095	1188	2142	2362	2467	1753	1235	1815	2399	1912	2256	1835	379	2349	2125	2175	1940	2249	2934	1683	2866	2578	636	2089	752		808	2364	1760	2799	1695
Seattle, Wash.	2618	2976	1228	2013	2300	2348	2078	1307	1749	2279	2274	2194	1839	1131	2305	2290	1940	1608	2574	2815	1638	2751	2465	174	2081	836	808		2245	1982	2684	1808
Spokane, Wash.	2340	2698	995	1735	2022	2070	1864	1089	1471	2001	2102	1916	1561	1205	2027	2012	1662	1330	2343	2537	1360	2473	2187	348	1803	712	882	278	1967	1726	2406	1544
Springfield, Ill.	592	1099	888	189	299	473	728	865	291	433	879	212	310	1899	275	370	263	480	758	906	412	826	546	2047	97	1324	2076	2039	377	494	751	507
Springfield, Mo.	652	1353	820	499	552	741	419	759	383	725	629	447	192	1636	475	281	573	594	636	1160	371	1080	800	1978	210	1254	1944	2009	666	183	1005	287
Toledo, Ohio	640	739	1176	244	210	116	1084	1218	549	62	1206	219	687	2276	301	654	319	637	986	578	681	514	228	2315	454	1612	2364	2245		850	447	884
Topeka, Kans.	863	1456	595	562	656	844	480	535	255	806	701	550	65	1531	585	508	598	508	863	1263	165	1183	903	1751	322	1028	1770	1802	750	223	1108	139
Tulsa, Okla.	772	1537	765	683	736	925	257	681	443	909	478	631	248	1452	659	401	757	695	647	1344	435	1264	984	1913	396	1172	1760	1982	850		1189	174
Washington, D. C.	608	429	1611	671	481	346	1319	1616	984	506	1375	558	1043	2631	582	867	758	1076	1078	233	1116	143	259	2754	793	2047	2799	2684	447	1189		1239
Wichita, Kans.	903	1587	583	696	787	975	365	509	392	940	608	681	190	1400	710	532	734	644	816	1394	298	1314	1034	1739	447	1003	1695	1808	884	174	1239	

Resorts

Resort	Address/Telephone	Waterfront	Boating	Fishing	Pool	Golf	Tennis	Horseback Riding	Skiing	Exercise	Children's Programs
EASTERN U.S. & CANADA											
The Balsams Grand Resort Hotel	Dixville Notch, NH 03576; 603/255-3400, 1-800/255-0800 (NH), 1-800/255-0600 (U.S. & Canada)	•	•		•	•	•	•		•	•
Boca Raton Resort & Club	501 E. Camino Real, Boca Raton, FL 33432; 407/395-3000, 1-800/327-0101	•	•		•	•	•			•	•
The Breakers	1 S. County Rd., Palm Beach, FL 33480; 407/655-6611, 1-800/833-3141	•			•	•				•	•
Le Château Montebello	392 Notre Dame, Montebello, Québec J0V1L0; 819/423-6341, 1-800/268-9411 (Can.), 1-800/828-7447 (U.S.)	•	•	•	•	•	•	•	•	•	•
The Cloister	Sea Island, GA 31561; 912/638-3611, 1-800/732-4752	•	•		•	•	•	•		•	•
Doral Resort & Country Club	4400 N.W. 87th Ave., Miami, FL 33178-2192; 305/592-2000, 1-800/22-DORAL				•	•	•			•	•
Grand Traverse Resort	6300 U.S. 31N, Acme, MI 49610-0404; 616/938-2100, 1-800/748-0303 (U.S.), 1-800/678-1308 (Can.)	•	•	•	•	•	•		X	•	•
The Greenbrier	U.S. 60 West, White Sulphur Springs, WV 24986; 304/536-1110, 1-800/624-6070				•	•	•	•	•	•	•
Grove Park Inn Resort	290 Macon Ave., Asheville, NC 28804; 704/252-2711, 1-800/438-5800				•	•	•			•	•
The Homestead	Hot Springs, VA 24445; 703/839-5500, 1-800/542-5734 (VA), 1-800/336-5771 (U.S.)				•	•	•	•	•	•	•
Marriott at Sawgrass	1000 TPC Blvd., Ponte Vedra Beach, FL 32082; 904/285-7777, 1-800/457-4653 (FL), 1-800/228-9290 (U.S.)		•	•	•	•	•			•	•
PGA National Resort & Spa	400 Ave. of the Champions, Palm Beach Gardens, FL 33418; 407/627-2000, 1-800/633-9150	•	•	•	•	•	•			•	
Resort at Longboat Key Club	301 Gulf of Mexico Dr., Longboat Key, FL 34228; 813/383-8821, 1-800/282-0113 (FL), 1-800/237-8821 (U.S.)	•	•	•	•	•	•			•	•
The Ritz-Carlton, Amelia Island	4750 Amelia Island Pkwy., Amelia Island, FL 32034; 904/277-1100, 1-800/241-3333	•	•	•	•	•	•			•	•
The Ritz-Carlton, Naples	280 Vanderbilt Beach Rd., Naples, FL 33963; 813/598-3300, 1-800/241-3333	•	•	•	•	•	•			•	•
The Ritz-Carlton, Palm Beach	100 S. Ocean Blvd., Manalapan, FL 33462; 407/533-6000, 1-800/241-3333	•	○	•	•	•	•			•	•
Sagamore	110 Sagamore Rd., Bolton Landing, NY 12814; 518/644-9400, 1-800/THE-OMNI	•	•		•	•	•	•		•	•
Topnotch Resort & Spa at Stowe	4000 Mountain Rd., P.O. Box 1458, Stowe, VT 05672; 802/253-8585, 1-800/451-8686				•	•	•	•	•	•	•
Turnberry Isle Resort & Club	19999 W. Country Club Dr., Aventura (N. Miami), FL 33180; 305/932-6200, 1-800/327-7028 (U.S.)	•	•	•	•	•	•			•	•
Westin Resort Hilton Head Island	2 Grasslawn Ave., Hilton Head Island, SC 29928; 803/681-4000, 1-800/933-3102	•	•	•	•	•	•			•	•
WESTERN U.S. & MEXICO											
The Arizona Biltmore	24th St. & Missouri, Phoenix, AZ 85016; 602/955-6600, 1-800/950-0086				•	•	•			•	•
The Boulders Resort	34631 N. Tom Darlington Dr., Carefree, AZ 85377; 602/488-9009, 1-800/553-1717				•	•	•	•		•	H
The Broadmoor	P.O. Box 1439 (1 Lake Ave.), Colorado Springs, CO 80906; 719/634-7711, 1-800/634-7711	•			•	•	•	•		•	•
Carmel Valley Ranch Resort	One Old Ranch Rd., Carmel, CA 93923; 408/625-9500, 1-800/4-CARMEL				•	•	•				
Flathead Lake Lodge	Flathead Lake Lodge Rd., Bigfork, MT 59911; 406/837-4391	•	•	•	•	•	•	•			•
Four Seasons Biltmore	1260 Channel Dr., Santa Barbara, CA 93108; 805/969-2261, 1-800/332-3442	•	•	•	•	•	•			•	•
Inn at Spanish Bay	2700 17 Mile Dr., Pebble Beach, CA 93953; 408/647-7500, 1-800/654-9300	•		•	•	•	•			•	•
Las Brisas	Carretera Escénica, Clemente Mejia 5255, 39868 Acapulco, Guerrero, Mexico; 74-841580 (Mex.), 1-800/228-3000 (U.S.)	•		•	•	•	•			•	
The Lodge at Pebble Beach	17 Mile Dr., P.O. Box 1128, Pebble Beach, CA 93953; 408/624-3811, 1-800/654-9300	•	•	•	•	•	•			•	H
Loews Ventana Canyon Resort	7000 N. Resort Dr., Tucson, AZ 85715; 602/299-2020, 1-800/234-5117				•	•	•	•		•	
Marriott's Camelback Inn Resort, Golf Club & Spa	5402 E. Lincoln Dr., Scottsdale, AZ 85253; 602/948-1700, 1-800/24-CAMEL				•	•	•			•	•
The Phoenician	6000 E. Camelback Rd., Scottsdale, AZ 85251; 602/941-8200, 1-800/888-8234				•	•	•			•	•
Quail Lodge Resort & Golf Club	8205 Valley Greens Dr., Carmel, CA 93923; 408/624-1581, 1-800/538-9516				•	•	•	•		•	•
Rancho Bernardo Inn	17550 Bernardo Oaks Dr., San Diego, CA 92128; 619/487-1611, 1-800/542-6096				•	•	•			•	•
The Ritz-Carlton, Laguna Niguel	33533 Ritz-Carlton Dr., Dana Point, CA 92629; 714/240-2000, 1-800/241-3333	•	•	•	•	•	•			•	•
The Ritz-Carlton, Rancho Mirage	68-900 Frank Sinatra Dr., Rancho Mirage, CA 92270; 619/321-8282, 1-800/241-3333				•	•	•	•		•	•
Stein Eriksen Lodge	7700 Stein Way, P.O. Box 3177, Park City, UT 84060; 801/649-3700, 1-800/453-1302 (U.S.)				•	•	•	•	•	•	H
Tall Timber	S.S.R. Box 90, Durango, CO 81301; 303/259-4813			•	•	•	•	•		•	
The Wigwam Resort & Country Club	300 E. Indian School Lane, Litchfield Park, AZ 85340; 602/935-3811, 1-800/327-0396				•	•	•	•		•	•

H = holidays only
X = cross-country

© 1995 Rand McNally & Company

Theme Parks

					Tickets						Facilities				
State/Province	Park Name	Address/Telephone	Season	Hours*	One-day	Multi-day	Season	Adult	Child	Senior	Locker Facilities	Baby Changing Facilities	Strollers	Picnic Facilities	Handicapped Access
California	Disneyland	1313 Harbor Blvd, Anaheim, CA 92803; 714/999-4000	Year-round	S 8AM-1AM W 10AM-varies	•	•		•	•	•	•	•	•	•	•
California	Paramount's Great America	PO Box 1776 (95052), Great America Pkwy, Santa Clara, CA 95054; 408/988-1776	Mar.-Oct.	10AM-varies	•			•	•	•	•	•	•	•	•
California	Knott's Berry Farm	8039 Beach Blvd, Buena Park, CA 90620; 714/220-5200	Year-round	S 9AM-varies W 10AM-varies	•			•	•	•	•	•	•	•	•
California	Marine World Africa USA	Marine World Pkwy, Vallejo, CA 94589 707/643-ORCA, 644-4000	Year-round	9:30AM-varies	•			•	•	•	•	•	•	•	•
California	Sea World of California	1720 S Shores Rd, Sea World Dr & I-5, San Diego, CA 92109; 619/226-3901, 222-6363	Year-round	10AM-varies	•			•	•	•	•	•	•	•	•
California	Six Flags Magic Mountain	Box 5500, Magic Mtn. Pkwy & I-5, Valencia, CA 91385; 805/255-4100, 255-4111	Mar. 28-Sept. 18	10AM-varies	•			•	•	•	•	•	•	•	•
Florida	Busch Gardens	PO Box 9158 (33674), 3605 Bougainvillea, Tampa, FL 33612; 813/987-5082	Year-round	varies	•			•	•	•	•	•	•	•	•
Florida	Sea World of Florida	7007 Sea World Dr, Orlando, FL 32821-8097; 407/351-3600	Year-round	9AM-varies	•			•	•	•	•	•	•	•	•
Florida	Universal Studios Florida	1000 Universal Studios Plaza, Orlando, FL 32819; 407/363-8000	Year-round	9AM-varies	•			•	•	•	•	•	•		•
Florida	Walt Disney World Resort	PO Box 10000, Lake Buena Vista, FL 32830-1000; 407/824-4321	Year-round	varies	•			•	•	•	•	•	•	•	•
Georgia	Six Flags Over Georgia	PO Box 43187, Atlanta, GA 30378; (7561 Six Flags Pkwy, Austell, 30001); 404/948-9290	Mar.-Oct.	10AM-varies	•			•	•	•	•	•	•	•	•
Illinois	Six Flags Great America	PO Box 1776, Grand Ave & I-94, Gurnee, IL 60031; 708/249-1776	Apr.-Oct.	10AM-varies	•			•	•	•	•	•	•	•	•
Minnesota	Valleyfair	One Valleyfair Dr, Shakopee, MN 55379; 612/445-7600	May-Sept.	10AM-varies	•			•	•	•	•	•	•	•	•
Missouri	Silver Dollar City	HCR 1 Box 791, West Hwy 76, Branson, Missouri 65616; 800/952-6626	May-Oct.	9:30AM-7PM	•			•	•	•	•	•	•	•	•
Missouri	Worlds of Fun	4545 Worlds of Fun Ave, Kansas City, MO 64161; 816/454-4545	Apr.-Oct.	10AM-varies	•			•	•	•	•	•	•	•	•
New Jersey	Six Flags Great Adventure	PO Box 120, Rte. 537, Jackson, NJ 08527; 908/928-1821	Mar. 26-Nov. 6	10AM-varies	•			•	•	•	•	•	•	•	•
New York	The Great Escape	PO Box 511, US 9, Lake George, NY 12845; 518/792-3500	Mem Day-Mid-Sept.	9:30AM-6PM	•			•	•		•		•	•	•
North Carolina	Paramount's Carowinds	PO Box 410289, Carowinds Blvd, Charlotte, NC 28241-0289; 800/888-4386, 704/588-2600	Mid Mar.-Mid Oct.	10AM-varies	•			•	•	•	•	•	•	•	•
Ohio	Cedar Point	PO Box 5006, Cedar Point, Sandusky, OH 44871-8006; 419/627-2350	May-Labor Day	9AM-varies	•	•		•	•	•	•	•	•		•
Ohio	Paramount's Kings Island	6300 Kings Island Dr, Kings Island, OH 45034; 513/573-5700	Mid-Apr.-Mid-Oct.	9AM-varies	•			•	•	•	•	•	•		•
Ohio	Sea World of Ohio	1100 Sea World Dr, Aurora (Cleveland), OH 44202-8700; 800/63-SHAMU, 216/562-8101	May-Sept.	10AM-varies	•			•	•	•		•	•	•	•
Ontario	Paramount Canada's Wonderland	PO Box 624, Vaughan, ON L6A 1S6; 416/832-7000	Mid May-Oct.	10AM-varies	•			•	•	•	•	•	•	•	•
Pennsylvania	Hersheypark	100 W Hersheypark Dr, Hershey, PA 17033; 800/HERSHEY	May-Sept.	10:30AM-varies	•			•	•	•	•	•	•	•	•
Pennsylvania	Kennywood	4800 Kennywood Blvd, West Mifflin, PA 15122; 412/461-0500	May-Labor Day	Noon-varies	•				•		•	•	•	•	•
Tennessee	Dollywood	1020 Dollywood Lane, Pigeon Forge, TN 37863-4101; 615/656-9448	May-Dec.	9AM-varies	•				•	•	•	•	•	•	•
Tennessee	Opryland	2802 Opryland Dr, Nashville, TN 37214; 615/889-6611	Mar.-Oct.	10AM-9PM	•			•	•	•	•	•	•	•	•
Texas	Six Flags AstroWorld	9001 Kirby Dr, Houston, TX 77054; 713/799-1234	May 23-Labor Day	10AM-varies	•			•	•	•	•	•	•	•	•
Texas	Sea World of Texas	10500 Sea World Drive, San Antonio, TX 78251; 210/523-3611	Mar.-Oct.	10AM-varies	•			•	•	•	•	•	•	•	•
Texas	Six Flags Over Texas	PO Box 191, I-30 at TX 360, Arlington, TX 76010; 817/640-8900	May-Aug.	10AM-varies	•			•	•	•	•	•	•	•	•
Virginia	Busch Gardens/ Williamsburg	1 Busch Gardens Blvd, Williamsburg, VA 23187-8785; 804/253-3350	Apr.-Oct.	10AM-varies	•			•	•	•	•	•	•	•	•
Virginia	Paramount's Kings Dominion	P.O. Box 2000, I-95 & State Rte. 30, Doswell, VA 23047; 804/876-5000	Mar.-Oct.	10:30AM-varies	•			•	•	•	•	•	•	•	•

*Hours subject to change. Phone before you visit. S Summer W Winter

Annual License Fees

State	Fishing/Hunting Information	Resident Fishing	Hunting	Resident Combination	Nonresident Fishing	Hunting	Nonresident Combination Fishing/Hunting
Alabama	Alabama Division of Game & Fish, 334/242-3465	$ 9.50	$16.00		$31.00	$177.00	
Alaska	Alaska Dept. of Fish & Game, 907/465-2376	15.00	25.00	40.00	50.00	85.00	
Arizona	Arizona Game & Fish Dept., 602/942-3000	12.00	18.00	34.00	38.00	85.50	137.50
Arkansas	Arkansas Game & Fish Commission, 1-800/364-GAME	10.50	25.00	35.50	25.00	185.00	
California	California Fish & Game, 916/227-2244	22.40	24.40		65.90	84.80	
Colorado	Colorado Division of Wildlife, 303/297-1192	20.25	15.25*	30.25	40.25	40.25*	
Connecticut	Connecticut Environmental Protection, 203/566-4409	15.00	10.00	21.00	25.00	42.00	54.00
Delaware	Delaware Division of Fish & Wildlife, 302/739-5841	8.50	12.50		15.00	45.00	
District of Columbia	D.C. Water & Fisheries Dept., 202/404-1151, ext. 3025	5.00	No hunting		7.50	No hunting	
Florida	Florida Game & Freshwater Fish Commission, 904/488-4676	12.00	11.00	22.00	30.00	150.00	
Georgia	Georgia Dept. of Natural Resources, 404/975-4230	9.00	10.00	18.00	24.00	59.00	
Hawaii	Hawaii Div. of Conservation & Resources Enforcement, 808/587-0077	3.75	10.00		7.50	95.00	
Idaho	Idaho Dept. of Fish & Game, 208/334-3700	16.00	7.00	21.00	51.00	101.00*	
Illinois	Illinois Dept. of Conservation, 217/782-2965	13.00	7.50	19.25	24.50	50.75	
Indiana	Indiana Division of Fish & Wildlife, 317/232-4080; ask for Public Affairs Unit	8.75	8.75	13.75	15.75	40.75	
Iowa	Iowa Dept. of Natural Resources, 515/281-5145	8.50	10.50	23.50	22.50	60.50*	
Kansas	Kansas Dept. of Wildlife & Parks, 316/672-5911	13.50	13.50	26.50	30.50	60.50	
Kentucky	Kentucky Dept. of Fish & Wildlife, 502/564-4762	12.50	12.50	20.00	30.00	95.00	
Louisiana	Louisiana Dept. of Wildlife & Fisheries, 504/765-2980	5.50	10.50		31.00	86.00	
Maine	Maine Dept. of Inland Fisheries & Wildlife, 207/287-2571	18.00†	18.00*†	34.00†	48.00†	83.00*†	119.00†
Maryland	Maryland Wildlife Division, 410/974-3195	10.00	24.50		20.00	120.50**	
Massachusetts	Massachusetts Division Fisheries & Wildlife, 508/792-7270	17.50	17.50	24.50	22.50	53.50*	
Michigan	Michigan Dept. of Natural Resources, 517/373-1230	9.85†	12.85†	19.70†	20.85†	100.35†	
Minnesota	Minnesota Dept. of Natural Resources, 612/296-6157	14.00	15.00*	29.00	28.50	61.00*	
Mississippi	Mississippi Dept. of Wildlife, Fisheries & Parks, 601/ 364-2124	8.00		17.00	25.00	225.00**	
Missouri	Missouri Dept. of Conservation, 314/751-4115, ext. 323	8.00	8.00*	14.00	25.00	50.00*	
Montana	Montana Dept. of Fish, Wildlife & Parks, 406/444-2535	17.00	**	64.00	50.00	**	
Nebraska	Nebraska Game & Parks Commission, 402/471-0641	11.50	8.50	19.50	25.00	40.00*	
Nevada	Nevada Dept. of Wildlife, 702/688-1558	15.50	20.50	34.00	45.50	100.50	
New Hampshire	New Hampshire Fish & Game, 603/271-3421	23.25	15.50	30.50	35.50	70.50	
New Jersey	New Jersey Division of Fish, Game & Wildlife, 609/292-2965	15.00	22.00	60.50	25.25	100.00	
New Mexico	New Mexico Dept. of Game & Fish, 505/827-7911	14.00	22.50	26.00	42.50	**	
New York	New York Dept. of Environmental Conservation, 518/457-3521	14.00	11.00*	31.00	35.00	100.00**	225.00
North Carolina	North Carolina Wildlife Resource Commission, 919/662-4370	15.00	15.00	40.00	30.00	80.00	130.00
North Dakota	North Dakota Game & Fish Dept., 701/221-6300	9.00	6.00*		25.00	75.00*	
Ohio	Ohio Division of Wildlife, 614/265-6300	15.00	15.00		24.00	91.00	
Oklahoma	Oklahoma Dept. of Wildlife Conservation, 405/521-3853	10.25	10.25	18.50	74.50	23.50	
Oregon	Oregon Dept. of Fish & Wildlife, 503/229-5410	17.50	15.00	29.50	40.50	53.00	
Pennsylvania	Pennsylvania Game Commission, fishing: 717/657-4518; hunting: 717/787-4250	12.50	12.75		25.50	80.75	
Rhode Island	Rhode Island Division of Fish, Wildlife, & Estuarine Resources, 401/ 789-3094	9.50	9.50*	15.50	20.50	20.50*	
South Carolina	South Carolina Wildlife & Marine Resources, 803/734-3843	10.00	12.00*	17.00	35.00	75.00*	
South Dakota	South Dakota Game, Fish & Parks Dept., 605/773-3485	14.00†	21.00†	30.00	35.00†	65.00†	
Tennessee	Tennessee Wildlife Resources Agency, 615/781-5480			16.50	51.00	156.00*	
Texas	Texas Parks & Wildlife Dept., 512/389-4800	13.00	25.00	25.00	20.00	200.00	
Utah	Utah State Division of Wildlife Resources, 801/596-8660	18.00	5.00**	25.00	40.00	5.00**	
Vermont	Vermont Fish & Wildlife Dept., 802/241-3700	18.00	12.00	26.00	35.00	75.00	95.00
Virginia	Virginia Dept. of Game & Inland Fisheries, 804/367-1000	12.50	12.50*		30.50	60.50*	
Washington	Washington Dept. of Wildlife, 360/753-5719	17.00†	16.00	30.00	48.00†	151.00	
West Virginia	West Virginia Dept. of Natural Resources, 304/558-2771	11.00†	11.00†	17.00†	25.00†	70.00†	
Wisconsin	Wisconsin Dept. of Natural Resources, 608/266-1877	12.00	90.00*		28.00	**	
Wyoming	Wyoming Game & Fish Dept., 307/777-4600	9.00	**		50.00	**	

©1995 Rand McNally & Company   *Annual fee for small game and/or birds.   **License fee depends on type of game.   †Does not include issuance fee.

Special Events

State/Province	Event	Information	Type of Celebration — Date(s)	Agricultural Fair/Livestock Show	Historical	Food	Parade	Arts & Crafts	Entertainment/Music	Sports
Alabama	City Stages	City Stages, P.O. Box 2266, Birmingham, AL 35201; 205/251-1272	Mid-June			•	•		•	
Alaska	Golden Days	Golden Days, Fairbanks C of C, 709 Second Ave., Fairbanks, AK 99701; 907/452-1105	Mid-July	•			•		•	•
Alberta	Edmonton Klondike Days Exposition	Edmonton Northlands, Box 1480, Edmonton, AB T5J 2N5; 403/471-7210	Late July		•				•	•
Arizona	Navajo Nation Tribal Fair	Fair Office, P.O. Box Drawer U, Window Rock, AZ 86515; 602/871-6702	Early Sept.	•	•		•		•	
Arkansas	Petit Jean Show '95	Petit Jean Show '95, c/o Museum of Automobiles, Route 3, Box 306, Morrilton, AR 72110; 501/727-5427	Mid-June							•
British Columbia	Benson & Hedges Inc., Symphony of Fire	Vancouver Fireworks Society, c/o Harbour Centre, 456 W. Cordova St., Vancouver, BC V6B 4K2; 604/688-1992	Late July–Early Aug.			•			•	
California	Snowfest	Snowfest, P.O. Box 7590, Tahoe City, CA 96145; 916/582-0489	Early Mar.	•						
Colorado	Cherry Creek Arts Festival	Festival Office, P.O. Box 6265, Denver, CO, 80206; 303/355-2787	Early July	•		•	•	•		
Connecticut	Norwalk Oyster Festival	Norwalk Seaport Assn., 132 Water St., South Norwalk, CT 06854; 203/838-9444	Mid-Sept.			•	•		•	•
Delaware	A Day in Old New Castle	Immanuel Episcopal Church, P.O. Box 47, New Castle, DE 19720; 302/328-2413	Mid-May			•	•		•	
District of Columbia	Pageant of Peace	National Park Service, 1100 Ohio Dr., SW, Washington, DC 20242; 202/619-7222	Mid-Dec.–Early Jan.			•			•	
Florida	Annual Maritime Festival	St. Augustine C of C, One Ribera St., St. Augustine, FL 32084; 904/829-5681	Early Oct.	•			•		•	
Georgia	Oktoberfest	Savannah Waterfront Assn., P.O. Box 572, Savannah, GA 31402; 912/234-0295	Early Oct.			•	•			
Hawaii	Aloha Festivals	Aloha Festivals, 750 Amana St., Suite 111, Honolulu, HI 96814; 808/944-8857	Mid-Sept.	•	•			•		
Idaho	Boise River Festival	Boise River Festival Office, 205 N. 10th, Suite 210, Boise, ID 83702; 208/383-7318	Late June	•	•			•	•	
Illinois	Lake Shelbyville Festival of Lights	Shelbyville County C of C, 244 E. Main St., Shelbyville, IL 62565; 800/874-3529	Late Oct.–Late Jan.			•				
Indiana	Feast of the Hunter's Moon	Tippecanoe County Historical Assn., 909 South St., Lafayette, IN 47901; 317/742-8411	Early Oct.		•	•		•	•	
Iowa	Nordic Fest	Nordic Fest, Inc., P.O. Box 364, Decorah, IA 52101; 319/382-9010	Late July			•	•	•	•	•
Kansas	Smoky Hill River Festival	Festival Office, P.O. Box 2181, Salina, KS 67402; 913/826-7410	Mid-June	•		•	•	•	•	
Kentucky	W.C. Handy Blues & Barbecue Festival	Henderson Tourist Commission, 2961 US 41 N, Henderson, KY 42420; 502/826-3128	Mid-June			•			•	•
Louisiana	French Quarter Festival	Festival Office, 1008 N. Peters, New Orleans, LA 70116; 504/522-5730	Mid-April	•		•			•	•
Maine	Schooner Days	Rockland/Thomaston Area C of C, P.O. Box 508, Rockland, ME 04841; 207/596-0376	Early July	•	•	•	•	•	•	•
Manitoba	Folklorama	Folklorama, 375 York Ave., Winnipeg, MB R3C 3J3; 800/665-0234 (U.S.) or 204/982-6221	Mid-Aug.	•	•	•	•		•	
Maryland	Autumn Glory Festival	Garrett County Promotion Council, Court House, 200 S. Third St., Oakland, MD 21550; 301/334-1948	Mid-Oct.			•	•	•	•	
Massachusetts	Boston Harborfest	Boston Harborfest, 45 School St., Boston, MA 02108-3204; 617/227-1528	Late June–Early July			•		•	•	•
Michigan	International Freedom Festival	The Parade Company, 9600 Mount Elliott, Detroit, MI 48312; 313/923-7400	Late June–Early July			•		•	•	•
Minnesota	Minnesota Renaissance Festival	Mid-America Festivals, 3525 145th St. W., Shakopee, MN 55379; 612/445-7361	Mid-Aug.–Late Sept.	•		•	•		•	•
Mississippi	Natchez Spring Pilgrimage	Natchez Pilgrimage Tours, P.O. Box 347, Natchez, MS 39121; 800/647-6742 or 601/446-6631	Early Mar.–Early Apr.							•
Missouri	Autumn Historic Folklife Festival	Hannibal Arts Council, P.O. Box 1202, Hannibal, MO 63401; 314/221-6545	Mid-Oct.			•	•		•	•
Montana	Custer's Last Stand Re-enactment	Hardin C of C, 200 N. Center Ave., Hardin, MT 59034; 406/665-1672	Late June	•						•

Type of Celebration	Agricultural Fair/Livestock Show	Historical	Food	Parade	Arts & Crafts	Entertainment/Music	Sports

State/Province	Event	Information	Date(s)	Agricultural Fair/Livestock Show	Historical	Food	Parade	Arts & Crafts	Entertainment/Music	Sports
Nebraska	Czech Festival	Wilber C of C, P.O. Box 1164, Wilber, NE 68465; 402/821-2732	Early Aug.		•		•	•	•	
Nevada	Hot August Nights	Hot August Nights, P.O. Box 819, Reno, NV 89504; 702/826-1955	Early Aug.		•			•		
New Brunswick	Festival by the Sea	Festival by the Sea, P.O. Box 6848, Station A, Saint John, NB E2L 4S3; 506/632-0086	Mid-Aug.		•					
Newfoundland	St. John's Regatta	St. John's Regatta Commission, P.O. Box 214, St. John's, NF A1C 5J2; 709/576-8511	Early Aug.	•					•	
New Hampshire	Annual Craftsmen's Fair	League of N.H. Craftsmen, 205 N. Main St., Concord, NH 03301; 603/224-1471	Mid-Aug.		•	•				
New Jersey	Victorian Week	Mid-Atlantic Center for the Arts, P.O. Box 340, Cape May, NJ 08204; 609/884-5404	Mid-Oct.		•	•			•	
New Mexico	¡Magnifico! Albuquerque Festival of Arts	¡Magnifico! P.O. Box 26866, Albuquerque, NM 87125; 800/284-2282	Late Apr.– Mid-May		•	•		•		
New York	I♥NY Catskills Summer Festival	Hunter Mountain Festivals, P.O. Box 295, Hunter, NY 12442; 518/263-3800	Late June– Early Sept.		•	•		•		
North Carolina	Celebration of Battle of Moores Creek Bridge	Moores Creek National Battlefield, P.O. Box 69, Currie, NC 28435; 910/283-5591	Late Feb.		•				•	
North Dakota	Medora Musical	Theodore Roosevelt-Medora Foundation, P.O. Box 198, Medora, ND 58645; 701/623-4444	Mid-June– Early Sept.		•					
Nova Scotia	Annapolis Valley Apple Blossom Festival	Festival, c/o Dorothy Butt, 37 Cornwallis St., Kentville, NS B4N 2E2; 902/678-8322	Late May		•		•	•		
Ohio	Circleville Pumpkin Show	Circleville Pumpkin Show, 216 N. Court St., Circleville, OH 43113; 614/474-7000 or 614/477-2240	Late Oct.		•	•				
Oklahoma	OK Mozart International Festival	OK Mozart, P.O. Box 2344, Bartlesville, OK 74005; 918/336-9900	Mid-June						•	
Ontario	Winterlude/ Bal de Neige	National Capital Commission, 161 Laurier Ave. W., Ottawa, ON K1P 6J6; 800/465-1867 or 613/239-5000	Early Feb.	•	•	•				
Oregon	Portland Rose Festival	Portland Rose Festival Association, 220 N.W. 2nd Ave., Portland, OR 97209; 503//227-2681	June	•	•			•	•	
Pennsylvania	Gettysburg Civil War Heritage Days	Gettysburg Travel Council, 35 Carlisle St., Gettysburg, PA 17325; 717/334-6274	Late June– Early July		•					•
Prince Edward Island	Summerside Highland Gathering	Highland Gathering, 619 Water St., E., Summerside, PEI C1N 4H8; 902/436-5377	Late June	•	•			•	•	
Québec	Québec Winter Carnival	Winter Carnival, 290 Joly, Québec City, PQ G1L 1N8; 418/626-3716	Early Feb.	•					•	
Rhode Island	JVC Jazz Festival-Newport	Newport Festivals, P.O. Box 605, Newport, RI 02840; 401/847-3700	Mid-Aug.		•					
Saskatchewan	Shakespeare on the Saskatchewan Festival	General Mgr., Nightcap Productions, Box 1646, Saskatoon, SK S7K 3R8; 306/653-2300	Early July– Late Aug.		•					
South Carolina	Sun Fun Festival	Myrtle Beach Area C of C, P.O. Box 2115, Myrtle Beach, SC 29578; 803/626-7444	Early June	•	•					
South Dakota	Sturgis Rally & Races	Sturgis Rally & Races, Inc., P.O. Box 189, Sturgis, SD 57785; 605/347-6570	Mid-Aug.	•				•		
Tennessee	Tennessee Valley Fair	Fair Office, P.O. Box 6066, Knoxville, TN 37914; 615/637-5840	Mid-Sept.		•			•		•
Texas	Charro Days	Charro Days, Inc., P.O. Box 3247, Brownsville, TX 78520; 210/542-4245 or 210/541-3141	Late Feb.	•	•		•		•	
Utah	Christmas Lights at Temple Square	Temple Square Public Relations, 50 W. North Temple, Salt Lake City, UT 84150; 801/240-4869	Late Nov.– Late Dec.		•					
Vermont	Stratton Arts Festival	Arts Festival, P.O. Box 576, Stratton Mountain, VT 05155; 802/297-3265	Mid-Sept.– Mid-Oct.		•	•				
Virginia	George Washington Birthday Celebration	Alexandria C & V Bureau, 221 King St., Alexandria, VA 22314; 703/838-4200 or 703/838-5005 (recording)	Mid-Feb.	•	•		•		•	
Washington	Autumn Leaf Festival	Leavenworth C of C, P.O. Box 327, Leavenworth, WA 98826; 509/548-5807	Late Sept.– Early-Oct.		•	•	•	•		
West Virginia	Winter Festival of Lights	Wheeling C & V Bureau, 1000 Boury Center, Wheeling, WV 26003; 800/828-3097 or 304/233-7709	Early Nov.– Late Feb.		•				•	
Wisconsin	Lumberjack World Championships	Lumberjack World Championship Foundation, P.O. Box 666, Hayward, WI 54843; 715/634-2484	Late July	•		•				
Wyoming	Green River Rendezvous	Museum of the Mountain Man, P.O. 909, Pinedale, WY 82941; 307/367-4101	Mid-July		•	•		•		

C&V: Convention & Visitors (Bureau)
C of C: Chamber of Commerce

National Park Areas

State/Province	Park Name	Campgrounds	Hotel, Motel, Lodge	Cabins	Handicap Access Rest Rooms	Restaurant, Snacks	Museum/Exhibit	Groceries, Ice	Bathhouses	Cross-Country Ski Trail	Snowmobile Route	Fishing	Boating	Swimming	Horseback Riding	Hiking	NPS Guided Tours
Alabama	Russell Cave Natl. Monument, Rte. 1, Box 175, Bridgeport, AL 35740						•									•	
	Tuskegee Institute Natl. Historic Site, P.O. Drawer 10, Tuskegee Institute, AL 36087	•					•									•	
Alaska	Denali Natl. Park and Preserve, P.O. Box 9, McKinley Park, AK 99755	•	•				•					•				•	•
	Glacier Bay Natl. Park and Preserve, P.O. Box 140, Gustavus, AK 99826	•	•				•	•						•		•	•
	Katmai Natl. Park and Preserve, P.O. Box 7, King Salmon, AK 99613	•	•				•	•				•				•	•
	Kenai Fjords Natl. Park, P.O. Box 1727, Seward, AK 99664	•	•				•	•	•								
	Lake Clark Natl. Park and Preserve, 4230 University Dr., Suite 311, Anchorage, AK 99508		•				•	•	•			•				•	•
	Sitka Natl. Historical Park, P.O. Box 738, Sitka, AK 99835	•	•				•									•	
	Wrangell-St. Elias Natl. Park and Preserve, P.O. Box 29, Glennallen, AK 99588		•	•			•	•	•							•	
Alberta	Banff Natl. Park, Box 900, Banff, AB T0L 0C0	•	•	•	•	•	•	•		•	•	•	•	•	•	•	•
	Jasper Natl. Park, Box 10, Jasper, AB T0E 1E0	•	•	•	•	•	•	•		•	•	•	•	•	•	•	•
	Waterton Lakes Natl. Park, Waterton Park, AB T0K 2M0	•	•	•	•	•	•	•		•		•	•	•	•	•	•
Arizona	Canyon de Chelly Natl. Monument, P.O. Box 588, Chinle, AZ 86503	•	•	•			•	•								•	•
	Casa Grande Ruins Natl. Monument, 1100 Ruins Dr., Coolidge, AZ 85228	•					•									•	
	Chiricahua Natl. Monument, Dos Cabezas Route, Box 6500, Willcox, AZ 85643	•	•				•									•	
	Grand Canyon Natl. Park, P.O. Box 129, Grand Canyon, AZ 86023	•	•	•			•						•			•	•
	Hubbell Trading Post Natl. Historic Site, P.O. Box 150, Ganado, AZ 86505-0150	•					•									•	
	Montezuma Castle Natl. Monument, P.O. Box 219, Camp Verde, AZ 86322						•									•	
	Navajo Natl. Monument, H.C. 71, Box 3, Tonalea, AZ 86044-9704	•		•			•									•	•
	Organ Pipe Cactus Natl. Monument, Rte. 1, Box 100, Ajo, AZ 85321	•	•				•									•	•
	Petrified Forest Natl. Park, P.O. Box 2217, Petrified Forest Natl. Park, AZ 86028		•				•	•								•	
	Pipe Spring Natl. Monument, Moccasin, AZ 86022	•					•									•	
	Saguaro Natl. Monument, 3693 South Old Spanish Trail, Tucson, AZ 85730	•	•				•									•	
	Tonto Natl. Monument, P.O. Box 707, Roosevelt, AZ 85545	•					•									•	
	Tumacacori Natl. Historical Park, P.O. Box 67, Tumacacori, AZ 85640	•					•									•	
	Tuzigoot Natl. Monument, P.O. Box 219, Camp Verde, AZ 86322						•									•	
	Walnut Canyon Natl. Monument, Walnut Canyon Rd., Flagstaff, AZ 86004-9705						•									•	
	Wupatki Natl. Monument, H.C. 33, Box 444A, Flagstaff, AZ 86004	•	•				•									•	
Arkansas	Fort Smith Natl. Historic Site, P.O. Box 1406, Fort Smith, AR 72902	•					•									•	
	Hot Springs Natl. Park, P.O. Box 1860, Hot Springs, AR 71902	•	•	•			•					•				•	
British Columbia	Mount Revelstoke/Glacier Natl. Park, Box 350, Revelstoke, BC V0E 2S0	•	•				•		•			•	•			•	•
California	Channel Islands Natl. Park, 1901 Spinnaker Dr., Ventura, CA 93001	•	•		•	•	•					•				•	•
	Death Valley Natl. Monument (Calif., Nev.), P.O. Box 579, Death Valley, CA 92328	•	•	•	•		•	•	•	•	•	•	•	•	•	•	•
	Devils Postpile Natl. Mon., c/o Sequoia and Kings Canyon Natl. Parks, Three Rivers, CA 93271	•	•	•	•		•										•
	Fort Point Natl. Historic Site, P.O. Box 29333, Presidio of San Francisco, CA 94129	•					•									•	
	Golden Gate Natl. Recreation Area, Fort Mason, Bldg. 201, San Francisco, CA 94123	•	•	•	•		•		•							•	•
	Joshua Tree Natl. Monument, 74485 National Monument Dr., Twentynine Palms, CA 92277	•	•		•		•									•	•
	Kings Canyon Natl. Park, Three Rivers, CA 93271	•	•	•			•	•				•			•	•	•
	Lassen Volcanic Natl. Park, Mineral, CA 96063	•	•	•	•	•	•	•			•	•		•	•	•	
	Lava Beds Natl. Monument, P.O. Box 867, Tulelake, CA 96134	•	•				•									•	
	Muir Woods Natl. Monument, Mill Valley, CA 94941	•	•				•									•	
	Pinnacles Natl. Monument, Paicines, CA 95043		•				•									•	
	Point Reyes Natl. Seashore, Point Reyes, CA 94956	•	•	•	•		•									•	•
	Redwood Natl. Park, 1111 Second St., Crescent City, CA 95531	•	•	•	•		•									•	•
	Santa Monica Mts. Natl. Recreation Area, 30401 Agoura Rd., Suite 100, Agoura Hills, CA 91301	•	•	•	•	•	•					•		•	•	•	•
	Sequoia Natl. Park, Three Rivers, CA 93271	•	•	•			•	•				•			•	•	•
	Whiskeytown-Shasta-Trinity Natl. Recreation Area, P.O. Box 188, Whiskeytown, CA 96095	•	•	•	•	•						•	•	•		•	
	Yosemite Natl. Park, P.O. Box 577, Yosemite Natl. Park, CA 95389	•	•	•	•	•	•	•		•	•	•	•	•	•	•	•
Colorado	Black Canyon of the Gunnison Natl. Monument, 2233 E. Main, Suite A, Montrose, CO 81401	•	•				•		•			•			•	•	•
	Colorado Natl. Monument, Fruita, CO 81521	•	•	•			•									•	•
	Curecanti Natl. Recreation Area, 102 Elk Creek, Gunnison, CO 81230	•	•		•	•	•	•	•			•	•	•		•	
	Dinosaur Natl. Monument (Colo., Utah), P.O. Box 210, Dinosaur, CO 81610	•	•				•	•				•				•	•

Activities, Facilities, and Accommodations in Selected National Park Service areas of the United States and Canada.

For information on National Park Service Areas not included in this list, write to:

U.S. Department of Interior
National Park Service
Office of Public Inquiries
P.O. Box 37127
Washington, D.C. 20013-7127

Canadian Parks Service
10 Wellington Street
Hull, PQ, Canada
K1A 0H3

State/Province	Park Name	NPS Guided Tours	Hiking	Horseback Riding	Swimming	Boating	Fishing	Snowmobile Route	Cross-Country Ski Trail	Bathhouses	Groceries, Ice	Museum/Exhibit	Restaurant, Snacks	Handicap Access Rest Rooms	Cabins	Hotel, Motel, Lodge	Campgrounds
		Activities								Facilities					Accommodations		
	Florissant Fossil Beds Natl. Monument, P.O. Box 185, Florissant, CO 80816	•	•						•			•		•			
	Great Sand Dunes Natl. Monument, 11500 Hwy. 150, Mosca, CO 81146	•	•				•				•	•		•			•
	Mesa Verde Natl. Park, Mesa Verde Natl. Park, CO 81321	•									•	•	•	•		•	•
	Rocky Mountain Natl. Park, Estes Park, CO 80517	•	•	•			•	•	•			•	•	•			•
District of Columbia	Ford's Theatre Natl. Historic Site, c/o NCP-Central, 900 Ohio Dr., SW, Washington, DC 20242	•										•					
	Frederick Douglass Natl. Historic Site, 1411 W St., SE, Washington, DC 20020	•										•		•			
	John F. Kennedy Center for the Performing Arts, Natl. Park Service, 2700 F. St., NW, Washington, DC 20566	•											•	•			
	Lincoln Memorial, c/o NCP-Central, 900 Ohio Dr., SW, Washington, DC 20242	•												•			
	Thomas Jefferson Memorial, c/o NCP-Central, 900 Ohio Dr., SW, Washington, DC 20242	•										•		•			
	Vietnam Veterans Memorial, c/o NCP-Central, 900 Ohio Dr., SW, Washington, DC 20242													•			
	Washington Monument, c/o NCP-Central, 900 Ohio Dr., SW, Washington, DC 20242	•										•	•	•			
	White House, c/o Nat'l Capital Region, 1100 Ohio Dr., SW, Washington, DC 20242											•		•			
Florida	Biscayne Natl. Park, P.O. Box 1369, Homestead, FL 33090	•	•		•	•	•					•		•			•
	Canaveral Natl. Seashore, 308 Julia St., Titusville, FL 32796-3521	•	•		•	•	•					•		•			
	Castillo de San Marcos Natl. Monument, 1 Castillo Dr. East, St. Augustine, FL 32084	•										•		•			
	Dry Tortugas Natl. Park, c/o Everglades Natl. Park, P.O. Box 279, Homestead, FL 33030				•	•	•					•		•			•
	Everglades Natl. Park, P.O. Box 279, Homestead, FL 33030	•	•			•	•				•	•	•	•	•	•	•
	Fort Matanzas Natl. Monument, c/o Castillo de San Marcos Natl. Monument, 1 Castillo Dr. East, St. Augustine, FL 32084	•			•		•					•					
	Gulf Islands Natl. Seashore, 1801 Gulf Breeze Pkwy., Gulf Breeze, FL 32561 (Also in Miss.)	•	•		•	•	•			•	•	•	•	•			•
Georgia	Andersonville Natl. Historic Site, Rte. 1, Box 800, Andersonville, GA 31711	•										•		•			
	Chattahoochee River Natl. Recreation Area, 1978 Island Ford Pkwy., Dunwoody, GA 30350	•	•			•	•						•	•			
	Cumberland Island Natl. Seashore, P.O. Box 806, St. Marys, GA 31558	•	•		•		•			•		•		•			•
	Fort Frederica Natl. Monument, Rte. 9, Box 286 C, St. Simons Island, GA 31522	•										•		•			
	Fort Pulaski Natl. Monument, P.O. Box 30757, Savannah, GA 31410	•	•			•	•					•		•			
	Martin Luther King, Jr., Natl. Historic Site, 522 Auburn Ave., NE, Atlanta, GA 30312	•										•					
	Ocmulgee Natl. Monument, 1207 Emery Hwy., Macon, GA 31201	•	•				•					•		•			
Hawaii	Haleakala Natl. Park, P.O. Box 369, Makawao, Maui, HI 96768		•	•								•		•	•		•
	Hawaii Volcanoes Natl. Park, Hawaii Natl. Park, HI 96718		•								•	•	•	•	•		•
	USS *Arizona* Memorial, 1 Arizona Memorial Place, Honolulu, HI 96818	•										•	•	•			
Idaho	Craters of the Moon Natl. Monument, P.O. Box 29, Arco, ID 83213	•	•						•			•		•			•
Illinois	Lincoln Home Natl. Historic Site, 413 S. Eighth St., Springfield, IL 62701	•										•		•			
Indiana	Indiana Dunes Natl. Lakeshore, 1100 N. Mineral Springs Rd., Porter, IN 46304-1299	•	•	•	•	•	•		•	•		•	•	•			
Iowa	Effigy Mounds Natl. Monument, 151 Hwy. 76, Harpers Ferry, IA 52146	•	•						•			•		•			
	Herbert Hoover Natl. Historic Site, P.O. Box 607, West Branch, IA 52358	•	•						•			•		•			
Kansas	Fort Larned Natl. Historic Site, Rte. 3, Larned, KS 67550	•					•					•		•			
	Fort Scott Natl. Historic Site, Old Fort Blvd., Fort Scott, KS 66701-1471	•										•		•			
Kentucky	Abraham Lincoln Birthplace Natl. Historic Site, 2995 Lincoln Farm Rd., Hodgenville, KY 42748		•									•		•			
	Cumberland Gap Natl. Historical Park, (Ky., Tenn., Va.) P.O. Box 1848, Middlesboro, KY 40965		•									•		•			•
	Mammoth Cave Natl. Park, Mammoth Cave, KY 42259	•	•	•		•	•				•	•	•	•	•	•	•
Louisiana	Jean Lafitte Natl. Historical Park & Pres., 365 Canal St., Suite 3080, New Orleans, LA 70130	•	•			•	•					•		•			
Maine	Acadia Natl. Park, P.O. Box 177, Bar Harbor, ME 04609		•	•	•	•	•	•	•		•	•	•	•		•	•
Manitoba	Riding Mountain Natl. Park, Wasagaming, MB R0J 2H0	•	•	•	•	•	•		•	•	•	•	•	•	•	•	•
Maryland	Assateague Island Natl. Seashore (Md., Va.), 7206 Natl. Seashore Lane, Berlin, MD 21811	•	•		•	•	•			•		•		•			•
	Chesapeake & Ohio Canal Natl. Hist. Park (W. Va., Md., Va.), P.O. Box 4, Sharpsburg, MD 21782	•	•	•		•	•					•	•	•			•
	Fort McHenry Natl. Mon. and Historic Shrine, end of E. Fort Avenue, Baltimore, MD 21230-5393	•										•		•			
Massachusetts	Adams Natl. Historic Site, 135 Adams St., P.O. Box 531, Quincy, MA 02269-0531	•										•					
	Boston Natl. Historical Park, Charlestown Navy Yard, Boston, MA 02129	•										•	•	•			
	Cape Cod Natl. Seashore, South Wellfleet, MA 02663	•	•	•	•		•			•		•	•	•		•	
	John Fitzgerald Kennedy Natl. Historic Site, 83 Beals St., Brookline, MA 02146	•										•					
	Minute Man Natl. Historical Park, P.O. Box 160, Concord, MA 01742	•										•		•			
	Salem Maritime Natl. Historic Site, Custom House, 174 Derby St., Salem, MA 01970	•	•			•	•					•	•				
Michigan	Isle Royale Natl. Park, 800 E. Lakeshore Dr., Houghton, MI 49931	•	•		•	•	•				•		•	•	•	•	•
	Pictured Rocks Natl. Lakeshore, P.O. Box 40, Munising, MI 49862		•		•	•	•	•	•			•		•			•

©1995 Rand McNally & Company

Activities, Facilities, and Accommodations in Selected National Park Service areas of the United States and Canada.

		Accommodations			Facilities					Activities							
State/Province	Park Name	Campgrounds	Hotel, Motel, Lodge	Cabins	Handicap Access Rest Rooms	Restaurant, Snacks	Museum/Exhibit	Groceries, Ice	Bathhouses	Cross-Country Ski Trail	Snowmobile Route	Fishing	Boating	Swimming	Horseback Riding	Hiking	NPS Guided Tours
	Sleeping Bear Dunes Natl. Lakeshore, 9922 Front St., P.O. Box 277, Empire, MI 49630	•	•		•	•	•		•	•		•	•				•
Minnesota	Grand Portage Natl. Monument, P.O. Box 668, Grand Marais, MN 55604	•	•				•		•			•		•			
	Pipestone Natl. Monument, P.O. Box 727, Pipestone, MN 56164						•					•		•			
	Voyageurs Natl. Park, 3131 Hwy. 53, International Falls, MN 56649	•	•		•	•	•	•	•			•	•	•		•	•
Mississippi	Natchez Trace Parkway (Miss., Ala., Tenn.), R.R. 1, NT-143, Tupelo, MS 38801		•	•	•	•	•					•	•		•		•
Missouri	George Washington Carver Natl. Monument, P.O. Box 38, Diamond, MO 64840	•					•					•		•			
	Harry S Truman Natl. Historic Site, 223 N. Main St., Independence, MO 64050	•					•					•		•			
	Jefferson Natl. Expansion Memorial, 11 North 4th St., St. Louis, MO 63102	•					•					•		•			
	Ozark Natl. Scenic Riverways, P.O. Box 490, Van Buren, MO 63965	•	•	•	•	•	•					•	•	•	•		
Montana	Bighorn Canyon Natl. Recreation Area (Mont., Wyo.), P.O. Box 458, Fort Smith, MT 59035	•	•		•	•	•	•				•	•	•			
	Glacier Natl. Park, West Glacier, MT 59936	•	•	•	•	•	•		•			•	•	•	•	•	•
	Little Bighorn Battlefield Natl. Monument, P.O. Box 39, Crow Agency, MT 59022	•					•					•		•			
Nebraska	Agate Fossil Beds Natl. Monument, P.O. Box 27, Gering, NE 69341		•			•						•		•			
	Scotts Bluff Natl. Monument, P.O. Box 27, Gering, NE 69341		•									•		•			
Nevada	Great Basin Natl. Park, Baker, NV 89311	•	•	•		•		•				•	•	•			•
	Lake Mead Natl. Recreation Area (Nev., Ariz.), 601 Nevada Hwy., Boulder City, NV 89005-2426		•		•	•	•			•	•	•	•	•		•	•
New Hampshire	Saint-Gaudens Natl. Historic Site, R.R. 3, Box 73, Cornish, NH 03745-9704	•	•				•					•		•			
New Jersey	Edison Natl. Historic Site, Main St. and Lakeside Ave., West Orange, NJ 07052	•					•					•		•			
New Mexico	Bandelier Natl. Monument, HCR 1, Box 1, Suite 15, Los Alamos, NM 87544	•	•			•	•					•		•			•
	Capulin Volcano Natl. Monument, P.O. Box 40, Capulin, NM 88414		•									•		•			
	Carlsbad Caverns Natl. Park, 3225 National Parks Hwy., Carlsbad, NM 88220	•	•		•	•	•					•		•			•
	Chaco Culture Natl. Historical Park, Star Route 4, Box 6500, Bloomfield, NM 87413	•	•				•					•		•			•
	El Malpais Natl. Monument, P.O. Box 939, Grants, NM 87020		•									•					
	El Morro Natl. Monument, Rte. 2, Box 43, Ramah, NM 87321-9603		•				•					•		•			•
	Gila Cliff Dwellings Natl. Monument, Rte. 11, Box 100, Silver City, NM 88061						•					•		•			
	Pecos Natl. Historical Park, P.O. Drawer 418, Pecos, NM 87522	•					•					•		•			
	Salinas Pueblo Missions Natl. Monument, P.O. Box 496, Mountainair, NM 87036	•					•					•		•			
	White Sands Natl. Monument, P.O. Box 1086, Holloman AFB, NM 88330	•	•				•					•		•			
New York	Castle Clinton Natl. Monument, c/o Manhattan Sites, NPS, 26 Wall St. New York, NY 10005	•					•					•		•			
	Fire Island Natl. Seashore, 120 Laurel St., Patchogue, NY 11772-3596	•	•		•	•	•					•	•	•	•	•	•
	Fort Stanwix Natl. Monument, 112 E. Park St., Rome, NY 13440	•					•					•		•			
	Gateway Natl. Recreation Area (N.Y., N.J.), Floyd Bennett Field, Bldg. 69, Brooklyn, NY 11234	•	•	•	•	•	•	•			•	•	•	•	•		•
	Home of Franklin D. Roosevelt Natl. Historic Site, 519 Albany Post Rd., Hyde Park, NY 12538						•					•		•			
	Sagamore Hill Natl. Historic Site, 20 Sagamore Hill Rd., Oyster Bay, NY 11771						•					•		•			
	Statue of Liberty Natl. Monument (N.Y., N.J.), Liberty Island, New York, NY 10004	•					•					•	•	•			
North Carolina	Blue Ridge Parkway (N.C., Va.), 200 BB&T Bldg., One Pack Square, Asheville, NC 28801	•	•	•		•	•	•	•			•	•	•	•	•	•
	Cape Hatteras Natl. Seashore, Rte. 1, Box 675, Manteo, NC 27954	•	•			•	•	•			•	•	•	•	•		•
	Cape Lookout Natl. Seashore, 131 Charles St., Harkers Island, NC 28531	•	•			•	•	•					•	•		•	
	Carl Sandburg Home Natl. Historic Site, 1928 Little River Rd., Flat Rock, NC 28731	•	•				•					•		•			
	Fort Raleigh Natl. Historic Site, c/o Cape Hatteras Natl. Seashore, Rte. 1, Box 675, Manteo, NC 27954	•					•					•		•			
North Dakota	Fort Union Trading Post Natl. Historic Site (ND, MT), R.R. 3, Box 71, Buford Route, Williston, ND 58801	•	•			•	•		•			•		•			
	Theodore Roosevelt Natl. Park, P.O. Box 7, Medora, ND 58645	•	•	•		•	•	•				•		•			•
Nova Scotia	Fortress of Louisbourg Natl. Historic Park, Box 160, Louisbourg, NS B0A, 1M0	•	•				•					•	•	•			
Ohio	Cuyahoga Valley Natl. Recreation Area, 15610 Vaughn Rd., Brecksville, OH 44141	•	•	•	•		•		•			•		•		•	
	Hopewell Culture Natl. Historical Park, 16062 State Route 104, Chillicothe, OH 45601-8694	•	•				•					•		•			
Oklahoma	Chickasaw Natl. Recreation Area, P.O. Box 201, Sulphur, OK 73086	•	•		•	•	•					•		•			•
Ontario	Georgian Bay Islands Natl. Park, P.O. Box 28, Honey Harbour, ON P0E 1E0	•	•		•	•	•		•	•		•		•			•
	St. Lawrence Islands Natl. Park, P.O. Box 469, RR 3, Mallorytown Landing, ON K0E 1R0	•	•		•	•	•			•		•		•			•
Oregon	Crater Lake Natl. Park, P.O. Box 7, Crater Lake, OR 97604	•	•				•	•	•	•	•	•	•	•	•	•	•
	Oregon Caves Natl. Monument, 19000 Caves Hwy., Cave Junction, OR 97523		•											•	•	•	
Pennsylvania	Delaware Water Gap Natl. Recreation Area (Pa., N.J.), Bushkill, PA 18324	•	•		•	•	•	•	•			•	•	•			
	Eisenhower National Historic Site, P.O. Box 1080, Gettysburg, PA 17325	•					•					•		•			
	Gettysburg Natl. Military Park, P.O. Box 1080, Gettysburg, PA 17325	•	•				•					•		•			
	Hopewell Furnace Natl. Historic Site, 2 Mark Bird Lane, Elverson, PA 19520		•				•					•		•			

Activities, Facilities, and Accommodations in Selected National Park Service areas of the United States and Canada.

State/Province	Park Name	Activities: NPS Guided Tours	Hiking	Horseback Riding	Swimming	Boating	Fishing	Snowmobile Route	Cross-Country Ski Trail	Facilities: Bathhouses	Groceries, Ice	Museum/Exhibit	Restaurant, Snacks	Handicap Access Rest Rooms	Accommodations: Cabins	Hotel, Motel, Lodge	Campgrounds
	Independence Natl. Historical Park, 313 Walnut St., Philadelphia, PA 19106	•										•	•	•			
	Upper Delaware Scenic and Recreational River (Pa., N.Y.), P.O. Box C, Narrowsburg, NY 12764	•		•	•	•	•				•	•	•	•	•	•	•
	Valley Forge Natl. Historical Park, P.O. Box 953, Valley Forge, PA 19481		•	•		•	•		•			•	•	•			
Prince Edward Island	Prince Edward Island Natl. Park, P.O. Box 487, Charlottetown, PE C1A 7L1	•	•		•	•	•		•	•		•	•	•			•
Québec	La Mauricie Natl. Park, 794 5th St., P.O. Box 758, Shawinigan, PQ G9N 6V9	•	•		•	•	•		•	•	•	•	•	•		•	•
South Carolina	Congaree Swamp Natl. Monument, 200 Caroline Sims Rd., Hopkins, SC 29061	•	•			•	•							•			
	Fort Sumter Natl. Monument, 1214 Middle St., Sullivans Island, SC 29482	•					•					•		•			
	Ninety Six Natl. Historic Site, P.O. Box 496, Ninety Six, SC 29666	•	•	•			•					•		•			
South Dakota	Badlands Natl. Park, P.O. Box 6, Interior, SD 57750	•	•									•	•		•		•
	Jewel Cave Natl. Monument, R.R. 1, Box 60AA, Custer, SD 57730	•										•		•			
	Mount Rushmore Natl. Memorial, P.O. Box 268, Keystone, SD 57751											•	•	•			
	Wind Cave Natl. Park, R.R. 1, Box 190, Hot Springs, SD 57747	•	•									•	•	•			•
Tennessee	Andrew Johnson Natl. Historic Site, P.O. Box 1088, Greeneville, TN 37744											•		•			
	Great Smoky Mountains Natl. Park (Tenn., N.C.), Gatlinburg, TN 37738	•	•	•			•				•	•		•	•	•	•
Texas	Amistad Natl. Recreation Area, P.O. Box 420367, Del Rio, TX 78842-0367		•	•	•	•	•				•		•	•			•
	Big Bend Natl. Park, P.O. Box 129, Big Bend Natl. Park, TX 79834	•	•	•		•	•				•	•	•	•		•	•
	Fort Davis Natl. Historic Site, P.O. Box 1456, Fort Davis, TX 79734		•									•		•			
	Guadalupe Mountains Natl. Park, H.C. 60, Box 400, Salt Flat, TX 79847	•	•									•		•			•
	Lake Meredith Recreation Area, P.O. Box 1460, Fritch, TX 79036				•	•	•				•	•					•
	Lyndon B. Johnson Natl. Historical Park, P.O. Box 329, Johnson City, TX 78636	•										•		•			
	Padre Island Natl. Seashore, 9405 S. Padre Island Dr., Corpus Christi, TX 78418-5597		•	•	•	•	•			•		•		•			•
	San Antonio Missions Natl. Historical Park, 2202 Roosevelt Ave., San Antonio, TX 78210	•	•									•					
Utah	Arches Natl. Park, P.O. Box 907, Moab, UT 84532	•	•									•		•			•
	Bryce Canyon Natl. Park, Bryce Canyon, UT 84717	•	•	•					•		•	•	•	•	•	•	•
	Canyonlands Natl. Park, 125 West 200 South, Moab, UT 84532	•	•	•		•								•			•
	Capitol Reef Natl. Park, H.C. 70, Box 15, Torrey, UT 84775	•	•	•			•					•		•			•
	Cedar Breaks Natl. Monument, 82 North 100 East St., Cedar City, UT 84720	•	•					•	•			•		•			•
	Glen Canyon Natl. Recreation Area (Utah, Ariz.), P.O. Box 1507, Page, AZ 86040	•	•		•	•	•				•	•	•	•		•	•
	Natural Bridges Natl. Monument, Box 1, Lake Powell, UT 84533		•									•					•
	Timpanogos Cave Natl. Monument, R.R. 3, Box 200, American Fork, UT 84003	•										•	•				
	Zion Natl. Park, Springdale, UT 84767-1099	•	•	•				•	•			•	•	•	•	•	•
Virginia	Appomattox Court House Natl. Historical Park, P.O. Box 218, Appomattox, VA 24522	•	•									•		•			
	Booker T. Washington Natl. Monument, Rte. 3, Box 310, Hardy, VA 24101	•	•									•					
	Colonial Natl. Historical Park, P.O. Box 210, Yorktown, VA 23690	•	•									•		•			
	George Washington Birthplace Natl. Mon., R.R. 1, Box 717, Washington's Birthplace, VA 22443	•	•				•					•		•			
	George Washington Memorial Pkwy. (Va., Md.), Turkey Run Park, McLean, VA 22101	•	•	•		•	•					•	•	•			
	Shenandoah Natl. Park, Rte. 4, Box 348, Luray, VA 22835	•	•	•			•		•		•	•	•	•	•	•	•
	Wolf Trap Farm for the Performing Arts, 1551 Trap Rd., Vienna, VA 22182	•											•	•			
Virgin Islands	Virgin Islands Natl. Park, 6010 Estate Nazareth, St. Thomas, VI 00802	•	•		•	•	•			•	•	•	•	•	•		•
Washington	Coulee Dam Natl. Recreation Area, 1008 Crest Dr., Coulee Dam, WA 99116	•			•	•	•			•		•		•			•
	Fort Vancouver Natl. Historic Site, 612 E. Reserve St., Vancouver, WA 98661-3897	•										•		•			
	Lake Chelan Natl. Recreation Area, 2105 Hwy. 20, Sedro Woolley, WA 98284	•	•	•		•	•		•		•		•		•	•	•
	Mount Rainier Natl. Park, Tahoma Woods, Star Route, Ashford, WA 98304	•	•				•	•	•		•	•	•	•		•	•
	North Cascades Natl. Park, 2105 Hwy. 20, Sedro Woolley, WA 98284		•	•			•										
	Olympic Natl. Park, 600 E. Park Ave., Port Angeles, WA 98362	•	•	•	•	•	•		•	•	•	•	•	•	•	•	•
	Ross Lake Natl. Recreation Area, 2105 Hwy. 20, Sedro Woolley, WA 98284	•	•	•		•	•	•			•		•	•	•		•
West Virginia	Harpers Ferry Natl. Historical Park (W.Va, Md., Va), P.O. Box 65, Harpers Ferry, WV 25425	•	•				•					•		•			
	New River Gorge Natl. River, P.O. Box 246, Glen Jean, WV 25846		•	•	•	•	•					•		•	•		•
Wisconsin	Apostle Islands Natl. Lakeshore, Rte. 1, Box 4, Bayfield, WI 54814	•	•		•	•	•					•		•			•
	St. Croix Natl. Scenic Riverway (Wis., Minn.), P.O. Box 708, St. Croix Falls, WI 54024		•		•	•	•		•		•	•	•	•	•	•	•
Wyoming	Devils Tower Natl. Monument, P.O. Box 8, Devils Tower, WY 82714	•	•				•					•		•			•
	Fort Laramie Natl. Historic Site, P.O. Box 86, Fort Laramie, WY 82212	•					•					•		•			
	Grand Teton Natl. Park, P.O. Drawer 170, Moose, WY 83012	•	•	•	•	•	•	•	•		•	•	•	•	•	•	•
	Yellowstone Natl. Park (Wyo., Idaho, Mont.), P.O. Box 168, Yellowstone Natl. Park, WY 82190	•	•	•		•	•	•	•		•	•	•	•	•	•	•

Tourism Information & Auto/RV Laws

State	Tourism Information	Studded Tires Permitted (Dates)	Child Restraints Required	Seat Belts Required	Auto Liability Insurance Mandatory	Overnight Off-road Parking Permitted	Brakes if Weight Over	Breakaway Brakes if Weight Over	Max Length of Car and Trailer Without Permit	Chains Required	Flares Required	Riding in Trailer Permitted
Alabama	Alabama Bureau of Tourism & Travel 800/ALABAMA; 205/242-4169	Prohibited	•	•	•	NO	3,000		65			
Alaska	Alaska Division of Tourism 907/465-2010	Oct. 1–Apr. 15 (d)	•	•	•	Designated areas only	(b)	(b)	48	•		
Arizona	Arizona Office of Tourism 800/842-8257; 602/542-TOUR	Oct. 1–May 1 (d)	•	•	•	Only if posted	3,000	3,000	40	•		
Arkansas	Arkansas Dept. of Parks & Tourism 800/NATURAL; 501/682-7777	Nov. 15–Apr. 15	•	•	•	Rest areas only	3,000	3,000	65	•	•	
California	California Office of Tourism 800/TO-CALIF; 916/322-1396	Nov. 1–Apr. 1	•	•		Designated areas only	(b)	(b)	65	•		• d
Colorado	Colorado Tourism Board 800/COLORADO	YES, with no restrictions	•	•	•	Unless posted	3,000	3,000	70	•		
Connecticut	Connecticut Tourism Division 800/CT-BOUND; 203/258-4355	Nov. 15–Apr. 30	•	•	•	Designated areas only	3,000	8,000	65	•		
Delaware	Delaware Tourism Office 800/441-8846; 302/739-4271	Oct.15–Apr. 15	•	•	•	Where posted	4,000	10,000	60			
District of Columbia	Washington DC Conv. & Visitors Assn. 202/789-7000	Oct. 15–Apr. 15	•	•		NO	3,000(a)	3,000	55	•		
Florida	Florida Division of Tourism 904/487-1462	Yes (d)	•	•		Only rest areas where posted	3,000	3,000	60	•	• d	
Georgia	Georgia Tourism Division 800/VISIT GA; 404/656-3590	Prohibited	•	•	•	Designated areas only			60			
Hawaii	Hawaii Visitors Bureau 808/923-1811	Prohibited	•	•	•	NO	3,000	3,000	45	•		
Idaho	Idaho Dept. of Commerce 800/635-7820; 208/334-2470	Oct. 1–Apr. 15	•	•	•	Designated areas only	1,500	1,500	75			
Illinois	Illinois Bureau of Tourism 800/223-0121; 312/814-4732	Prohibited	•	•	•	YES	3,000	5,000	60			• d
Indiana	Indiana Tourism Division 800/782-3775; 317/232-8860	Oct. 1–May 1	•	•	•	NO	(b)	3,000	60	•	•	•
Iowa	Iowa Division of Tourism 800/345-IOWA; 515/242-4705	Nov. 1–Apr. 1	•	•		Not at rest areas	3,000	3,000	60	•	•	•
Kansas	Kansas Travel & Tourism Development 800/2 KANSAS; 913/296-2009	Nov. 1–Apr. 15 (d)	•	•	•	YES	(h,d)		65	• d		
Kentucky	Kentucky Dept. of Travel Development 800/225-TRIP, ext. 67; 502/564-4930	Prohibited	•	•	•	Designated areas only					•	
Louisiana	Louisiana Office of Tourism 800/33GUMBO; 504/342-8119	Prohibited	•	•	•	NO	3,000	3,000	65	•	•	
Maine	Maine Publicity Bureau 800/533-9595; 207/582-9300	Oct. 1–May 1	•		•	NO	3,000	3,000	65	•		
Maryland	Maryland Office of Tourism & Development 800/543-1036; 410/333-6611	Prohibited	•	•	•	Where posted	3,000	3,000	35	•		
Massachusetts	Massachusetts Office of Travel & Tourism 800/447-MASS; 617/727-3201	Nov. 1–Apr. 1	•	•	•	Overnight rest areas only	10,000	10,000	66	•	• d	
Michigan	Michigan Travel Bureau 800/5432-YES; 517/373-0670	Yes (d)	•	•	•	Unless posted	5,500		59			•
Minnesota	Minnesota Office of Tourism 800/657-3700; 612/296-5029	Prohibited	•	•	•	As posted	3,000	6,000	65		•	•
Mississippi	Mississippi Tourism Division 800/647-2290; 601/359-3297	Prohibited	•	•		NO	2,000	2,000	50			•
Missouri	Missouri Div. of Tourism 800/877-1234; 314/751-4133	Nov. 1–Mar. 31	•	•	•	YES			55			•

Cars Towing Trailers
- Riding in Trailer Permitted
- Flares Required
- Chains Required
- Max Length of Car and Trailer Without Permit
- Breakaway Brakes if Weight Over
- Brakes if Weight Over
- Overnight Off-road Parking Permitted

Automobiles
- Auto Liability Insurance Mandatory
- Seat Belts Required
- Child Restraints Required
- Studded Tires Permitted (Dates)

(a) Or, when trailer exceeds 40% of weight of towing vehicle   (b) Equipment required   (c) Must be able to stop within legal distance
(d) Possible exceptions. Refer to state laws   (e) Studded tires permitted if studs don't project more than ¹⁄₁₆" when compressed

Cars Towing Trailers
- Riding in Trailer Permitted
- Flares Required
- Chains Required
- Max Length of Car and Trailer Without Permit
- Breakaway Brakes if Weight Over
- Brakes if Weight Over
- Overnight Off-road Parking Permitted

Automobiles
- Auto Liability Insurance Mandatory
- Seat Belts Required
- Child Restraints Required
- Studded Tires Permitted (Dates)

State	Tourism Information	Studded Tires Permitted (Dates)	Child Restraints Required	Seat Belts Required	Auto Liability Insurance Mandatory	Overnight Off-road Parking Permitted	Brakes if Weight Over	Breakaway Brakes if Weight Over	Max Length of Car and Trailer Without Permit	Chains Required	Flares Required	Riding in Trailer Permitted
Montana	Travel Montana 800/541-1447; 406/444-2654	Oct. 1–May 31	•	•	•	YES	3,000	3,000	65	• (d)	• (d)	• (d)
Nebraska	Nebraska Travel & Tourism 800/228-4307; 402/471-3796	Nov. 1–Apr. 1	•	•	•	Not in rest areas	3,000	3,000	65	• (d)	• (d)	•
Nevada	Nevada Commission on Tourism 800/NEVADA-8	Oct. 1–Apr. 30	•	•	•	YES	3,000(d)	3,000	70			• (d)
New Hampshire	New Hampshire Travel & Tourism 800/944-1167; 603/271-2666	YES, with no restrictions	•			Not in rest areas	3,000(d)	3,000	48	•		• (d)
New Jersey	New Jersey Division of Travel & Tourism 800/JERSEY-7; 609/292-2470	Nov. 15–Apr. 1	•	•	•	Rest areas only	3,000(a)		50			•
New Mexico	New Mexico Dept. of Tourism 800/545-2040	YES, with no restrictions	•	•	•	NO			40	•		•
New York	New York Tourism Division 800/CALL-NYS; 518/474-4116	Oct. 15–May 1	•	•	•	YES	1,000		65	•		
North Carolina	North Carolina Travel & Tourism Div. 800/VISIT-NC; 919/733-4171	YES, with no restrictions	•	•	•	NO	4,000		60			•
North Dakota	North Dakota Tourism 800/435-5663; 701/224-2525	Oct. 15–Apr. 15	•	•	•	NO (d)			75	•		• (d)
Ohio	Ohio Div. of Travel & Tourism 800/BUCKEYE	Nov. 1–Apr. 15	•	•	•	Where posted	2,000	2,000	65	•	•	
Oklahoma	Oklahoma Tourism & Recreation Dept. 800/652-6552; 405/521-2409	Nov. 1–Apr. 1	•	•	•	Only at rest areas	3,000	3,000	70	•	•	•
Oregon	Oregon Tourism Division 800/547-7842	Nov. 1–Apr. 15	•	•	•	NO			50	•		• (d)
Pennsylvania	Pennsylvania Bureau of Travel Marketing 800/VISIT-PA	Nov. 1–Apr. 15	•	•		As posted	(b)	3,000	60	•	•	
Rhode Island	Rhode Island Division of Tourism 800/556-2484; 401/277-2601	Nov. 15–Mar. 31	•	•	•	Designated rest areas only	(b)	3,000				•
South Carolina	South Carolina Division of Tourism 800/346-3634; 803/734-0235	YES (e)	•	•	•	NO(d)	3,000	3,000	15 for trailer			• (d)
South Dakota	South Dakota Tourism 800/SDAKOTA; 605/773-3301	Oct. 1–Apr. 30	•	•	•	Designated areas only	3,000	3,000	80	•	• (d)	• (d)
Tennessee	Tennessee Tourist Dev. 615/741-2158	Oct. 15–Apr. 15	•	•		NO	3,000(d)		65	•		
Texas	Texas Dept. of Commerce 800/8888 TEX; 800/452-9292; 512/462-9191	Rubber studs permitted	•	•	•	YES	4,500	3,000	65	•	• (d)	• (d)
Utah	Utah Travel Council 801/538-1030	Oct. 15–Mar. 31	•	•	•	Designated areas only	750		65	•		
Vermont	Vermont Dept. of Travel & Tourism 802/828-3236 (in state)	YES (d)	•	•	•	NO	3,000(a)	3,000	48			
Virginia	Virginia Division of Tourism 800/932-5827; 804/786-4484	Oct. 15–Apr. 15	•	•		NO	3,000(a)	(b)	60	•		(d)
Washington	Washington Tourism Development Div. 800/544-1800; 206/586-2102	Nov. 1–Apr. 1	•	•	•	NO	3,000	3,000	48 for trailer		•	•
West Virginia	West Virginia Div. of Tourism & Parks 800/CALL-WVA; 304/558-2286	Nov. 1–Apr. 15	•	•	•	Not on Interstate	(b)	3,000	55	•	•	•
Wisconsin	Wisconsin Division of Tourism 800/432-TRIP; 608/266-2161	Prohibited	•	•		NO	3,000	3,000	65	•		
Wyoming	Wyoming Div. of Tourism 800/CALL-WYO; 307/777-7777	YES, with no restrictions	•	•	•	Designated areas only	(b,c)		85			• (d)

(a) Or, when trailer exceeds 40% of weight of towing vehicle   (b) Equipment required   (c) Must be able to stop within legal distance
(d) Possible exceptions. Refer to state laws   (e) Studded tires permitted if studs don't project more than 1/16″ when compressed

Canada Travel Information

Canadian Citizens Visiting the United States

Canadian nationals, and aliens having a common nationality with nationals of Canada, are not required to present passports or visas to visit for a period of six months or less, except when arriving from a visit outside the Western Hemisphere. However, such persons should carry evidence of their citizenship. Visitors entering for a period of more than six months and less than one year are required to furnish valid passports.

This article does not attempt to cover all regulations or requirements. Obtain further information from the nearest United States or Canadian immigration office in the U.S., Canada, or a port of entry.

United States Citizens Visiting Canada

Passports are not required of native-born United States citizens to enter Canada. They should carry identifying papers such as a passport, certified birth certificate plus a photo ID, or a voter's registration card plus a photo ID with the same address to show proof of their U.S. citizenship at the port of entry. Naturalized citizens must carry their naturalization certificate. Aliens who reside in the U.S. must have their Alien Registration Receipt Card.

Automobiles will be admitted for touring in Canada without payment of any duty or fee for any period up to twelve months. Vehicle Registration Cards and evidence of valid insurance should be carried. Any necessary permits are issued at any port of entry.

Returning motorists must report for inspection. Each U.S. resident, after 48 hours, may bring back articles for personal use valued at up to $400.00 free of duty, provided same has not been claimed in the preceding 30 days.

Hunting in Canada

Hunting is controlled by federal, provincial, and territorial laws. Hunting licenses obtained from each province or territory in which they plan to hunt are required of all non-residents. An additional permit, the federal migratory game bird hunting permit, is also required for those planning to hunt migratory game birds. Obtain the permit from any post office. Information:

Environment Canada
Canadian Wildlife Service
Ottawa, Ontario, Canada K1A 0H3
819/997-2957

Many of Canada's provincial parks and reserves forbid the entry of any type of weapon. Each province can provide you with regulations. Export permits are required to take out all unprocessed wildlife from the Northwest Territories.

Some species such as black bear require a CITES export permit.

The importation of firearms into Canada is strictly controlled. You may bring a hunting rifle or shotgun and up to 200 rounds of ammunition into Canada for sporting or competition use, but you must be at least 18 years of age. No hand guns or automatic weapons are allowed. For specific rules and further information:

Revenue Canada
Inspection & Control Division
Connaught Building, Mackenzie Ave.
Ottawa, Ontario, Canada K1A 0L5
613/954-7129

Fishing in Canada

Fishing is likewise controlled by federal, provincial, and territorial laws. Fishing licenses obtained from each province or territory in which they plan to fish are required of all non-residents. British Columbia also requires tidal-waters sports fishing licenses.

Fishing in national wildlife areas is controlled by Environment Canada (see address above). Special fishing permits are required to fish in all national parks. These permits are available at any national park site for a small fee. For information on rules in national parks write to:

Parks Canada
Depatment of Canadian Heritage
Ottawa, Ontario, Canada K1A 0M5

Hunting & Fishing in the United States

Hunting and fishing are controlled by the respective states. Information regarding licenses, seasons, fees, and specific rules may be obtained by contacting the individual states in which you plan to hunt or fish. (See Fishing/Hunting chart for addresses of state agencies.)

Firearms and Ammunition

Firearms and ammunition are subject to restrictions and import permits approved by the Bureau of Alcohol, Tobacco and Firearms (ATF). In order to import these items, applications to import may be made only by or through a licensed importer, dealer, or manufacturer. Items prohibited by the National Firearms Act will not be admitted unless specifically authorized.

If a person is returning with firearms or ammunition that was previously taken out of the U.S. by that person, no import permit is required upon presentation of proof of such action. To facilitate reentry, have the items registered, before departing from the United States, at any Customs office or ATF field office. No more than three nonautomatic firearms and 1,000 cartridges will be registered for any one person.

For further information, contact:
Alcohol, Tobacco & Firearms
Department of the Treasury
Washington, D.C. 20226

CANADIAN TOURISM OFFICES

If you wish to obtain information about particular areas, places of interest, activities, or events, contact the provincial or territorial offices directly.

Travel **Alberta**
10155-102 Street, 3rd Floor
Edmonton, AB T5J 4L6
800/661-8888 (N. Am. exc. AK)
403/427-4321

Tourism **British Columbia**
Parliament Buildings
Victoria, BC V8V 1X4
800/663-6000 (N. Am.)
604/663-6000

Travel **Manitoba**
7-155 Carlton St., Dept. RI5
Winnipeg, MB R3C 2H8
800/665-0040 ext. RI5 (N. Am. exc. AK, HI)
204/945-3777 ext. RI5

New Brunswick Tourism
P.O. Box 12345
Fredericton, NB E3B 5C3
800/561-0123 (N. Am.)

Newfoundland and **Labrador** Tourism
& Culture
P.O. Box 8730
St. John's, NF A1B 4K2
800/563-6353 (N. Am.)
709/729-2830

Northwest Territories Dept. of Economic
Development & Tourism
Box 1320
Yellow Knife, NT X1A 2L9
800/661-0788 (N. Am.)
403/873-7200

Nova Scotia Economic Development & Tourism
Box 130
Halifax, NS B3J 2M7
800/565-0000 (N. Am. exc. AK, HI)
902/424-4247

Ontario Travel
Queen's Park
Toronto, ON M7A 2E5
800/ONTARIO (N. Am. exc. AK)
416/314-0944

Prince Edward Island Visitor Services
P.O. Box 940
Charlottetown, PE C1A 7M5
800/463-4PEI (4734) (N. Am. exc. AK, HI)
902/368-4444

Tourisme **Québec**
P.O. Box 979
Montréal, PQ H3C 2W3
800/363-7777 (N. Am. exc. AK)
514/873-2015

Tourism **Saskatchewan**
1919 Saskatchewan Dr.
Regina, SK S4P 3V7
800/667-7191 (N. Am.)

Tourism **Yukon**
Box 2703
Whitehorse, YT Y1A 2C6
403/667-5340

General information about Canada is available from Canadian Consulates in major U.S. cities.

Road Condition "Hot Lines"

United States

Call the following numbers for road conditions and
road construction.

Alabama
conditions: (205) 242-4378 24 hours

Alaska
conditions: (907) 243-7675, #2 (recording)
construction: (907) 243-7675, #8 (recording)

Arizona
conditions: (602) 252-1010, ext. 7623 (recording)
construction: (602) 255-7386 weekdays;
   (602) 779-2711 (Northern AZ)

Arkansas
conditions: (501) 569-2374 (recording)
construction: (501) 569-2227 weekdays

California
conditions: (916) 445-7623

Colorado
conditions: (303) 639-1234 (recording)
construction: (303) 757-9228 weekdays;
   (303) 573-ROAD (recording)

Connecticut
conditions & construction: (800) 443-6817 (CT only)
conditions: (203) 594-2650 weekdays

Delaware
conditions: (800) 652-5600 (CT only)
construction: (302) 739-6677 (recording)

District of Columbia
conditions: (202) 936-1111 (recording)
construction: (202) 767-8539 weekdays

Florida
no central source

Georgia
conditions: (404) 656-5267 weekdays

Hawaii
construction: (808) 536-6566 (recording)

Idaho
conditions: (208) 336-6600 (recording)
construction: (208) 334-8888 (recording)

Illinois
conditions: (312) 368-4636 (recording);
   (217) 782-5730 (recording in winter)
construction: (800) 452-4368 (May–Oct.)

Indiana
conditions–north: (317) 232-8300 (recording)
conditions–south: (317) 232-8298 (recording)
construction: (317) 232-5115 weekdays

Iowa
conditions: (515) 288-1047 (recording)
construction: (515) 239-1471 weekdays

Kansas
conditions & construction: (913) 296-3102
   24 hours

Kentucky
conditions: (502) 564-4556 weekdays
construction: (502) 564-4780 weekdays

Louisiana
conditions: (504) 379-1541 weekdays

Maine
conditions Nov.–Apr. (24 hrs); May–Oct.
   weekdays: (207) 287-3427
construction: (207) 287-2672 weekdays

Maryland
conditions: (800) 543-2515 (MD only);
   (401) 333-1215
construction: (800) 323-6742 (MD only)

Massachusetts
conditions: (617) 374-1234 weekdays

Michigan
no central source

Minnesota
conditions: (800) 542-0220 (recording);
   (612) 296-3076 (recording)

Mississippi
conditions: (601) 987-1212 24 hours

Missouri
no central source

Montana
conditions: (800) 332-6171 (recording);
   (406) 444-6339 (recording)

Nebraska
conditions & construction: (402) 479-4512
   weekdays
conditions (winter): (402) 471-4533 (recording);
   (800) 906-9069 (NE only)

Nevada
conditions & construction recordings for:
South–Las Vegas: (702) 486-3116
Northwest–Reno: (702) 793-1313
Northeast–Elko: (702) 738-8888

New Hampshire
conditions: (603) 485-9526 24 hours

New Jersey
conditions & construction for the Turnpike:
   (908) 247-0900 24 hours
conditions & construction for the Garden State
   Parkway: (908) 727-5929 (recording)

New Mexico
conditions: (505) 827-5118 weekdays
   (then recording)
construction: (800) 432-4269 (NM only) weekdays;
   (505) 827-5118 weekdays

New York
conditions for New York Thruway: (800) 847-8929

North Carolina
conditions: (919) 549-5100, category 7623
   (recording)

North Dakota
conditions: (800) 472-2686 (ND only) (recording);
   (701) 224-2898 (recording)
construction: (701) 224-4418 weekdays

Ohio
conditions: (614) 466-7170 weekdays

Oklahoma
conditions: (405) 425-2385
construction: (405) 521-2554 weekdays

Oregon
conditions: (800) 976-7277 (OR only);
   (503) 889-3999 (recording)

Pennsylvania
conditions & construction for the Turnpike:
   (800) 331-3414 (PA only); (717) 939-9551,
   ext. 5550 weekdays
construction: (717) 939-9551 ext. 5550 weekdays
conditions for the Interstate (814) 355-7545
   weekdays

Rhode Island
conditions: (401) 738-1211 (recording)
construction: (401) 277-2468 weekdays

South Carolina
conditions & construction: (803) 737-1030
   24 hours

South Dakota
conditions & construction: (605) 773-3536
   24 hours

Tennessee
no central source

Texas
conditions: (800) 452-9292 weekdays

Utah
conditions & construction: (801) 964-6000
   (recording)

Vermont
conditions: (802) 828-2468 weekdays

Virginia
conditions & construction: (800) 367-ROAD
   (VA only)
conditions (winter): (804) 786-3181 weekdays

Washington
conditions: Mountain Pass Report (Nov.–Apr.)
   (206) 434-PASS
construction: (206) 705-7075 weekdays

West Virginia
conditions: (304) 558-2889 (recording);

Wisconsin
conditions (winter) & construction (summer):
   (800) 762-3947 (recording);

Wyoming
conditions: (307) 635-9966 (recording)
construction: (307) 777-4437 weekdays

Canada

Alberta
conditions: (403) 246-5853 (recording)

British Columbia
conditions: (604) 525-4997 (recording)

Manitoba
conditions: (204) 945-3704 (recording)

New Brunswick
conditions: (800) 561-4063; (800) 661-2527
   (recording)

Newfoundland
conditions: (weekdays; 24-hour operation
   in winter)
Clarenville: (709) 466-7953
Deer Lake: (709) 635-2162
Grand Falls: (709) 292-4300
St. John's: (709) 729-2391

Northwest Territories
Hwy 1–7 conditions: (403) 874-2208 (recording)
Hwy 8 conditions: (403) 979-2678

Nova Scotia
conditions: (902) 424-3933 (recording)

Ontario:
conditions: (800) ONTARIO Canada and U.S.
   except AK (recording)

Prince Edward Island
conditions: (902) 368-4770 weekdays

Québec
conditions: (514) 873-4121 (recording)

Saskatchewan
conditions: (306) 787-7623

Yukon
conditions: (403) 667-8215 (recording)

Keys:
weekdays = normal business hours
24 hours = a person answers 24 hours/7 days
   per week
recording = available at all times unless noted

Intercity Toll Road Information

State	Road Name	Location	Miles	Auto Toll	Auto/2-Axle Trailer
Delaware	Delaware Route 1	Dover to Smyrna	18.0	$1.00	$3.00
	Kennedy Memorial Highway	Md. State Line to N.J. State Line	16.8	1.25	2.50
Florida	Airport Expressway (Miami)	NW 42nd Av. to I-95	3.6	.25	.75
	Bee Line Main	Orlando Airport Plaza to FL 520	20	1.00	2.50
	Bee Line East	FL 520 to Cape Canaveral	22	.20	.40
	Bee Line West	I-4 to Co 436 (Semoran Blvd.)	11	.50	1.00
	Central Florida GreenWay	International Dr. to US 17/92	46	4.50	10.50
	Crosstown Expressway (Tampa)	Gandy Blvd. (US 92) to I-75	13.9	1.25	2.50
	Dolphin Expressway	Le Jeune Road to I-95	4.1	.25	.75
	Don Shula Expressway	Homestead Ext. (Florida's Tpk.) to Killian Pkwy.	2.1	.25	.75
	East-West Expressway (Orlando)	Colonial Dr. (east) to Colonial Dr. (west)	22	2.50	6.00
	Everglades Parkway (Alligator Alley)	Naples to Andytown	78	1.50	4.00
	Florida's Turnpike	I-75 to Miami	260	12.50	26.00
	Florida's Tpk. (Homestead Ext.)	Miramar to Florida City	48	2.00	6.00
	Gratigny Parkway	Palmetto Expwy. to NW 27th Av.	5.2	.25	.75
	Sawgrass Expressway	I-75 to I-95	23	1.50	3.10
	Veterans Expressway	Dale Mabry Hwy. to Courtney Campbell Causeway	4.7	1.25	3.50
Illinois	Chicago Skyway	I-94, Chicago, to Ind. State Line	7.3	2.00	3.50
	East-West Tollway	I-88, Chicago, to Rock Falls	97	2.70	5.40
	North-South Tollway	I-290, Addison, to I-55, Bolingbrook	17.5	1.00	2.00
	Northwest Tollway	Des Plaines to South Beloit	76	2.00	4.00
	Tri-State Tollway	Ind. State Line to Wis. State Line	83	2.40	4.80
Indiana	Indiana Toll Road	Ohio State Line to Ill. State Line	157	4.65	5.35
Kansas	Kansas Turnpike	Kansas City to Okla. State Line	236	7.00	9.50
Kentucky	Audubon Parkway	Pennyrile Parkway to Owensboro	24.6	.50	1.10
	Cumberland Parkway	Bowling Green to Somerset	88.5	2.00	4.00
	Daniel Boone Parkway	London to Hazard	59.1	1.40	2.80
	Green River Parkway	Owensboro to Bowling Green	70.7	1.50	3.00
Maine	Maine Turnpike	York to Augusta	100	3.25	4.75
Massachusetts	Massachusetts Turnpike	Boston to N.Y. State Line	135	5.60	6.70
New Hampshire	F.E. Everett Turnpike	Nashua to Concord	44.7	1.50	2.50
	New Hampshire Turnpike	Portsmouth to Seabrook	16.1	1.00	1.50
	Spaulding Turnpike	Portsmouth to Milton, N.H.	33.2	1.00	2.00
New Jersey	Atlantic City Expressway	Turnersville to Atlantic City	44	1.25	3.75
	Garden State Parkway	Montvale to Cape May	173	4.20	8.40
	New Jersey Turnpike	Del. Mem. Br. to Washington Br.	142	4.60	15.60
New York	New York Thruway –				
	Eastbound	Pa. State Line to N.Y.C.	496	18.10	33.30
	Westbound	N.Y.C. to Pa. State Line	496	15.60	28.80
	Berkshire Section	Selkirk to Mass. Turnpike	24	1.20	2.15
	New England Section	N.Y.C. to Conn. State Line	15	1.00	2.00
	Niagara Section	Buffalo to Niagara Falls	21	1.50	3.00
Ohio	J. W. Shockness y Ohio Tpk.	Pa. State Line to Ind. State Line	241	4.90	7.50
Oklahoma	Cherokee Turnpike	US 412 to US 59	32.8	1.75	3.50
	Chickasaw Turnpike	Ada to Sulphur	27.1	.50	1.00
	Cimarron Turnpike	I-35 to Tulsa	59.2	2.00	4.00
	Creek Turnpike	US 75 to US 64	7.6	.50	1.00
	H.E. Bailey Turnpike	Oklahoma City to Texas State Line	86.4	3.25	6.25
	Indian Nation Turnpike	Henryetta to Hugo	105.2	3.75	6.75
	Kilpatrick Turnpike	Oklahoma City, OK 74 to I-35/44	9.5	.75	1.50
	Muskogee Turnpike	Tulsa to Webber Falls	53.1	2.00	3.75
	Turner Turnpike	Oklahoma City to Tulsa	86	2.75	5.75
	Will Rogers Turnpike	Tulsa to Mo. State Line	88.5	2.75	5.75
Pennsylvania	Amos K. Hutchinson Bypass (PA Tpk. 66)	US 119 to US 22	13	1.00	2.00
	James E. Ross Hwy. (PA Tpk. 60)	Beaver Falls to New Castle (Northbound)	17	1.00	1.45
		New Castle to Beaver Falls (Southbound)	17	.50	.75
	James J. Manderino Hwy. (PA Tpk. 43)	I-70 to US 40	5	.50	1.00
	Pennsylvania Turnpike	N.J. State Line to Ohio State Line	359	14.70	21.70
	Pa. Turnpike (N.E. Sect.)	Norristown to Scranton	110	3.60	5.40
Texas	Dallas North Tollway	I-35E, Dallas, to FM 544, Dallas	17.2	1.00	2.00
	Hardy Toll Road	I-45, Houston, to I-610, Houston	21.6	2.00	6.00
	Sam Houston Tollway	US 59, Houston, to I-45	28	2.50	6.00
Virginia	Chesapeake Bay Bridge & Tunnel	US 13, Norfolk/VA Beach to Eastern Shore	17	10.00	16.00
	Downtown Expressway (Richmond)	I-195 to I-95	3.4	.35	.55
	Dulles Toll Road	VA 123 to VA 28	13	.85	1.70
	Powhite Parkway	I-195 to VA 150	2.1	.35	.55
	Powhite Parkway Extension	VA 150 to Old Hundred Rd.	10.4	.75	1.25
	Virginia Bch. – Norfolk Expy.	US 60, Virginia Bch., to I-64, Norfolk	12.1	.25	.40
West Virginia	West Virginia Turnpike	Charleston to Princeton	88	3.75	6.00

Car Rental Companies
Toll-free Numbers

The following selected list of car rental companies provides toll-free "800" numbers for making reservations in Canada and the United States. Although these numbers were in effect at press time, the Guide cannot be responsible should any of these numbers change. Many companies do not have toll-free reservation numbers; consult your local telephone directory for a regional listing.

Toll-free Numbers

AAPEX Courtesy Car Rental
Lighthouse Point, FL 33064
3400 N. Federal Hwy.
(800) 327-9106 Cont'l USA, except FL, and Canada

Agency Rent-A-Car
Solon, OH 44139
Corporate Office
30000 Aurora Rd.
(800) 321-1972 Cont'l USA and Canada

Alamo Rent-A-Car
Ft. Lauderdale, FL 33335
Corporate Office
P.O. Box 22776
(800) 327-9633 USA and Canada

Allstate Car Rental
Las Vegas, NV 89119
McCarran Int'l Airport
5175 Rent-A-Car Rd.
(800) 634-6186 USA and Canada

Altra Auto Rental
Solon, OH 44139
Corporate Office
30000 Aurora Rd.
(800) 232-9555 Cont'l USA (except OH) and Canada
(800) 321-1972 OH

Avis-Reservations Center
Garden City, NY 11530-9795
900 Old Country Rd.
(800) 331-1212 Domestic Reservations
(800) 331-1084 Int'l Reservations, including Canada

Avon Rent-A-Car
Los Angeles, CA 90045
9220 S. Sepulveda Blvd.
(800) 421-6808 USA and Canada

Aztec Car Rental
San Diego, CA 92101
2401 Pacific Hwy.
(800) 231-0400 USA and Canada

Brooks Car Rental
Las Vegas, NV 89109
3041 Las Vegas Blvd. South
(800) 634-6721 USA

Budget Rent-A-Car
Lisle, IL 60532
4225 Naperville Rd.
(800) 527-0700 Cont'l USA, HI and Canada
(800) 472-3352 Int'l Reservations

Dollar Systems
El Segundo, CA 90045-5359
100 N. Sepulveda Blvd., 6th Floor
(800) 800-4000 USA and Canada

Enterprise Rent-A-Car
St. Louis, MO 63124
8850 Ladue Rd.
(800) 325-8007 Cont'l USA and Canada

Freedom Rent-A-Car
Las Vegas, NV 89109
3763 Las Vegas Blvd. South
(800) 331-0777 Cont'l USA and Canada

Hertz Corporation
Oklahoma City, OK 73120
Worldwide Reservation Center
10401 N. Pennsylvania
(800) 654-3131 USA
(800) 654-3001 International

Interamerican Car Rental
Miami, FL 33126
1790 NW Le Jeune Rd.
(800) 327-1278 Cont'l USA

National Car Rental
Minneapolis, MN 55435
7700 France Ave. South
(800) CAR-RENT USA and Canada
(800) CAR-EURO International

Payless Car Rental Int'l Inc.
St. Petersburg, FL 33784-0669
P.O. Box 60669
(800) PAYLESS USA and Canada

Practical Rent-A-Car
Las Vegas, NV 89109
3763 Las Vegas Blvd. South
(800) 233-1663 Cont'l USA and Canada

Sears Rent-A-Car
Lisle, IL 60532
4225 Naperville Rd.,
(800) 527-0770 USA and Int'l Reservations

Thrifty Rent-A-Car
Tulsa, OK 74153-0250
P.O. Box 35250
(800) 367-2277 USA and Canada

U-SAVE Auto Rental, Inc.
Hanover, MD 21076
7525 Connelley Dr., Suite A
(800) 272-U-SAV USA and Canada

Value Rent-A-Car
Boca-Raton, FL 33431
P.O. Box 5040
(800) GO-VALUE Cont'l USA and Canada

Limousine Services

The following limousine service companies provide toll-free "800" numbers for making reservations throughout the United States.

Carey International
Washington, DC 20016
4530 Wisconsin Ave., NW, 5th Floor
(800) 336-4646 USA (except DC)
(800) 336-4747 Canada

Dav-El Limousine
New York, NY 10011
West 23rd St. & 12th Ave.
(800) 922-0343 USA and Canada

Hotel/Motel Toll-free Numbers

This selected list is a handy guide for hotel/motel toll-free reservation numbers. You can save time and money by using these toll-free numbers for continental USA and Canada. Although these "800" numbers were in effect at press time, the Atlas cannot be responsible should any of these numbers change. Many establishments do not have toll-free reservation numbers. Consult your local phone directory for a regional listing.

Adam's Mark Hotels
(800) 444-ADAM Cont'l USA and Canada

Best Western International, Inc.
(800) 528-1234 USA and Canada

Budgetel Inns
(800) 4 BUDGET USA and Canada

Budget Host
(800) BUD-HOST

Canadian Pacific Hotels
(800) 828-7447 USA
(800) 268-9411 Canada

Clarion Hotels
(800) CLARION USA

Colony Hotels & Resorts
(800) 777-1700 USA and Canada

Comfort Inns
(800) 228-5150 USA and Canada

Country Hearth Inn
(800) 848-5767 Cont'l USA

Courtyard by Marriott
(800) 321-2211 USA and Canada

Days Inn
(800) 325-2525 USA and Canada

Delta Inns & Resorts
(800) 877-1133 USA
(800) 268-1133 Canada

Doubletree Hotels
(800) 222-TREE Cont'l USA, AK, and Canada

Downtowner (See Hospitality International entry)

Drury Inn
(800) 325-8300 Cont'l USA, HI and Canada

Econo Lodges
(800) 55-ECONO USA and Canada

Embassy Suites
(800) EMBASSY USA
(800) 458-5848 Canada

Exel Inns of America
(800) 356-8013 Cont'l USA and Canada

Fairfield Inns by Marriott
(800) 228-2800 Cont'l USA and Canada

Fairmont Hotels
(800) 527-4727 USA and Canada

Forte Hotels
(800) 225-5843 USA and Canada

Four Seasons Hotels & Resorts
(800) 332-3442 USA
(800) 268-6282 Canada

Friendship Inns
(800) 453-4511 Cont'l USA and Canada

Guest Quarters Suite Hotel
(800) 424-2900 USA and Canada

Hampton Inns
(800) HAMPTON Cont'l USA and Canada

Harley Hotels
(800) 321-2323 Cont'l USA and Canada

Helmsley Hotels
(800) 221-4982 USA and Canada

Hilton Hotels
(800) HILTONS USA and Canada

Holiday Inns, Inc.
(800) HOLIDAY USA and Canada

Homewood Suites
(800) CALL-HOM(E) USA and Canada

Hospitality International/Downtowner/ Master Hosts/Passport Motor Inns/Red Carpet/Scottish Inns
(800) 251-1962 Cont'l USA and Canada

Howard Johnson Lodges
(800) 654-2000 USA and Canada

Hyatt Hotels Corp.
(800) 233-1234 USA (except NE) and Canada
(800) 228-3366 NE

Inter-Continental Hotels
(800) 327-0200 USA and Canada

Journey's End Hotels
(800) 668-4200 USA and Canada

Knights Inn
(800) 843-5644 Cont'l USA and Canada

LK Motels
(800) 282-5711 Cont'l USA and Canada

La Quinta Motor Inns, Inc.
(800) 531-5900 Cont'l USA and Canada

Loews Hotels
(800) 23-LOEWS USA and Canada

Luxbury Hotels
(800) CLASS-4-U USA and Canada

Marriott Hotels & Resorts
(800) 228-9290 Cont'l USA, HI, and Canada

Master Hosts (See Hospitality International entry)

Meany Tower Hotels
(800) 648-6440 Cont'l USA

Meridien Hotels
(800) 543-4300 Cont'l USA and Canada

Omni Hotels
(800) THE-OMNI USA and Canada

Passport Motor Inns
(See Hospitality International entry)

Preferred Hotels
(800) 323-7500 USA and Canada

Quality Inns
(800) 228-5151 USA and Canada

Radisson Hotels International
(800) 333-3333 USA and Canada

Ramada Inns, Inc.
(800) 228-2828 USA and Canada

Red Carpet (See Hospitality International entry)

Red Lion Inns
(800) 547-8010 USA and Canada

Red Roof Inns
(800) THE ROOF USA and Canada

Regent International Hotels
(800) 545-4000 USA and Canada

Renaissance Hotels
(800) 228-9898 USA and Canada

Residence Inn by Marriott
(800) 331-3131 USA and Canada

The Ritz-Carlton
(800) 241-3333 USA and Canada

Rodeway Inns International
(800) 228-2000 USA and Canada

Scottish Inns
(See Hospitality International entry)

Sheraton Hotels & Motor Inns
(800) 325-3535 USA and Canada

Shoney's Inn
(800) 222-2222 USA
(800) 233-4667 Canada

Signature Inn
(800) 822-5252 USA and Canada

Sonesta Hotels
(800) SONESTA USA and Canada

Stouffer Hotels & Resorts
(800) 468-3571 USA and Canada

Super 8 Motels, Inc.
(800) 800-8000 USA and Canada

Susse Chalet
(800) 5-CHALET USA and Canada

Thriftlodge Hotels
(800) 525-9055 USA and Canada

Travelodge Hotels
(800) 578-7878 USA and Canada

Vagabond Inns, Inc.
(800) 522-1555 USA and Canada

WestCoast Hotels, Inc.
(800) 426-0670 USA and Canada

Westin Hotels & Resorts
(800) 228-3000 USA and Canada

Wyndham Hotels & Resorts
(800) 822-4200 USA
(800) 631-4200 Canada

Clear Channel Radio Stations

Stations listed are American and Canadian AM unlimited time stations designated to operate with 50 kilowatt power and to render service over an extended area. A clear channel station's signals are generally protected for a distance of up to 750 miles at night.

United States

Alaska
650, KYAK, Anchorage
750, KFQD, Anchorage
820, KCBF, Fairbanks

Arizona
1580, KCWW, Tempe
660, KTNN, Window Rock

Arkansas
1090, KAAY, Little Rock

California
940, KFRE, Fresno
640, KFI, Los Angeles
1020, KTNQ, Los Angeles
1070, KNX, Los Angeles
1140, KRAK, Sacramento
1530, KFBK, Sacramento
760, KFMB, San Diego
680, KNBR, San Francisco
740, KCBS, San Francisco
810, KGO, San Francisco
1100, KFAX, San Francisco
1580, KBLA, Santa Monica

Colorado
850, KOA, Denver

Connecticut
1080, WTIC, Hartford

District of Columbia
1500, WTOP, Washington

Florida
710, WAQI, Miami
740, WWNZ, Orlando
540, WTGO, Pine Hills

Georgia
750, WSB, Atlanta

Hawaii
870, KAIM, Honolulu

Idaho
670, KBOI, Boise

Illinois
670, WMAQ, Chicago
720, WGN, Chicago
780, WBBM, Chicago
890, WLS, Chicago
1000, WMVP, Chicago

Indiana
1190, WOWO, Fort Wayne

Iowa
1040, WHO, Des Moines
1540, KXEL, Waterloo

Kentucky
840, WHAS, Louisville

Louisiana
870, WWL, New Orleans
1130, KWKH, Shreveport

Maryland
1090, WBAL, Baltimore

Massachusetts
680, WRKO, Boston
850, WHDH, Boston
1030, WBZ, Boston
1510, WSSU, Boston

Michigan
760, WJR, Detroit

Minnesota
830, WCCO, Minneapolis
1500, KSTP, St. Paul

Missouri
1120, KMOX, St. Louis

Nebraska
880, KRVN, Lexington
1110, KFAB, Omaha

Nevada
720, KDWN, Las Vegas
780, KROW, Reno

New Mexico
770, KKOB, Albuquerque
1020, KCKN, Roswell

New York
1540, WPTR, Albany
1520, WWKB, Buffalo
660, WFAN, New York
710, WOR, New York
770, WABC, New York
880, WCBS, New York
1010, WINS, New York
1050, WEVD, New York
1130, WBBR, New York
1560, WQEW, New York
1180, WHAM, Rochester
810, WGY, Schenectady

North Carolina
1110, WBT, Charlotte
680, WPTF, Raleigh

Ohio
700, WLW, Cincinnati
1530, WCKY, Cincinnati
1100, WWWE, Cleveland
1220, WKNR, Cleveland

Oklahoma
1520, KOMA, Oklahoma City
1170, KVOO, Tulsa

Oregon
1120, KPNW, Eugene
1190, KEX, Portland

Pennsylvania
1060, KYW, Philadelphia
1210, WOGL, Philadelphia
1020, KDKA, Pittsburgh

Tennessee
650, WSM, Nashville
1510, WLAC, Nashville

Texas
1080, KRLD, Dallas
820, WBAP, Fort Worth
740, KTRH, Houston
1200, WOAI, San Antonio

Utah
1160, KSL, Salt Lake City

Virginia
1140, WRVA, Richmond

Washington
710, KIRO, Seattle
1000, KOMO, Seattle
1090, KING, Seattle
1510, KGA, Spokane

West Virginia
1170, WWVA, Wheeling

Wyoming
1030, KTWO, Casper

Canada

Alberta
1010, CBR, Calgary

British Columbia
1130, CKWK, Vancouver

Manitoba
990, CBW, Winnipeg

New Brunswick
1070, CBA, Moncton

Newfoundland
640, CBN, St. John's

Ontario
860, CJBC, Toronto
1550, CBE, Windsor

Québec
1580, CBJ, Chicoutimi
690, CBF, Montréal
940, CBM, Montréal

Saskatchewan
540, CBK, Regina

Climate USA

State	City	Winter (Dec–Feb)				Spring (Mar–May)				Summer (June–Aug)				Fall (Sept–Nov)			
		Maximum Normal Daily Temp. (F)	Minimum Normal Daily Temp. (F)	Total Precipitation (In)	Total Days with Precipitation	Maximum Normal Daily Temp. (F)	Minimum Normal Daily Temp. (F)	Total Precipitation (In)	Total Days with Precipitation	Maximum Normal Daily Temp. (F)	Minimum Normal Daily Temp. (F)	Total Precipitation (In)	Total Days with Precipitation	Maximum Normal Daily Temp. (F)	Minimum Normal Daily Temp. (F)	Total Precipitation (In)	Total Days with Precipitation
Alabama	Birmingham	56	38	16	32	75	53	15	35	90	70	13	29	76	54	10	23
	Mobile	63	43	15	32	77	57	17	26	90	72	22	42	78	58	12	24
Alaska	Juneau	31	20	12	56	46	31	10	52	63	46	12	50	47	36	19	63
Arizona	Phoenix	67	39	2	12	84	52	1	6	103	74	2	10	87	57	2	8
	Tucson	65	39	2	12	81	50	1	7	97	71	5	21	83	56	2	11
Arkansas	Little Rock	52	31	13	28	71	49	15	30	90	68	10	23	74	49	10	21
California	Los Angeles	67	48	8	16	71	53	3	11	81	62	.07	2	78	58	2	6
	Sacramento	55	39	10	28	72	46	4	17	91	57	.2	2	76	50	3	11
	San Diego	65	47	5	19	68	54	2	14	74	63	.13	1	73	58	2	8
	San Francisco	57	47	12	31	61	49	5	19	64	53	.2	3	67	54	4	13
Colorado	Denver	45	18	2	17	60	34	6	29	84	56	5	27	66	37	3	16
Connecticut	Hartford	35	18	10	33	58	36	11	33	82	59	11	31	63	41	11	29
Delaware	Wilmington	42	25	9	30	62	42	10	34	84	64	11	28	67	47	9	25
District of Columbia	Washington	45	29	8	28	66	45	10	32	86	67	12	28	69	50	9	23
Florida	Jacksonville	66	45	9	24	79	57	10	24	89	71	21	41	79	61	14	27
	Miami	76	59	6	19	82	67	12	23	89	75	23	48	84	70	20	40
	Tampa	71	51	7	19	82	61	8	19	90	73	23	45	83	65	11	25
Georgia	Atlanta	53	34	13	32	70	50	14	30	86	68	12	31	72	52	9	21
Hawaii	Honolulu	81	66	10	30	83	69	6	28	87	73	2	21	86	71	6	25
Idaho	Boise	40	24	4	34	61	37	3	25	85	55	1	11	64	39	2	20
Illinois	Chicago	33	17	5	32	58	37	9	37	81	59	10	29	63	41	7	29
	Peoria	34	18	5	27	60	40	11	35	84	63	11	27	64	43	8	24
Indiana	Indianapolis	38	22	8	34	61	41	11	37	84	63	11	28	65	44	8	26
	South Bend	33	19	7	43	57	37	10	38	81	60	11	29	62	42	9	31
Iowa	Des Moines	31	15	3	23	58	38	9	31	83	63	11	28	62	42	7	22
Kansas	Wichita	44	24	3	16	67	44	8	26	90	68	12	24	70	47	7	19
Kentucky	Lexington	43	26	11	36	65	44	13	37	85	64	12	31	67	46	8	27
	Louisville	44	26	10	34	65	44	13	36	86	65	11	29	68	46	9	25
Louisiana	New Orleans	64	45	14	29	78	58	14	24	90	72	17	38	79	60	12	22
	Shreveport	59	39	12	27	76	55	14	27	92	72	9	22	78	56	9	21
Maine	Portland	33	13	11	33	52	32	10	35	77	54	8	28	59	38	11	29
Maryland	Baltimore	43	26	9	28	64	42	10	33	85	64	12	28	68	47	9	23
Massachusetts	Boston	37	24	11	35	56	41	11	34	79	62	9	30	62	48	11	29
Michigan	Detroit	34	21	6	38	56	38	8	37	81	61	9	29	62	45	7	29
	Grand Rapids	32	18	6	44	56	35	9	37	81	58	9	29	61	41	9	34
	Sault Ste. Marie	24	8	6	53	47	28	7	35	73	50	10	33	52	36	10	43
Minnesota	Duluth	21	2	3	34	47	27	8	34	73	52	12	35	51	34	7	31
	Minneapolis-St. Paul	24	7	2	25	53	33	7	31	80	59	11	32	57	37	6	25
Mississippi	Jackson	60	37	14	30	77	52	15	29	92	69	11	29	79	52	9	23
Missouri	Kansas City	43	26	4	21	64	45	10	30	89	69	11	26	70	48	8	21
	St. Louis	42	25	6	26	65	45	11	33	87	67	11	26	68	48	8	25
Montana	Billings	35	16	2	22	54	33	5	29	81	55	4	24	59	37	3	19
	Great Falls	33	14	3	24	54	31	5	30	80	51	5	26	58	35	3	20
Nebraska	Lincoln	36	15	3	17	61	39	8	29	86	64	12	25	65	42	6	20
	Omaha	36	16	2	20	62	39	9	30	86	64	13	28	66	42	6	19
Nevada	Las Vegas	58	34	1	8	78	50	1	6	101	72	1	7	80	53	1	6
	Reno	47	21	3	18	64	31	2	15	87	45	1	8	69	31	1	10

*Includes the liquid water equivalent of snowfall–(10" of snow equals approx. 1" of liquid water).

State	City	Winter (Dec–Feb)				Spring (Mar–May)				Summer (June–Aug)				Fall (Sept–Nov)			
		Maximum Normal Daily Temp. (F)	Minimum Normal Daily Temp. (F)	Total Precipitation (In)	Total Days with Precipitation	Maximum Normal Daily Temp. (F)	Minimum Normal Daily Temp. (F)	Total Precipitation (In)	Total Days with Precipitation	Maximum Normal Daily Temp. (F)	Minimum Normal Daily Temp. (F)	Total Precipitation (In)	Total Days with Precipitation	Maximum Normal Daily Temp. (F)	Minimum Normal Daily Temp. (F)	Total Precipitation (In)	Total Days with Precipitation
New Hampshire	Concord	33	12	8	32	56	32	9	34	80	54	9	31	61	37	10	29
New Jersey	Atlantic City	43	24	10	31	61	50	10	32	82	62	12	31	66	46	10	29
	Newark	40	25	9	32	60	42	11	34	83	65	11	29	66	48	10	26
New Mexico	Albuquerque	49	25	1	12	70	41	1	11	90	63	3	22	71	44	2	14
New York	Albany	32	15	7	35	57	35	8	37	81	58	9	31	61	40	8	31
	Buffalo	31	19	8	57	52	36	9	42	77	59	9	31	59	43	10	38
	New York	40	27	9	31	60	43	10	34	83	66	11	30	66	50	10	25
	Rochester	33	18	7	52	55	35	8	40	80	58	8	30	61	42	8	37
	Syracuse	33	18	8	54	52	36	9	44	80	59	10	33	61	43	9	39
North Carolina	Charlotte	53	32	11	30	72	48	11	30	87	67	12	31	72	51	9	21
	Greensboro	50	29	10	30	70	46	10	30	86	65	13	34	71	47	9	23
	Raleigh	52	30	10	29	71	46	10	29	87	65	14	30	72	48	9	23
North Dakota	Bismarck	23	2	1	23	52	29	4	26	81	55	8	29	57	32	3	19
Ohio	Cincinnati	41	26	9	34	64	44	12	37	85	64	11	31	67	47	8	28
	Cleveland	35	22	7	46	57	38	10	43	80	59	10	31	62	44	8	35
	Columbus	38	21	7	38	62	39	11	40	83	60	11	31	65	42	7	28
Oklahoma	Oklahoma City	50	28	4	16	70	48	11	25	91	69	9	22	73	50	7	18
	Tulsa	50	29	5	20	70	48	12	28	91	69	11	23	73	50	9	19
Oregon	Portland	46	34	16	54	60	41	8	43	76	54	3	18	63	44	11	39
Pennsylvania	Harrisburg	39	24	8	32	63	41	10	36	85	63	10	29	66	45	8	24
	Philadelphia	42	26	9	30	63	42	10	33	85	64	12	28	67	47	9	25
	Pittsburgh	37	22	8	47	60	39	11	41	81	59	10	33	63	43	7	32
Rhode Island	Providence	38	22	11	33	56	38	11	34	79	60	9	30	63	44	11	27
South Carolina	Columbia	58	34	10	29	76	51	11	28	91	69	15	33	76	52	9	21
	Greenville	53	34	12	30	71	49	12	32	87	68	12	32	72	51	10	25
South Dakota	Rapid City	37	13	1	20	56	31	6	31	83	56	7	29	62	35	2	16
	Sioux Falls	27	7	2	18	56	34	7	27	83	59	15	29	60	36	9	20
Tennessee	Chattanooga	51	31	16	34	72	47	13	32	89	66	22	33	73	48	10	24
	Knoxville	50	33	14	35	71	48	12	35	87	67	11	31	71	50	9	27
	Memphis	51	33	14	29	71	51	15	30	90	70	10	25	73	51	9	21
	Nashville	49	30	14	33	70	48	13	34	89	67	10	28	72	49	9	24
Texas	Corpus Christi	68	48	5	22	81	62	6	17	93	75	8	17	83	63	9	22
	Dallas	58	36	6	19	75	53	11	24	94	73	7	16	78	56	8	19
	El Paso	59	32	1	11	78	49	1	6	94	68	3	18	77	49	2	12
	Houston	65	43	11	26	79	58	11	25	93	72	13	28	82	58	13	24
	San Antonio	64	42	5	23	80	58	7	23	95	73	7	15	81	59	8	19
Utah	Salt Lake City	40	21	4	28	62	36	5	28	88	57	3	15	65	39	3	18
Vermont	Burlington	28	10	6	41	52	32	7	38	79	56	11	37	58	39	9	37
Virginia	Norfolk	50	33	10	29	67	48	9	31	85	68	15	30	70	53	10	24
	Richmond	49	28	9	28	69	45	9	31	87	65	14	31	71	48	10	23
Washington	Seattle-Tacoma	47	36	14	55	59	42	7	41	73	54	3	20	61	46	10	36
	Spokane	35	23	6	43	56	36	4	29	80	53	2	17	57	38	4	26
West Virginia	Charleston	45	26	10	44	66	43	11	42	84	62	12	35	68	45	8	30
	Parkersburg	43	26	9	40	65	43	11	39	84	63	12	33	67	46	7	29
Wisconsin	Madison	28	11	4	28	54	33	8	33	79	57	11	31	58	38	7	27
	Milwaukee	30	14	4	31	53	34	8	36	78	57	10	30	59	40	7	28
Wyoming	Cheyenne	40	17	1	17	55	30	5	30	80	52	6	32	61	34	2	18

*Includes the liquid water equivalent of snowfall–(10" of snow equals approx. 1" of liquid water).

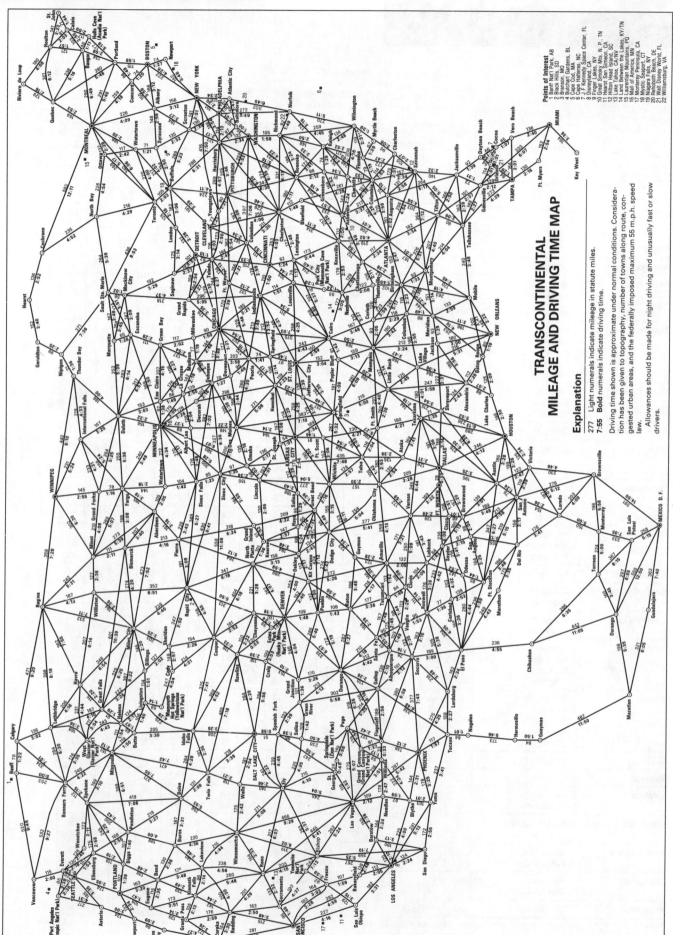

TRANSCONTINENTAL
MILEAGE AND DRIVING TIME MAP

Explanation

277 Light numerals indicate mileage in statute miles.

7:55 **Bold** numerals indicate driving time.

Driving time shown is approximate under normal conditions. Consideration has been given to topography, number of towns along route, congested urban areas, and the federally imposed maximum 55 m.p.h. speed law.

Allowances should be made for night driving and unusually fast or slow drivers.

Points of Interest
1 Banff Nat'l Park, AB
2 Black Hills, SD
3 Branson, MO
4 Butchart Gardens, BL
5 Cape Cod, MA
6 Cape Hatteras, NC
7 J. F. Kennedy Space Center, FL
8 Disneyland, CA
9 Finger Lakes, NY
10 Great Smoky Mts. N. P., TN
11 Hearst San Simeon, CA
12 Hilton Head Island, SC
13 Lake Tahoe, CA/NV
14 Land Between the Lakes, KY/TN
15 Laurentian Mountains, PQ
16 Mall of America, MN
17 Monterey Peninsula, CA
18 Mystic Seaport, CT
19 Niagara Falls, NY
20 Rehoboth Beach, DE
21 Walt Disney World, FL
22 Williamsburg, VA

Major Cities Guide

Use the information in this section to acquaint yourself with 89 major cities, including principal United States cities, Mexico City, and the Canadian cities of Calgary, Edmonton, Montréal, Ottawa, Toronto, Vancouver, and Winnipeg. The following information is provided:

Population Statistics. Populations of cities, figure followed by (1990C), and their metropolitan areas (figure preceded by an asterisk) are the census figures for the year cited. The U.S. metropolitan areas have been defined by the Bureau of the Census and the U.S. Office of Management and Budget to include one or more major cities and their counties. A metropolitan area may also include additional counties that have strong economic and social ties to the central county or counties.

Weather Information. Altitude figures, and average January and July temperatures, provide weather information in brief for each city.

Telephone Numbers / Time Zones. Telephone area codes and time zone information help orient you to each new location. Local time and weather numbers are provided where applicable.

Airport Transportation / Maps. Flyers will find useful information in each city listing under Airport Transportation: the distance from each city's airport(s) to downtown, and the type of transportation service that is available.

In addition, detailed airport maps locating parking facilities, terminals and airlines for 41 of the largest cities are included in this section.

Hotels and Restaurants. A listing of selected hotels/motels and restaurants is provided for each city; establishments chosen for their broad appeal to the traveling public.

Attractions. Use the selected list of important local attractions to plan a visit to attractions of special interest. Addresses and telephone numbers are provided for each attraction listed.

City Maps. Get around each city easily by using the large-scale maps for each city to identify streets, suburbs, freeways, hospitals, airports, points of interest, and more.

Tourist Information Sources. Each city listing also provides information sources for that city. Addresses and/or phone numbers are given for local convention and visitors bureaus, offices of tourism, and chambers of commerce.

Whether you call or write ahead for information, or stop in once you're in town, these offices can provide special information about their city that will enhance your visit and make the most of your sightseeing time.

Contents/Cities and Airports

Major North American Cities

MAJOR CITY DATA Populations of cities, figure followed by (1990C), and their metropolitan areas (figure preceded by an asterisk) are the final census figures for the year cited. A population figure followed by (1991E) is an estimate for the year cited. Metropolitan areas have been defined by the Bureau of the Census and the U.S. Office of Management and Budget to include one or more major cities and their counties. A metropolitan area may also include additional counties that have strong economic and social ties to the central county or counties

Directory entries for hotels located within approximately two miles of the airport are preceded by the ✈ symbol.

All information in the major city guide has been checked for accuracy at the time of publication. Since changes do occur, the publisher cannot be responsible for any variations from the information printed.

Albany, New York

Population: (*874,304) 101,082 (1990C)
Altitude: 150 feet

Average Temp.: Jan., 22°F.; July, 72°F.
Telephone Area Code: 518
Time: (none)   **Weather:** 476-1111
Time Zone: Eastern

AIRPORT TRANSPORTATION:

Nine miles to downtown Albany. Taxicab, bus, and airport limousine service.

SELECTED HOTELS:

✈ Albany Marriott, 189 Wolf Rd., 458-8444; FAX 458-7365
✈ Best Western—Albany Airport, 200 Wolf Rd., 458-1000; FAX 458-2807
✈ Comfort Inn, 1606 Central Ave., 869-5327; FAX 456-8971
✈ Days Inn, 16 Wolf Rd., 459-3600; FAX 459-3677
✈ The Desmond, 660 Albany-Shaker Rd., 869-8100; FAX 869-7659
✈ Holiday Inn Turf on Wolf Rd., 205 Wolf Rd., 458-7250; FAX 458-7377
Omni Hotel, 10 Eyck Plaza, State & Lodge sts., 462-6611; FAX 462-2091
Ramada Inn, 1228 Western Ave., 489-2981; FAX 489-8967
Travelers Motor Inn, 1630 Central Ave., 456-0222; FAX 452-1376

SELECTED RESTAURANTS:

Signature Food: American
Anthony's Park Plaza, 27 Elk, 434-2711
Chaucer's, Rtes. 9 & 146, Clifton Park, 383-3660
Grimaldi's Ristorante, 1553 Central Ave., 869-0634
Jack's Oyster House, 42 State, 465-8854
Jade Fountain, 1652 Western Ave., 869-9585
La Serre, 14 Green St., 463-6056
L'Auberge, 351 Broadway, 465-1111
Ogden's, 42 Howard St., 463-6605
Olde Dater Tavern, 130 Meyer Rd., Clifton Park, 877-7225
Scrimshaw, in The Desmond, 869-8100

SELECTED ATTRACTIONS:

Albany Urban Cultural Park/Henry Hudson Planetarium (visitor center), Broadway & Clinton, 434-5132 or -6311
Crailo State Historic Site (museum of Dutch life in the Upper Hudson Valley), 9½ Riverside Ave., Rensselaer, 463-8738
Empire State Plaza Art Collection (modern American art), Underground Concourse and Outdoor Plaza Level of Empire State Plaza, 473-7521

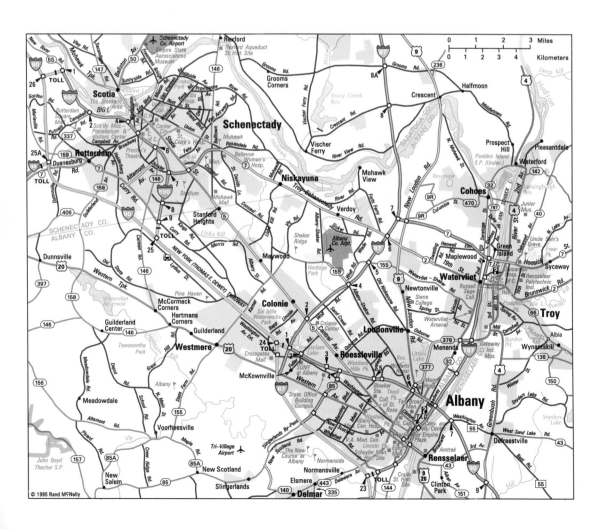

© 1995 Rand McNally

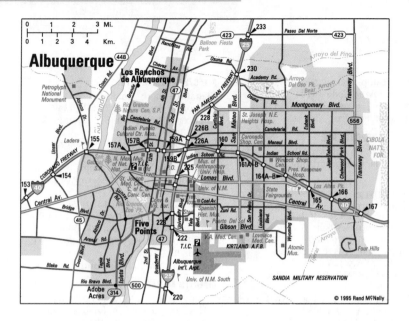

SELECTED ATTRACTIONS:

The Albuquerque Museum, 2000
Mountain Rd. NW, 242-4600

Cliffs Amusement Park, 4800 Osuna Rd.
NE, 881-9373

Indian Pueblo Cultural Center, 2401 12th
St. NW, 843-7270

National Atomic Museum, Wyoming Ave.,
Kirtland Air Force Base, 845-4636

New Mexico Museum of Natural History
& Science, 1801 Mountain Rd. NW,
841-8837

Petroglyph National Monument, 6900
Unser Blvd. NW, 839-4429, 766-8375, or
873-6620

Rio Grande Nature Center State Park, 2901
Candelaria NW, 344-7240

Sandia Peak Aerial Tramway, #10
Tramway Loop NE, 298-8518

Turquoise Trail Scenic and Historic Area,
North NM 14, the "Scenic Route" to
Santa Fe, and NM 536 to the Sandia
Crest, 281-5233

INFORMATION SOURCES:

Albuquerque Convention & Visitors
Bureau
121 Tijeras NE
P.O. Box 26866
Albuquerque, New Mexico 87125-6866
(505) 842-9918; (800) 284-2282 or
733-9918 (convention services)

Greater Albuquerque Chamber of
Commerce
Albuquerque Convention Center 2nd &
Marquette Albuquerque, New Mexico
87102 (505) 764-3700

Anchorage, Alaska

Population: (*226,338) 226,338 (1990C)
Altitude: 118 feet
Average Temp.: Jan., 12°F.; July, 58°F.
Telephone Area Code: 907
Time: 844 **Weather:** 936-2525
Time Zone: Alaskan (one hour earlier
than Pacific time)

AIRPORT TRANSPORTATION:

Six miles to downtown Anchorage.
Taxicab, airport limousine, and bus
service.

SELECTED HOTELS:

Anchorage Hilton Hotel, 500 W. Third
Ave., 272-7411; FAX 265-7140

Best Western Barratt Inn, 4616 Spenard
Rd., 243-3131; FAX 249-4917

Comfort Inn Heritage Suites, 111 W.
Warehouse Ave., 277-6887;
FAX 274-9830

Days Inn Anchorage, 321 E. Fifth Ave.,
276-7226; FAX 265-5145

Hotel Captain Cook, Fifth Ave. & K St.,
276-6000; FAX 278-5366

Inlet Tower Suites, 1200 L St., 276-0110;
FAX 258-4914

Ramada Northern Lights Inn, 598 W.
Northern Lights Blvd., 561-5200;
FAX 563-8217

Regal Alaskan Hotel, 4800 Spenard Rd.,
243-2300; FAX 243-8815

Sheraton Anchorage Hotel, 401 E. Sixth
Ave., 276-8700; FAX 276-7561

WestCoast International Inn, 3333
International Airport Rd., 243-2233;
FAX 248-3796

First Church in Albany, 110 N. Pearl St.
463-4449

Five Rivers Environmental Education
Center, Game Farm Rd., Delmar,
475-0291

Historic Cherry Hill (five-generation Van
Rensselaer home), 523½ S. Pearl St.,
434-4791

New York State Capitol (tours), 474-2418

New York State Museum, Empire State
Plaza, 474-5877

Schuyler Mansion State Historic Site, 32
Catherine St., 434-0834

Shaker Heritage Society (1848 Shaker
meeting house), Albany Shaker Rd.,
adjacent to the airport & within the Ann
Lee Health Facility Complex, 456-7890

INFORMATION SOURCES:

Albany Cty. Convention & Visitors Bureau
52 S. Pearl Albany,
New York 12207
(518) 434-1217; (800) 258-3582

Albany-Colonie Regional Chamber of
Commerce
540 Broadway Albany,
New York 12207 (518) 434-1214

Albuquerque, New Mexico

Population: (*480,577) 384,736 (1990C)
Altitude: 4,958 feet
Average Temp.: Jan., 35°F.; July, 79°F.
Telephone Area Code: 505
Time: 247-1611 **Weather:** 821-1111
Time Zone: Mountain

AIRPORT TRANSPORTATION:

Eight miles to downtown Albuquerque.
Taxicab, limousine, and city bus service.

SELECTED HOTELS:

Albuquerque Doubletree, 201 Marquette
Ave. NW, 247-3344; FAX 247-7025

Albuquerque Hilton Hotel, 1901
University Blvd. NE, 884-2500; FAX
889-9118

Albuquerque Marriott Hotel, 2101

Louisiana Blvd. NE, 881-6800;
FAX 888-2982

✈ Best Western Airport Inn, 2400 Yale
Blvd. SE, 242-7022; FAX 243-0620

✈ Best Western Fred Harvey—
Albuquerque Int'l Airport, 2910 Yale
Blvd. SE, 843-7000; FAX 843-6307

Best Western Winrock Inn, 18 Winrock
Center, 883-5252; FAX 889-3206

Holiday Inn—Midtown, 2020 Menaul
Blvd. NE, 884-2511; FAX 884-5720

Holiday Inn Pyramid—Journal Center,
5151 San Francisco Rd., 821-3333;
FAX 828-0230

Hyatt Regency, 330 Tijeras NW, 842-1234;
FAX 766-6710

La Posada, 125 2nd St. NW, 242-9090;
FAX 242-8664

Quality Hotel 4 Seasons, 2500 Carlisle NE,
888-3311; FAX 881-7452

Ramada Hotel Classic, 6815 Menaul Blvd.
NE, 881-0000; FAX 881-3736

Ramada Inn East, 25 Hotel Circle NE,
296-5472; FAX 291-9428

Sheraton—Old Town, 800 Rio Grande
Blvd. NW, 843-6300; FAX 842-9863

SELECTED RESTAURANTS:

Signature Food: green chile

The Cooperage, 7220 Lomas Blvd. NE,
255-1657

El Pinto, 10500 4th St. NW, 898-1771

Garduños, 10551 Montgomery NE,
298-5000

La Cascada, in the Albuquerque
Doubletree Hotel, 247-3344

Luna Mansion, Jct. US 85 & NM 6,
865-7333

Nicole's, in the Albuquerque Marriott
Hotel, 881-6800

Prairie Star, on Jemez Canyon Rd.,
867-3327

Rancher's Club, in the Albuquerque
Hilton Hotel, 884-2500

Seagull Street, 5410 Academy NE,
821-0020

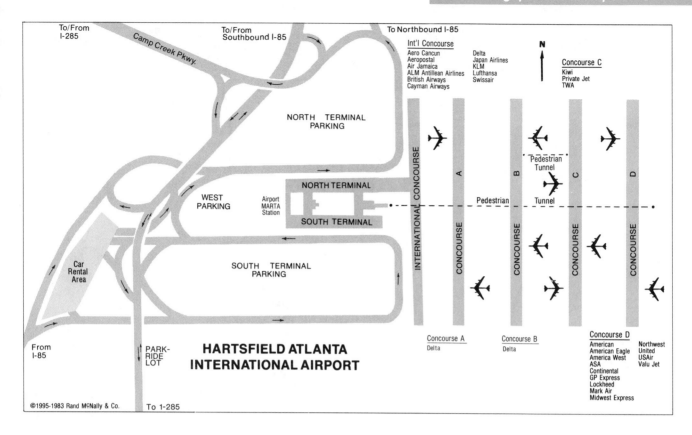

Hartsfield Atlanta International Airport map. Labels include: To/From I-285, Camp Creek Pkwy., To/From Southbound I-85, To Northbound I-85, NORTH TERMINAL PARKING, WEST PARKING, Airport MARTA Station, NORTH TERMINAL, SOUTH TERMINAL, SOUTH TERMINAL PARKING, Car Rental Area, From I-85, PARK-RIDE LOT, INTERNATIONAL CONCOURSE, CONCOURSE A, CONCOURSE B, CONCOURSE C, CONCOURSE D, Pedestrian Tunnel, HARTSFIELD ATLANTA INTERNATIONAL AIRPORT. ©1995-1983 Rand McNally & Co. To I-285

Int'l Concourse: Aero Cancun, Aeropostal, Air Jamaica, ALM Antillean Airlines, British Airways, Cayman Airways, Delta, Japan Airlines, KLM, Lufthansa, Swissair

Concourse C: Kiwi, Private Jet, TWA

Concourse A: Delta

Concourse B: Delta

Concourse D: American, American Eagle, America West, ASA, Continental, GP Express, Lockheed, Mark Air, Midwest Express, Northwest, United, USAir, Valu Jet

---

Westmark Anchorage Hotel, 720 W. Fifth Ave., 276-7676; FAX 276-3615
Westmark Inn Third Avenue (open May through Sept.), 115 E. Third Ave., 272-9403; FAX 272-3879

SELECTED RESTAURANTS:

Signature Food: salmon
Atlasta Deli, 36th & Arctic, 563-3354
The Bistro, in the Sheraton Anchorage Hotel, 276-8700
Corsair, 944 W. Fifth Ave., 278-4502
Crow's Nest, in the Hotel Captain Cook, 276-6000
La Mex, 900 W. Sixth Ave., 274-7678
Quarter Deck, in the Hotel Captain Cook, 276-6000
Romano's, 2415 C St., 276-0888
Tony Roma's, 1430 E. Tudor Rd., 561-7427

SELECTED ATTRACTIONS:

Alaska Zoo, 4731 O'Malley Rd., 346-2133
Anchorage Museum of History and Art, 121 W. Seventh Ave., 343-4326
Earthquake Park, about 4 mi. from downtown on W. Northern Lights Blvd.
Elmendorf Air Force Base Wildlife Museum, enter Boniface Gate, about 3 mi. from downtown off Glenn Hwy., 552-2282
Imaginarium, 725 W. Fifth Ave., 276-3179
Oscar Anderson House/Elderberry Park, 420 M St., 274-2336
Portage Glacier/Begich, Boggs Visitor Center, 783-2326 or -3242
TAHETA Art & Culture Group Co-op (Alaska Native arts & crafts studios), 605 A St., 272-5829

INFORMATION SOURCES:

Anchorage Convention & Visitors Bureau
1600 A St., Suite 200
Anchorage Alaska 99501-5162
(907) 276-4118
Log Cabin Visitor Center
Fourth St. & F St.
Anchorage, Alaska 99501
(907) 274-3531
Anchorage Chamber of Commerce
441 W. 5th Ave., Suite 300
Anchorage, Alaska 99501
(907) 272-7588

Atlanta, Georgia

Population: (*2,833,511) 394,017 (1990C)
Altitude: 1,050 feet
Average Temp.: Jan., 52°F.; July, 85°F.
Telephone Area Code: 404
Time and Weather: 603-3333
Time Zone: Eastern

AIRPORT TRANSPORTATION:

Eight miles to downtown Atlanta.
Taxicab and limousine bus service.

SELECTED HOTELS:

Atlanta Hilton & Towers, 255 Courtland St. NE, 659-2000; FAX 522-8926
Atlanta Marriott Marquis, 265 Peachtree Center Ave., 521-0000; FAX 586-6299
Atlanta Marriott Perimeter Center, 246 Perimeter Center Pkwy., 394-6500; FAX 394-4338
Best Western American Hotel, 160 Spring St. NW, 688-8600; FAX 658-9458
Colony Square Hotel, 188 14th St. NE, 892-6000; FAX 872-9192

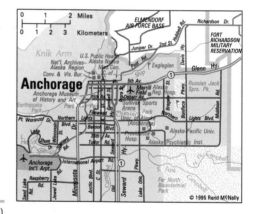

Map of Anchorage area showing Elmendorf Air Force Base, Fort Richardson Military Reservation, Knik Arm, Anchorage, International Airport, and surrounding roads. © 1995 Rand McNally

✈ Holiday Inn—Airport North, 1380 Virginia Ave., 762-8411; FAX 767-4963
✈ Howard Johnson's North—Atlanta Airport, 1377 Virginia Ave., 762-5111; FAX 762-1277
Hyatt Regency—Atlanta, 265 Peachtree St. NE, 577-1234; FAX 588-4137
Omni Hotel at CNN Center, 100 CNN Center, 659-0000; FAX 525-5050
Radisson Atlanta, 165 Courtland St., 659-6500; FAX 524-1259
Ritz-Carlton Atlanta, 181 Peachtree St. NE, 659-0400; FAX 688-0400
Stone Mountain Inn, U.S. 78, Stone Mountain Pk., 469-3311; FAX 498-5691
✈ Stouffer Concourse Hotel, One Hartsfield Center, 209-9999; FAX 209-7031
The Westin Peachtree Plaza, 210 Peachtree St. NW, 659-1400; FAX 589-7424

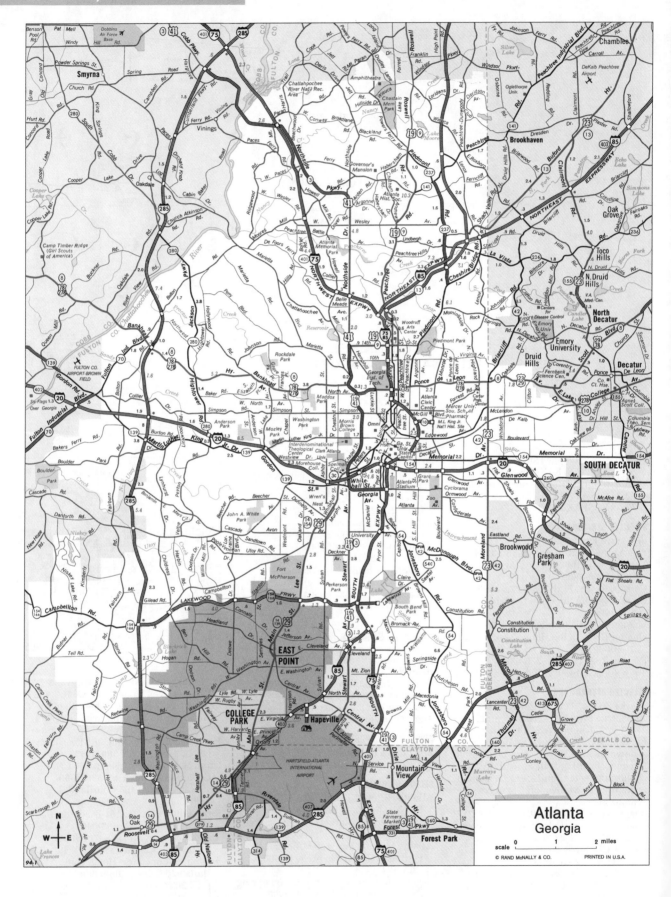

Atlanta
Georgia

scale 0   1   2 miles

© RAND McNALLY & CO.   PRINTED IN U.S.A.

SELECTED RESTAURANTS:

Signature Food: fried green tomatoes; grits

The Abbey, Piedmont & North aves., 876-8831

Avanzare, in the Hyatt Regency—Atlanta, 577-1234

Bugatti's, in the Omni Hotel at CNN Center, 659-0000

Coach And Six, 1776 Peachtree St. NW, 872-6666

The Crab House, Piedmont at North Ave. (2nd Level—Rio), 872-0011

Daily's, 17 International Blvd., 681-3303

Dante's Down the Hatch, 3380 Peachtree Rd., 266-1600

Garden Terrace, in the Atlanta Hilton & Towers, 659-2000

La Grotta, 2637 Peachtree Rd. NE, 231-1368

The Mansion, Piedmont at North Ave., 876-0727

Nikolai's Roof Restaurant, in the Atlanta Hilton & Towers, 659-2000

Pano's and Paul's, 1232 W. Paces Ferry Rd. NW, 261-3662

Terrace Garden Inn, 3405 Lenox Rd. NE, 261-9250

SELECTED ATTRACTIONS:

Carter Presidential Center, 1 Copenhill Ave., 420-5110

Fernbank Museum of Natural History, 767 Clifton Rd. NE, 378-0127

High Museum of Art, 1280 Peachtree St. NE, 892-3600

Martin Luther King Jr. Historic District, Auburn Ave. between Jackson & Randolph sts., (birth home) 331-3920, (Ebenezer Baptist Church) 688-7263, (grave & exhibit center) 524-1956

The Road to Tara Museum, 659 Peachtree St. (inside the Georgian Terrace), 814-4000

Stone Mountain Park, US 78, Stone Mountain, 498-5600

Underground Atlanta, Peachtree St. at Alabama St.

World of Coca-Cola Pavilion, 55 Martin Luther King Jr. Dr., 676-5151

INFORMATION SOURCES:

Atlanta Convention & Visitors Bureau Suite 2000, 233 Peachtree St. NE Atlanta, Georgia 30303 (404) 521-6600

Atlanta Chamber of Commerce 235 International Blvd. NW Atlanta, Georgia 30303 (404) 880-9000

Austin, Texas

Population: (*781,572) 465,622 (1990C)
Altitude: 501 feet
Average Temp.: Jan., 50°F.; July, 85°F.
Telephone Area Code: 512
Time: 476-7744 **Weather:** 476-7736
Time Zone: Central

AIRPORT TRANSPORTATION:

Seven-and-a-half miles to downtown Austin.

Taxicab, hotel shuttle, limousine and bus service.

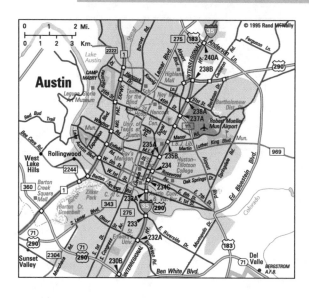

SELECTED HOTELS:

✦ Austin Doubletree Hotel, 6505 I-35N, 454-3737; FAX 454-6915

Austin Marriott at the Capitol, 701 E. 11th St., 478-1111; FAX 478-3700

Four Seasons Hotel—Austin, 98 San Jacinto Blvd., 478-4500; FAX 478-3117

Guest Quarters, 303 W. 15th St., 478-7000; FAX 478-5103

Hawthorn Suites—Northwest, 8888 Tallwood Dr., 343-0008; FAX 343-6532

Holiday Inn, 6911 N. I-35, 459-4251; FAX 459-9274

Howard Johnson's North Plaza, 7800 Interregional Hwy. 35 N., 836-8520; FAX 837-0897

Hyatt Regency—Austin, 208 Barton Springs, 477-1234; FAX 480-2069

Radisson Hotel, 111 E. Cesar Chavez St., 478-9611; FAX 473-8399

✦ Red Lion Hotel, 6121 I-35N at US 290, 323-5466; FAX 453-1945

Wyndham Hotel Southpark, 4140 Governor's Row, 448-2222; FAX 442-8028

SELECTED RESTAURANTS:

Signature Food: Tex-Mex

County Line on the Hill, 6500 W. Bee Caves, 327-1742

County Line on the Lake, 5204 FM 2222, 346-3664

Foothills of Austin, in the Hyatt Regency—Austin, 477-1234

Green Pastures, 811 W. Live Oak, 444-4747

Matt's El Rancho, 2613 S. Lamar, 462-9333

Oasis Cantina Del Lago, 6550 Comanche Trail, 266-2442

Threadgill's, 6416 N. Lamar, 451-5440

SELECTED ATTRACTIONS:

Barker Texas History Center, Sid Richardson Hall, Unit 2, University of Texas campus, 495-4515

Elisabet Ney Museum (sculpture), 304 E. 44th St., 458-2255

French Legation Museum, 802 San Marcos at E. 7th St., 472-8180

Governor's Mansion, 1010 Colorado, 463-5518

Huntington Art Gallery, in the Harry Ransom Center, West 21st & Guadalupe, 471-7324

Lyndon B. Johnson Presidential Library and Museum, 2313 Red River, University of Texas campus, 482-5136

Neill-Cochran House (ca 1855), 2310 San Gabriel, 478-2335

O. Henry Home and Museum, 409 E. Fifth St., 472-1903

State Capitol Building, 11th & Congress, 463-0063

Texas Memorial Museum, 2400 Trinity, University of Texas campus, 471-1605

Treaty Oak, 503 Baylor

INFORMATION SOURCE:

Austin Convention & Visitors Bureau 201 E. 2nd Austin, Texas 78701 Convention Info. (512) 474-5171 Visitor Info. (512) 478-0098; (800) 926-2282

Baltimore, Maryland

Population: (*2,382,172) 736,014 (1990C)
Altitude: Sea level to 32 feet
Average Temp.: Jan., 37°F.; July, 79°F.
Telephone Area Code: 410
Time: 844-2525 **Weather:** 936-1212
Time Zone: Eastern

AIRPORT TRANSPORTATION:

Ten miles to downtown Baltimore.

Taxicab and limousine bus service.

SELECTED HOTELS:

Baltimore Ramada Hotel, 1701 Belmont Ave. at Security Blvd., 265-1100; FAX 944-2325

Brookshire Hotel, 120 E. Lombard St., 625-1300; FAX 625-0912

Cross Keys Inn, 5100 Falls Rd., 532-6900; FAX 532-2403

Harbor Court Hotel, 550 Light St., 234-0550; FAX 659-5925

Holiday Inn at the Harbor, 301 W. Lombard St., 685-3500; FAX 727-6169

✦ Holiday Inn—BWI Airport, 890 Elkridge Landing Rd., 859-8400; FAX 684-6778

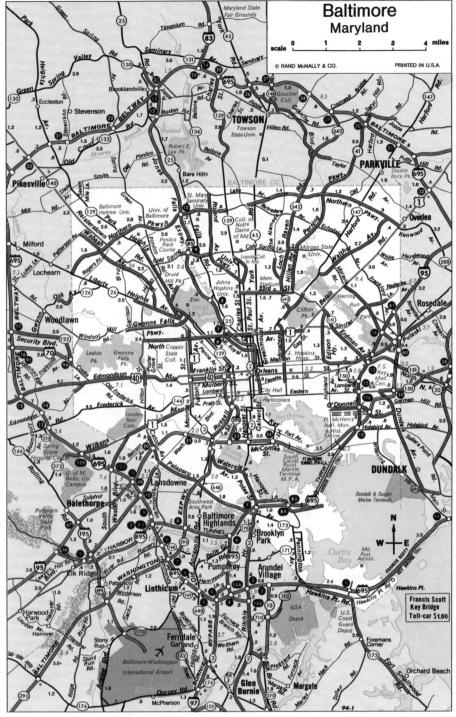

Baltimore
Maryland

scale 0 1 2 3 4 miles

© RAND McNALLY & CO.   PRINTED IN U.S.A.

Citronelle's, in the Latham Hotel, 727-7101

Crossroads, in Cross Keys Inn Hotel, 532-6900

Hampton's, in the Harbor Court Hotel, 234-0550

Haussner's Restaurant, 3242 Eastern Ave., 327-8365

Obrycki's Crab House, 1727 E. Pratt St., 732-6399

The Prime Rib, N. Calvert & Chase sts., 539-1804

Tio Pepe Restaurant, 10 E. Franklin St., 539-4675

SELECTED ATTRACTIONS:

B&O Railroad Museum, 901 W. Pratt St., 752-2490

Baltimore Museum of Art, Art Museum Dr., off the 3100 block of N. Charles St., 396-7101

Baltimore Zoo, Druid Hill Park, 366-5466

Fort McHenry National Monument and Historic Shrine, Foot of E. Fort Ave., 962-4290

Harbor Cruises, Ltd., 301 Light St., 727-3113

Harborplace/The Gallery at Harborplace, Light & Pratt sts. and Calvert & Pratt sts., 332-4191

Lexington Market, Lexington & Eutaw sts., 685-6169

Maryland Science Center, 601 Light St., 685-5225

National Aquarium in Baltimore, Pier 3, 501 E. Pratt St., 576-3800

Top of the World Trade Center, 401 E. Pratt St., 837-4515

INFORMATION SOURCE:

Baltimore Area Convention and Visitors Association
100 Light St., 12th Floor
Baltimore, Maryland 21202
(410) 659-7300; (800) 343-3468
Visitor Info. Center (410) 837-4636;
(800) 282-6632 (except Baltimore)

Baton Rouge, Louisiana

Population: (*528,264) 219,531 (1990C)
Altitude: 58 feet
Average Temp.: Jan., 51°F.; July, 82°F.
Telephone Area Code: 504
Time: 387-5411   **Weather:** 357-9743
Time Zone: Central

AIRPORT TRANSPORTATION:

Seven miles to downtown Baton Rouge.
Taxicab, hotel van, and limousine service.

SELECTED HOTELS:

Baton Rouge Hilton, 5500 Hilton Ave., 924-5000; FAX 925-1330

Bellemont Hotel, 7370 Airline Hwy., 357-8612; FAX 357-4974

Courtyard By Marriott, 2421 S. Acadian Thruway, 924-6400; FAX 923-3041

Holiday Inn—East, I-10 & Siegen Ln., 293-6880; FAX 293-6880

Holiday Inn—South, 9940 Airline Hwy., 924-7021; FAX 924-7021, Ext. 1987

Quality Inn—Baton Rouge, 10920 Mead Rd., 293-9370; FAX 293-8889

Ramada Hotel, 1480 Nicholson Dr., 387-1111; FAX 387-1111

Sheraton Baton Rouge, 4728 Constitution Ave., 925-2244; FAX 927-6925

Hyatt Regency Baltimore, 300 Light St., 528-1234; FAX 685-3362

Latham Hotel, 612 Cathedral St., 727-7101; FAX 789-3312

Marriott Inner Harbor, Pratt & Eutaw sts., 962-0202; FAX 962-0202

Marriott's Hunt Valley Inn, 245 Shawan Rd., Hunt Valley, 785-7000; FAX 785-0341

Omni Inner Harbor Hotel, 101 W. Fayette St., 752-1100; FAX 752-0832

Pikesville Hilton Inn, 1726 Reisterstown Rd. at the Beltway, Pikesville, 653-1100; FAX 484-4138

✈ Sheraton International Hotel, 7032 Elm Rd., at Baltimore/Washington Internat'l Airport, 859-3300; FAX 859-0565

Stouffer Harborplace Hotel, 202 E. Pratt St., 547-1200; FAX 539-5780

SELECTED RESTAURANTS:

Signature Food: steamed hardshell crabs

Bamboo House, Harborplace—Pratt St. Pavilion, 625-1191

Chiapparelli's, 237 S. High St., 837-0309

SELECTED RESTAURANTS:

Signature Food: gumbo; jambalaya; seafood

Chalet Brandt, 7655 Old Hammond Hwy., 927-6040

The Chinese Restaurant, 1710 Nicholson Dr., 387-9443

Don's Seafood & Steakhouse, 6823 Airline Hwy., 357-0601

Giamanco's, 4624 Government St., 928-5045

Juban's, 3739 Perkins Rd., 346-8422

Maison LaCour, 11025 N. Harrell's Ferry Rd., 275-3755

Mike Anderson's Seafood, 1031 W. Lee Dr., 766-3728

Mulate's Cajun Restaurant, 8322 Bluebonnet Rd., 767-4794

Ruth's Chris Steak House, 4836 Constitution, 925-0163

SELECTED ATTRACTIONS:

Louisiana Arts & Science Center/Riverside Museum, North Blvd. & River Rd., 344-5272

Louisiana State Capitol, at the north end of 4th St., 342-7317

Louisiana State University (museums, Indian mounds), south of downtown between Highland Rd. and Nicholson Dr., 388-3202

LSU Rural Life Museum (19th-century buildings), Essen Ln. at I-10, 765-2437

McGee's Atchafalaya Basin Tours, the Atchafalaya Swamps in Henderson, (318) 228-8519

Magnolia Mound Plantation, 2161 Nicholson Dr., 343-4955

Nottoway Plantation, LA Hwy. 1, White Castle, 346-8263

Oak Alley Plantation, LA Hwy. 18, Vacherie, 265-2151

Old State Capitol—Louisiana Center for Political & Governmental History, North Blvd. at River Rd., 342-0500

U.S.S. *Kidd*/Nautical Historic Center, Government St. at River Rd., 342-1942

INFORMATION SOURCES:

Baton Rouge Area Convention & Visitors Bureau
  730 North Boulevard
  Baton Rouge, Louisiana 70802
  (504) 383-1825; (800) LA-ROUGE
The Greater Baton Rouge Chamber of Commerce
  564 Laurel St.
  P.O. Box 3217
  Baton Rouge, Louisiana 70821
  (504) 381-7125

Billings, Montana

Population: (*113,419) 81,151 (1990C)
Altitude: 3,124 feet
Average Temp.: Jan., 22°F.; July, 72°F.
Telephone Area Code: 406
Time: 1-976-7651 **Weather:** 657-6988
Time Zone: Mountain

AIRPORT TRANSPORTATION:

Three miles to downtown Billings.
Taxicab and limousine service.

SELECTED HOTELS:

Airport Metra Inn, 403 Main, 245-6611; FAX (none)

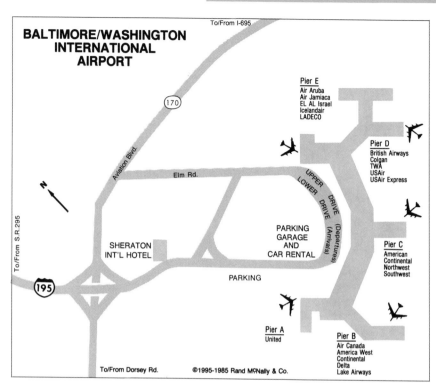
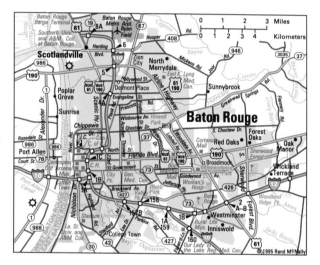
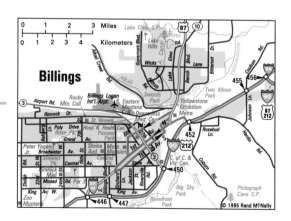

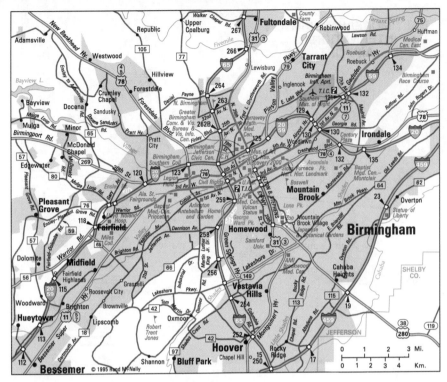

Ramada Inn Central, 300 N. 10th St.,
  328-8560; FAX 323-5819
The Tutwiler, Park Place at 21st St. N.,
  322-2100; FAX 325-1183
Wynfrey Hotel, 1000 Riverchase Galleria,
  987-1600; FAX 988-4597

SELECTED RESTAURANTS:
Signature Food: barbeque
Bombay Cafe, 2839 7th Ave. S., 322-1930
Christian's Classic Cuisine, in the
  Tutwiler Hotel, 323-9822
GG in the Park, 3625 8th Ave. S.,
  254-3506
Highlands: A Bar & Grill, 2011 11th Ave.
  S., 939-1400
John's Restaurant, 112 N. 21st St.,
  322-6014
Michael's Sirloin Room, 431 20th St. S.,
  322-0419
Winston's, in the Wynfrey Hotel,
  987-1600

SELECTED ATTRACTIONS:
Alabama Jazz Hall of Fame, 17th St. & 4th
  Ave. N., 254-2720
Alabama Sports Hall of Fame Museum,
  22nd St. & Civic Center Blvd., 323-6665
Arlington Antebellum Home and Gardens,
  331 Cotton Ave. SW, 780-5656
Birmingham Civil Rights Institute, 520
  16th St. N., 328-9696
Birmingham Museum of Art, 2000 8th
  Ave. N., 254-2565
Birmingham Zoo, 2630 Cahaba Rd.,
  879-0408
Botanical and Japanese Gardens, 2612
  Lane Park Rd., 879-1227
Discovery 2000/Discovery Place
  (children's museum), 1320 22nd St. S.,
  939-1177; Red Mountain Museum
  (earth's history), 1422 22nd St. S.,
  933-4153
Five Points South (historic district), 11th
  Ave. S. & 20th St. Riverchase Galleria,
  U.S. 31 South at I-459, 985-3039
Sloss Furnaces (iron-making), beside First
  Ave.—north viaduct on 32nd St.,
  324-1911
Vulcan Park (panoramic view from
  world's largest cast iron statue), Valley
  Ave. at Highway 31 S., 328-6198

INFORMATION SOURCES:
Greater Birmingham Convention and
Visitors Bureau
  2200 Ninth Ave. N.
  Birmingham, Alabama 35203
  (205) 252-9825; (205) 458-8000;
  (800) 458-8085
Birmingham Area Chamber of Commerce
  2027 First Ave. N.
  North Birmingham, Alabama 35203
  (205) 323-5461

✈ Best Western Ponderosa Inn, 2511 1st
  Ave. N., 259-5511; FAX 245-8004
Billings Inn, 880 N. 29th St., 252-6800;
  FAX 252-6800
Billings Plaza Holiday Inn, 5500 Midland
  Rd., 248-7701; FAX 248-8954
Elliott Inn, 1345 Mullowney, 252-2584;
  FAX 252-2584
✈ Radisson Northern Hotel, Broadway &
  1st Ave. N., 245-5121; FAX 259-9862
Ramada Inn, 1223 Mullowney Ln.,
  248-7151; FAX 248-1695
✈ Sheraton Hotel, 27 N. 27th St.,
  252-7400; FAX 252-2401
War Bonnet Inn, I-90 & S. 27th St.,
  248-7761; FAX 248-7761

SELECTED RESTAURANTS:
Signature Food: Montana beef
Bert and Ernie's, 2824 2nd Ave. N.,
  248-4313
The Cattle Company, 300 S. 24th, Rimrock
  Mall, 656-9090
The Granary, 1500 Poly Dr., 259-3488
Jake's, 2701 1st Ave. N., 259-9375
Juliano's, 2912 7th Ave. N., 248-6400
Miyajima Gardens, 5364 Midland Rd.,
  245-8240
The Rex, 2401 Montana Ave., 245-7477

SELECTED ATTRACTIONS:
Boothill Cemetery, east end of Black Otter
  Trail, 252-4016
Moss Mansion, 914 Division St., 256-5100
Peter Yegen Museum, at Logan Field
  Airport, 256-6811
Pictograph Caves, 5 miles southeast of
  Billings off I-90, 252-4016
Pompeys Pillar, 28 miles northeast of
  Billings, 657-6262
Western Heritage Center, 2822 Montana
  Ave., 256-6809

Zoo Montana, 2100 S. Shiloh Rd.,
  652-8100

INFORMATION SOURCE:
Billings Convention & Visitors Bureau
Billings Area Chamber of Commerce
  815 S. 27th St.
  P.O. Box 31177
  Billings, Montana 59107
  (406) 245-4111

Birmingham, Alabama

Population: (*907,810) 265,968 (1990C)
Altitude: 601 feet
Average Temp.: Jan., 46°F.; July, 82°F.
Telephone Area Code: 205
Time: 979-8463   **Weather:** 945-7000
Time Zone: Central

AIRPORT TRANSPORTATION:
Five miles to downtown Birmingham.
Taxicab, bus and limousine service.

SELECTED HOTELS:
✈ Holiday Inn—Airport, 5000 10th Ave.
  N., 591-6900; FAX 591-2093
Holiday Inn—East, 7941 Crestwood Blvd.,
  956-8211; FAX 956-1234
Holiday Inn—Galleria Area, 1548
  Montgomery Hwy., 822-4350;
  FAX 822-0350
Holiday Inn Homewood, 260 Oxmoor Rd.,
  942-2041; FAX 290-9309
Mountain Brook Inn, 2800 US 280,
  870-3100; FAX 870-5938
Radisson Birmingham, 808 S. 20th St. at
  University Blvd., 933-9000;
  FAX 933-0920
✈ Ramada Inn—Airport, 5216 Airport
  Hwy., 591-7900; FAX 592-6476

Boise, Idaho

Population: (*205,775) 125,738 (1990C)
Altitude: 2,726 feet
Average Temp.: Jan., 29°F.; July, 75°F.
Telephone Area Code: 208
Time: 1-844-8463   **Weather:** 342-6569
Time Zone: Mountain

AIRPORT TRANSPORTATION:
Three miles to downtown Boise.

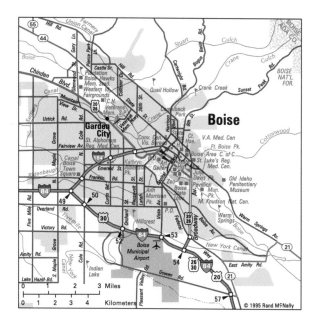

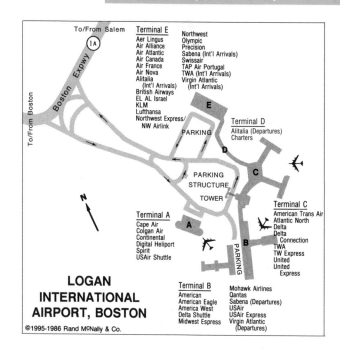

LOGAN INTERNATIONAL AIRPORT, BOSTON
©1995-1986 Rand McNally & Co.

Taxicab, bus, airport limousine, and hotel shuttle service.

SELECTED HOTELS:

✈ Best Western Airport Motor Inn, 2660 Airport Way, 384-5000; FAX 384-5566

The Boisean, 1300 S. Capitol, 343-3645; FAX 343-4823

Doubletree Hotel, 475 Park Center Blvd., 345-2002; FAX 345-8354

✈ Holiday Inn, 3300 Vista Ave., 344-8365; FAX 343-9635

Plaza Suite Hotel, 409 S. Cole, 375-7666; FAX 376-3608

Red Lion Inn—Downtowner, 1800 Fairview Ave., 344-7691; FAX 336-3652

Red Lion Inn—Riverside, 29th & Cinden Blvd., 343-1871; FAX 344-1079

Residence Inn, 1401 Lusk Ave., 344-1200; FAX 384-5354

Shilo Inn—Riverside, 3031 Main, 344-3521; FAX 384-1217

University Inn, 2360 University Dr., 345-7170; FAX 345-5118

SELECTED RESTAURANTS:

Signature Food: potato

Angell's, 999 W. Main St., 342-4900

Boise Black Angus, 3101 Main, 345-7600

Charthouse, 2288 N. Garden St., 336-9370

The Gamekeeper, in the Owyhee Plaza Hotel, 1109 Main, 343-4611

Lock, Stock & Barrel, 4705 Emerald, 336-4266

Murphy's Oyster Bar & Grill, 1555 Broadway Ave., 344-3691

Pacific Grill & Smokehouse, in the Red Lion Inn—Riverside, 343-1871

Peter Schatt's, in the Idanha Hotel, 10th & Main, 336-9100

The Renaissance, 311 Main St., 344-6776

The Sandpiper, 1100 W. Jefferson, 344-8911

SELECTED ATTRACTIONS:

Boise Art Museum, 670 S. Julia Davis Dr., 345-8330

Boise Greenbelt (20-mile riverfront pathway), Willow Lane Athletic Complex to Lucky Peak Dam, 384-4240

The Discovery Center (hands-on science center), 131 Myrtle St., 343-9895

Idaho Botanical Garden, 2355 Old Penitentiary Rd., 343-8649

Idaho Historical Museum, 610 N. Julia Davis Dr., 334-2120

Les Bois Park (horse racing), 5610 Glenwood, 376-RACE

Old Idaho Territorial Penitentiary, 2445 Old Penitentiary Rd., 334-2844

World Center for Birds of Prey, Exit 50 off I-84 at the end of S. Cole Rd., 362-8687

Zoo Boise, in Julia Davis Park, 384-4260

INFORMATION SOURCES:

Boise Convention and Visitors Bureau
168 N. 9th, Suite 200
Boise, Idaho 83702
(208) 344-7777; (800) 635-5240

Boise Area Chamber of Commerce
300 N. 6th
P.O. Box 2368
Boise, Idaho 83701
(208) 344-5515

Boston, Massachusetts

Population: (*2,870,669) 574,283 (1990C)
Altitude: Sea level to 330 feet
Average Temp.: Jan., 29°F.; July, 72°F.
Telephone Area Code: 617
Time: 637-1234   **Weather:** 936-1234
Time Zone: Eastern

AIRPORT TRANSPORTATION:

Three miles to downtown Boston.
Taxicab and limousine bus service.

SELECTED HOTELS:

Boston Harbor Hotel, 70 Rowes Wharf, 439-7000; FAX 330-9450

Boston Marriott, Copley Place, 110 Huntington Ave., 236-5800; FAX 236-5885

The Colonnade, 120 Huntington Ave., 424-7000; FAX 424-1717

Copley Plaza Hotel, 138 St. James Ave., 267-5300; FAX 267-7668

The Four Seasons, 200 Boylston St., 338-4400; FAX 423-0154

✈ Harborside Hyatt, 101 Harborside Dr., 568-1234; FAX 567-8856

Hotel Meridien, 250 Franklin St., 451-1900; FAX 423-2844

✈ Logan Airport Hilton, 75 Service Rd., Logan International Airport, 569-9300; FAX 569-3981

Omni Parker House, 60 School St., 227-8600; FAX 742-5729

The Ritz-Carlton, Boston, 15 Arlington St., 536-5700; FAX 536-1335

Sheraton Boston Hotel & Towers, Prudential Center, 39 Dalton St., 236-2000; FAX 236-1702

Westin Hotel Copley Place, 10 Huntington Ave., 262-9600; FAX 424-7483

SELECTED RESTAURANTS:

Signature Food: lobster

Anthony's Pier 4, 140 Northern Ave., 423-6363

The Cafe Budapest, 90 Exeter St., 266-1979

Caffe Lampara, 916 Commonwealth Ave., 566-0300

Copley's Restaurant, in the Copley Plaza Hotel, 267-5300

The Dining Room, in The Ritz-Carlton, Boston, 536-5700

Felicia's, 145A Richmond St., up one flight, 523-9885

Hampshire House, 84 Beacon St., 227-9600

Jimmy's Harborside Restaurant, 242 Northern Ave., 423-1000

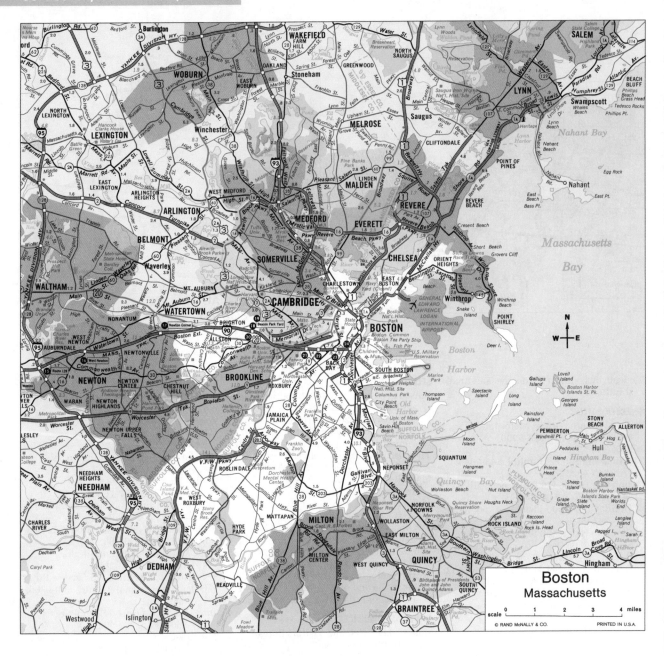

Boston
Massachusetts

scale 0 1 2 3 4 miles

© RAND McNALLY & CO.   PRINTED IN U.S.A.

Julien, in the Hotel Meridien, 451-1900
Legal Seafoods, in the Boston Park Plaza
    Hotel, 35 Columbus Ave., 426-4444
Locke–Ober Cafe, 3 Winter Pl., 542-1340
Maison Robert, Old City Hall, 45 School
    St., 227-3370

SELECTED ATTRACTIONS:

Boston Tea Party Ship & Museum,
    Congress St. Bridge, 338-1773
Cheers!/Bull & Finch Pub, 84 Beacon St.,
    227-9605
Children's Museum, Museum Wharf, 300
    Congress St., 426-8855
Faneuil Hall/Faneuil Hall Marketplace,
    Congress & North sts., 523-1300
The Freedom Trail (self-guided walking
    tour), 242-5642
Harvard University, Cambridge, 495-1000
John F. Kennedy Library & Museum,
    Columbia Point, Dorchester, 929-4523

Museum of Fine Arts, 465 Huntington
    Ave., 267-9300
Museum of Science, 1 Science Park,
    723-2500
New England Aquarium, Central Wharf,
    973-5200

INFORMATION SOURCES:

Greater Boston Convention & Visitors
Bureau, Inc.
    Prudential Tower, Suite 400, Box 490
    Boston, Massachusetts 02199
    (617) 536-4100
Greater Boston Chamber of Commerce
    1 Beacon St., 4th Floor
    Boston, Massachusetts 02108
    (617) 227-4500

Buffalo, New York

Population: (*968,532) 328,123 (1990C)
Altitude: 600 feet
Average Temp.: Jan., 26°F.; July, 71°F.
Telephone Area Code: 716
Time: 844-1717 **Weather:** 632-1319
Time Zone: Eastern

AIRPORT TRANSPORTATION:

Nine miles to downtown Buffalo.
Taxicab and limousine bus service.

SELECTED HOTELS:

Best Western Inn—Downtown, 510
    Delaware Ave., 886-8333;
    FAX 884-3070
Buffalo Hilton, 120 Church St., 845-5100;
    FAX 845-5377
Buffalo Marriott Inn, 1340 Millersport
    Hwy., Amherst, 689-6900;
    FAX 689-0483

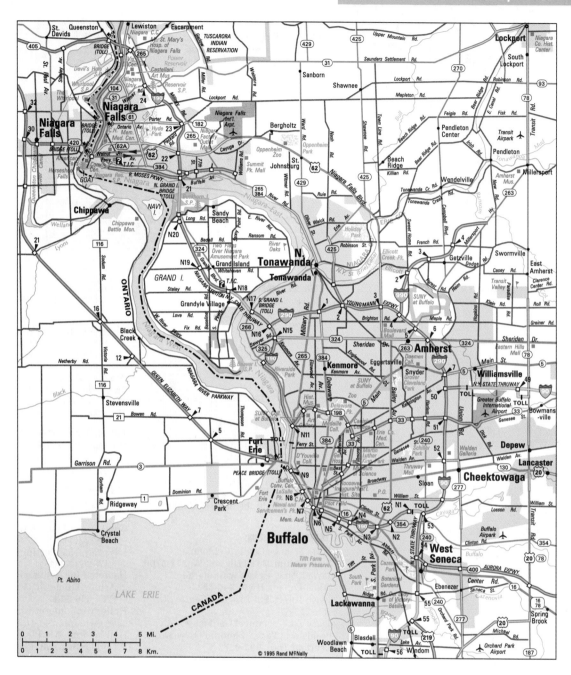

© 1995 Rand McNally

✈ Holiday Inn—Buffalo Airport, 4600 Genesee St., Cheektowaga, 634-6969; FAX 634-0920

Holiday Inn—Buffalo Downtown, 620 Delaware Ave., 886-2121; FAX 886-7942

✈ Holiday Inn Express, 6700 Transit Rd., Williamsville, 634-7500; FAX 634-7502

Holiday Inn—Gateway, 601 Dingens St., 896-2900; FAX 896-3765

Hyatt Regency, Pearl & W. Huron, 856-1234; FAX 852-6157

✈ Quality Inn, 4217 Genesee St., Cheektowaga, 633-5500; FAX 633-4231

✈ Radisson Hotel & Suites, 4243 Genesee St., 634-2300; FAX 632-2387

✈ Ramada Inn—Buffalo Airport, 6643 Transit Rd., 634-2700; FAX 634-1644

Sheraton Buffalo, 2040 Walden Ave., Cheektowaga, 681-2400; FAX 681-8067

Williamsville Inn, 5447 Main St., Williamsville, 634-1111; FAX 631-3367

SELECTED RESTAURANTS:

Signature Food: buffalo chicken wings; beef on wick

Asa Ransom House, 10529 Main St., Clarence, 759-2315

Daffodil's Restaurant, 930 Maple Rd., Williamsville, 688-5413

E.B. Greens, in the Hyatt Regency, 856-1234

Grill Ninety-One, 91 Niagara St., 856-8373

Justine's, in the Buffalo Hilton, 845-5100

Lord Chumley's, 481 Delaware Ave., 886-2220

Old Red Mill Inn, 8326 Main St., Clarence, 633-7878

Park Lane at the Circle, Delaware Ave. at Gates Circle, 883-3344

SELECTED ATTRACTIONS:

Albright-Knox Art Gallery, 1285 Elmwood Ave., 882-8700

Buffalo and Erie County Historical Society, 25 Nottingham Ct., 873-9644

Buffalo Museum of Science, 1020 Humboldt Pkwy., 896-5201

Buffalo Naval & Servicemen's Park, 1 Naval Park Cove, 847-1773

Buffalo Zoo, on Parkside Ave. across from Delaware Park, 837-3900

Historic Theatre District, 856-3150

Miss Buffalo-Niagara Clipper, 856-6696

The Original American Kazoo Company, 8703 South Main St., Eden, 992-3960

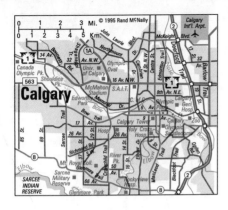

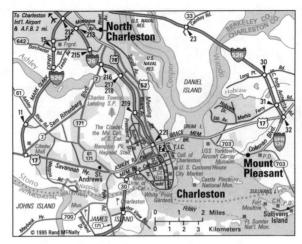

INFORMATION SOURCES:

Greater Buffalo Convention and Visitors
Bureau
107 Delaware Ave.
Buffalo, New York 14202
(716) 852-0511; (800) 283-3256
Greater Buffalo Partnership
Main Place Tower, 3rd Floor
Buffalo, New York 14202
(716) 852-7100

Calgary, Alberta, Canada

Population: (*671,326) 636,104 (1986C)
Altitude: 3,439 feet
Average Temp.: Jan., 13°F.; July, 62°F.
Telephone Area Code: 403
Time and Weather: 263-3333
Time Zone: Mountain

AIRPORT TRANSPORTATION:

Ten miles to downtown Calgary.
Taxicab, airporter shuttle, and limousine
service.

SELECTED HOTELS:

Best Western Hospitality Inn, 135
Southland Dr. SE, 278-5050;
FAX 278-5050
Blackfoot Inn, 5940 Blackfoot Trail SE,
252-2253; FAX 252-3574
✈ Chateau Airport Hotel, 2001 Airport
Rd. NE, 291-2600; FAX 291-3419
Delta Bow Valley, 209 4th Ave. SE,
266-1980; FAX 298-5060
The Palliser, 9th Ave. & 1st St. SW,
262-1234; FAX 260-1260
✈ Port O'Call Inn, 1935 McKnight Blvd.
NE, 291-4600; FAX 250-6827
Prince Royal Inn, 618 5th Ave. SW,
263-0520; FAX 298-4888
Radisson Plaza, 110 9th Ave. SE,
266-7331; FAX 262-8442
Ramada Downtown, 708 8th Ave. SW,
263-7600; FAX 237-6127
Sandman Hotel, 888 7th Ave. SW,
237-8626; FAX 290-1238
Sheraton Cavalier, 2620 32nd Ave. NE,
291-0107; FAX 291-2834
The Westin Hotel, 320 4th Ave. SW,
266-1611; FAX 233-7471

SELECTED RESTAURANTS:

Signature Food: beef
Atrium Terrace, in the Chateau Airport
Hotel, 291-2600
Caesar's, 512 4th Ave. SW, 264-1222
Hy's Steak House, 316 4th Ave. SW,
263-2222
Japanese Village, 302 4th Ave. SW,
262-2738
La Caille on the Bow, 805 1st Ave. SW,
262-5554
La Chaumiere, 121 17th Ave. SE,
228-5690
La Dolce Vita, 916 1st Ave. NE, 263-3445
Owl's Nest Dining Room, in the Westin
Hotel, 266-1611
Panorama Room, atop the Calgary Tower,
101 9th Ave. SW, 266-7171
Rimrock Room, in the Palliser Hotel,
262-1234

SELECTED ATTRACTIONS:

Alberta Science Centre/Centennial
Planetarium, 701 11th St. SW, 221-3700
Calaway Park, 10 km. west on the
Trans-Canada Hwy., 240-3822
Calgary Tower, 9th Ave. and Centre St.,
266-7171
Calgary Zoo, Botanical Gardens &
Prehistoric Park, Memorial Dr. west of
Deerfoot Trail, 232-9372
Canada Olympic Park, on the western city
limits along the Trans-Canada Hwy.,
286-2632
Eau Claire Market & IMAX Theatre, 2nd
St. & 2nd Ave., 264-6450
Fort Calgary, 750 9th Ave. SE, 290-1875
Glenbow Museum, 130 9th Ave. SE,
264-8300
Heritage Park (turn-of-the-century village),
Heritage Dr. & 14th St. SW,
259-1900
Inglewood Bird Sanctuary, south of 9th
Ave. & Sanctuary Rd. SE, 269-6688
Museum of Movie Art, #9 3600 21st St.
NE, 250-7588

INFORMATION SOURCES:

Calgary Convention & Visitors Bureau
237 8th Ave. SE
Calgary, Alberta, Canada T2G 0K8
(403) 263-8510; (800) 661-1678
The Calgary Chamber of Commerce
517 Centre St. S.

Calgary, Alberta, Canada T2G 2C4
(403) 263-7435

Charleston, South Carolina

Population: (*506,875) 80,414 (1990C)
Altitude: 118 feet
Average Temp.: Jan., 49°F.; July, 80°F.
Telephone Area Code: 803
Time: 572-8463   **Weather:** 744-3207
Time Zone: Eastern

AIRPORT TRANSPORTATION:

Twelve miles to downtown Charleston.
Taxicab, airport shuttle, hotel van, and
limousine service.

SELECTED HOTELS:

Charleston Marriott, I-26 at Montague
Ave., 747-1900; FAX 744-2530
Comfort Inn Riverview, 144 Bee St.,
577-2224; FAX 577-9001
Days Inn, 155 Meeting St., 722-8411;
FAX 763-5361
Hampton Inn—Riverview Hotel, 11
Ashley Pointe Dr., 556-5200;
FAX 556-5200
Holiday Inn International Airport, I-26 &
W. Aviation, 744-1621; FAX 744-0942
Holiday Inn—Mt. Pleasant, 250 Johnnie
Dodds Blvd., Mt. Pleasant, 884-6000;
FAX 881-1786
Holiday Inn—Riverview, 301 Savannah
Hwy., 556-7100; FAX 556-6176
Mills House Hotel, 115 Meeting St.,
577-2400; FAX 722-2112
Northwoods Atrium Inn—Best Western,
7401 Northwoods Blvd., 572-2200;
FAX 863-8316
Omni Hotel at Charleston Place, 130
Market St., 722-4900; FAX 722-0728
Quality Inn Heart of Charleston, 125
Calhoun St., 722-3391; FAX 577-0361
✈ Radisson, I-26 & E. Aviation Ave.,
744-2501; FAX 744-2501, ext. 11
✈ Ramada Inn Airport, I-26 & W.
Montague, 744-8281; FAX 744-6230
Sheraton Charleston, 170 Lockwood Dr.,
723-3000; FAX 723-3000, ext. 1595

SELECTED RESTAURANTS:

Signature Food: she crab soup; benne
wafers

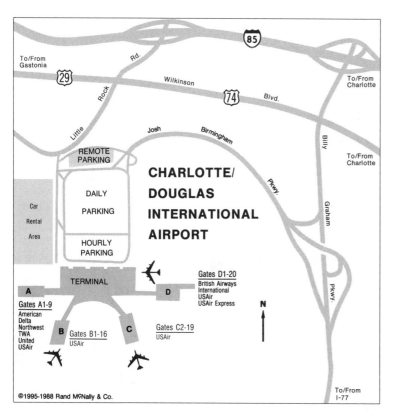

Barbadoes Room, in the Mills House
Hotel, 577-2400
Carolina's, 10 Exchange St., 724-3800
French Quarter, in the Lodge Alley Inn,
195 E. Bay St., 722-1611
Henry's, 54 North Market St., 723-4363
Marianne, 235 Meeting St., 722-7196
The Palmetto Cafe, in the Omni Hotel at
Charleston Place, 722-4900
Papillon, 32 North Market St., 723-3614
Shem Creek Bar & Grill, 508 Mill St., Mt.
Pleasant, 884-8102

SELECTED ATTRACTIONS:

Audubon Swamp Garden, 10 mi. NW on
SC Hwy. 61, enter at Magnolia
Plantation, 571-1266
The Charleston Museum, 360 Meeting St.,
722-2996
Charles Towne Landing 1670, 1500 Old
Town Rd., 852-4200
Drayton Hall Plantation, 3380 Ashley
River Rd., 766-0188
Fort Moultrie, 1214 Middle St. on
Sullivan's Island, 883-3123
Fort Sumter tours, City Marina, 722-1691
Gibbes Museum of Art, 135 Meeting St.,
722-2706
Heyward-Washington House, 87 Church
St., 722-0354
Magnolia Plantation & Gardens, 10 mi.
NW on SC Hwy. 61, 571-1266
Patriots Point Naval & Maritime Museum,
40 Patriots Point Rd., Mt. Pleasant,
884-2727

INFORMATION SOURCES:

Charleston Trident Convention & Visitors
Bureau
81 Mary St.
P.O. Box 975
Charleston, South Carolina 29402
(803) 853-8000; (800) 868-8118
Visitor Reception & Transportation Center
375 Meeting St.
Charleston, South Carolina 29402-0975
(803) 853-8000
Charleston Trident Chamber of Commerce
81 Mary St.
Charleston, South Carolina 29402-0975
(803) 577-2510

Charlotte, North Carolina

Population: (*1,162,093) 395,934 (1990C)
Altitude: 700 feet
Average Temp.: Jan., 52°F.; July, 88°F.
Telephone Area Code: 704
Time: 375-6711   **Weather:** 359-8466
Time Zone: Eastern

AIRPORT TRANSPORTATION:

Ten miles to downtown Charlotte.
Taxicab and hotel shuttle service.

SELECTED HOTELS:

Adam's Mark Charlotte, 555 S. McDowell
St., 372-4100; FAX 348-4645
Charlotte Marriott City Center, 100 W.
Trade, 333-9000; FAX 342-3419
✈ Comfort Inn Airport, I-85 at Little Rock
Rd., 394-4111; FAX 394-4111
Embassy Suites, 4800 S. Tryon, 527-8400;
FAX 527-7035
Holiday Inn, 3501 E. Independence Blvd.,
537-1010; FAX 531-2439

© RAND McNALLY & CO.   PRINTED IN U.S.A.

Chicago
Illinois

scale 0 1 2 3 4 5 miles

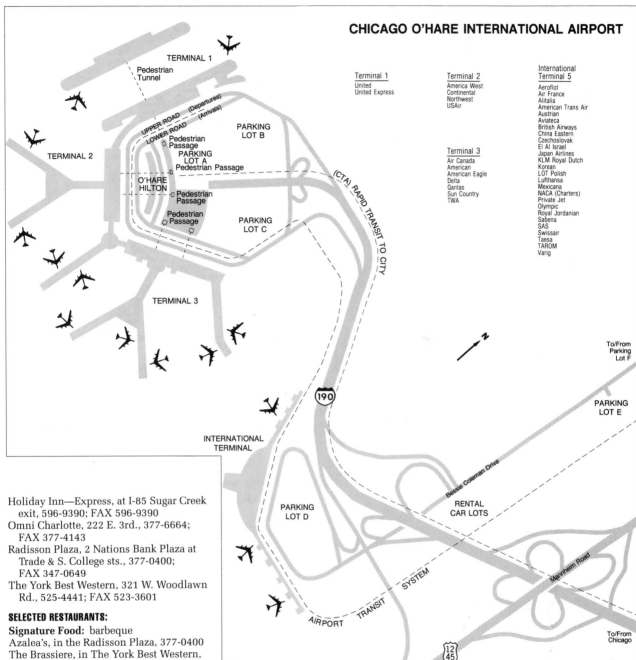

CHICAGO O'HARE INTERNATIONAL AIRPORT

Terminal 1
United
United Express

Terminal 2
America West
Continental
Northwest
USAir

Terminal 3
Air Canada
American
American Eagle
Delta
Qantas
Sun Country
TWA

International Terminal 5
Aeroflot
Air France
Alitalia
American Trans Air
Austrian
Aviateca
British Airways
China Eastern
Czechoslovak
El Al Israel
Japan Airlines
KLM Royal Dutch
Korean
LOT Polish
Lufthansa
Mexicana
NACA (Charters)
Private Jet
Olympic
Royal Jordanian
Sabena
SAS
Swissair
Taesa
TAROM
Varig

©1995 Rand McNally & Co.

Holiday Inn—Express, at I-85 Sugar Creek
exit, 596-9390; FAX 596-9390
Omni Charlotte, 222 E. 3rd., 377-6664;
FAX 377-4143
Radisson Plaza, 2 Nations Bank Plaza at
Trade & S. College sts., 377-0400;
FAX 347-0649
The York Best Western, 321 W. Woodlawn
Rd., 525-4441; FAX 523-3601

SELECTED RESTAURANTS:
Signature Food: barbeque
Azalea's, in the Radisson Plaza, 377-0400
The Brassiere, in The York Best Western,
525-4441
Epicurean, 1324 East Blvd., 377-4529
Hereford Barn Steak House, 4320 N. I-85
Service Rd., 596-0854
The Lamplighter, 1065 E. Morehead,
372-5343
Ranch House, 5614 Wilkinson Blvd.,
399-5411
Slug's 30th Edition, 2 First Union Tower,
30th Floor, 372-7778

SELECTED ATTRACTIONS:
Charlotte Motor Speedway, Concord,
455-3200
Discovery Place (science museum,
Omnimax Theatre & Planetarium), 301
N. Tryon St., 372-6261
Mint Museum of Art, 2730 Randolph Rd.,
337-2000
Paramount's Carowinds (theme park), I-77
& Carowinds Blvd. (exit 90), 588-2606

Reed Gold Mine State Historic Site, 20 mi.
east of Charlotte off Albemarle Rd. (NC
Hwy. 24/27), 786-8337

INFORMATION SOURCES:
Charlotte Convention & Visitors Bureau
122 E. Stonewall St.
Charlotte, North Carolina 28202-1838
(704) 334-2282; (800) 231-4636
Charlotte Chamber of Commerce
129 W. Trade St.
P. O. Box 32785
Charlotte, North Carolina 28232
(704) 378-1300

Chicago, Illinois

Population: (*6,069,974) 2,783,726
(1990C)
Altitude: 596 feet
Average Temp.: Jan., 27°F.; July, 75°F.
Telephone Area Code: 312
Time and Weather: 976-2100
Time Zone: Central

AIRPORT TRANSPORTATION:
Nineteen miles from O'Hare to downtown
Chicago; 10 miles from Midway to
downtown Chicago.
Taxicab, bus/rapid transit, and limousine
bus service from both airports.

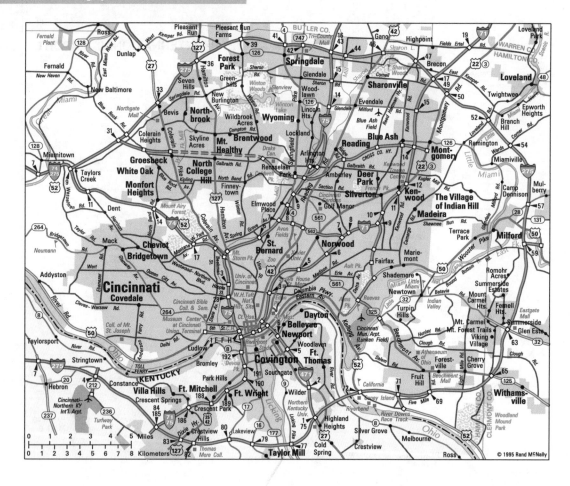

© 1995 Rand McNally

SELECTED HOTELS:

Barclay Chicago, 166 E. Superior, 787-6000; FAX 787-4331

Chicago Marriott Hotel, 540 N. Michigan Ave., 836-0100; FAX 836-6139

Courtyard by Marriott, 30 E. Hubbard, 329-2500; FAX 329-0293

Days Inn, 644 N. Lake Shore Dr., 943-9200; FAX 649-5580

The Drake, 140 E. Walton Place, 787-2200; FAX 951-5803

Fairmont Hotel, 200 N. Columbus Dr. at Illinois Center, 565-8000; FAX 856-1032

Holiday Inn, 350 N. Orleans St., 836-5000; FAX 222-9508

Hotel Nikko, 320 N. Dearborn, 744-1900; FAX 527-2650

Hyatt Regency Chicago, 151 E. Wacker Dr., 565-1234; FAX 565-2966

Hyatt Regency O'Hare, 9300 W. Bryn Mawr Ave., Rosemont, (708) 696-1234; FAX (708) 698-0139

Marriott O'Hare, 8535 W. Higgins Rd., 693-4444; FAX 714-4294

Omni Ambassador East Hotel, 1301 N. State St., 787-7200; FAX 787-4760

✈ Omni O'Hare Hilton, at O'Hare International Airport, 686-8000; FAX 601-2873

Palmer House Hilton, 17 E. Monroe St., 726-7500; FAX 263-2556

Park Hyatt, Chicago Ave. & Rush St., 280-2222; FAX 280-1963

Radisson Plaza Ambassador West Hotel, 1300 N. State St., 787-7900; FAX 787-2067

The Ritz-Carlton, Chicago, 160 E. Pearson St., 266-1000; FAX 266-1194

Swissotel Chicago, 323 E. Wacker, 565-0565; FAX 565-0540

Tremont Hotel, 100 E. Chestnut St., 751-1900; FAX 280-2111

The Westin Hotel, 909 N. Michigan Ave., 943-7200; FAX 943-9347

SELECTED RESTAURANTS:

Signature Food: deep dish pizza; Chicago-style hotdog

Biggs Restaurant, 1150 N. Dearborn, 787-0900

Cape Cod Room, in the Drake Hotel, 787-2200

Cricket's, in the Tremont Hotel, 751-1900

Gordon Restaurant, 500 N. Clark, 467-9780

House of Hunan, 535 N. Michigan Ave., 329-9494

Jaxx, in the Park Hyatt, 280-2222

Lawry's The Prime Rib, 100 E. Ontario St., 787-5000

Michael Jordan's—The Restaurant, 500 N. La Salle, 644-3865

Nick's Fishmarket, 1 First National Plaza, Monroe at Dearborn, 621-0200

Pizzeria Uno, 29 E. Ohio St., 321-1000

Planet Hollywood, 633 N. Wells, 266-7827

The Pump Room, in the Omni Ambassador East Hotel, 266-0360

Signature Room at the Ninety-Fifth, 172 E. Chestnut, in the John Hancock Center, 787-9596

Su Casa, 49 E. Ontario St., 943-4041

SELECTED ATTRACTIONS:

Adler Planetarium, 1300 S. Lake Shore Dr., 322-0300

Art Institute, Michigan Ave. at Adams St., 443-3600

Brookfield Zoo, 31st St. & 1st Ave., Brookfield, (708) 485-0263

Field Museum of Natural History, Roosevelt Rd. at S. Lake Shore Dr., 922-9410

John G. Shedd Aquarium, 1200 S. Lake Shore Dr., 939-2426

Lincoln Park Zoo, Fullerton Ave. & N. Lake Shore Dr., 294-4660

The "Magnificent Mile" (shopping), N. Michigan Ave.

Museum of Science & Industry, 57th St. & S. Lake Shore Dr., 684-1414

Sears Tower, 233 S. Wacker Dr., 875-9696

Spertus Museum of Judaica, 618 S. Michigan Ave., 922-9012

INFORMATION SOURCES:

Chicago Convention & Tourism Bureau, Inc.
McCormick Place-on-the-Lake
2301 S. Lake Shore Dr.
Chicago, Illinois 60616-1497
(312) 567-8500

Chicago Office of Tourism
78 E. Washington
Chicago, Illinois 60602
(312) 744-2400

Chicagoland Chamber of Commerce
200 N. LaSalle St., 6th Floor
Chicago, Illinois 60611
(312) 494-6700

Cincinnati, Ohio

Population: (*1,452,645 364,040 (1990C)
Altitude: 683 feet
Average Temp.: Jan., 35°F.; July, 78°F.
Telephone Area Code: 513
Time: 721-1700 **Weather:** 241-1010
Time Zone: Eastern

AIRPORT TRANSPORTATION:

Thirteen miles to downtown Cincinnati.
Taxicab and limousine bus service.

SELECTED HOTELS:

Cincinnati Marriott, 11320 Chester Rd.,
   772-1720; FAX 772-6466
The Clarion Hotel of Cincinnati, 141 W.
   6th St., 352-2100; FAX 352-2148
Days Inn Central, 8001 Reading Rd.,
   821-5110; FAX 821-8689
Harley Hotel of Cincinnati, 8020
   Montgomery Rd., 793-4300;
   FAX 793-1413
Holiday Inn, 3855 Hauk Rd., 563-8330;
   FAX 563-9679
Howard Johnson Plaza, 11440 Chester
   Rd., 771-3400; FAX 771-6340
Hyatt Regency, 151 W. 5th St., 579-1234;
   FAX 579-0107
Imperial House—West, 5510 Rybolt Rd.,
   574-6000; FAX 574-6566
Omni Netherland Plaza, 35 W. 5th St.,
   421-9100; FAX 421-4291
Terrace Hotel, 15 W. 6th St., 381-4000;
   FAX 381-5158
The Vernon Manor Hotel, 400 Oak St.,
   281-3300; FAX 281-8933
The Westin Hotel, 5th & Vine, 621-7700;
   FAX 852-5670

SELECTED RESTAURANTS:

Signature Food: chili
Celestial, 1071 Celestial St., 241-4455
La Normandie Grill, 118 E. 6th St.,
   721-2761
Maisonette, 114 E. 6th St., 721-2260
The Orchids at the Palm Court, in the
   Omni Netherland Plaza, 421-9100
The Palace Restaurant, in the
   Cincinnatian Hotel, 601 Vine St.,
   381-3000
The Terrace Garden, in the Terrace Hilton,
   381-4000
Windjammer, 11330 Chester Rd.,
   Sharonville, 771-3777

SELECTED ATTRACTIONS:

Children's Museum of Cincinnati,
   Longworth Hall, 700 Pete Rose Way,
   421-5437
Cincinnati Fire Museum, 315 W. Court St.,
   621-5553
Cincinnati Zoo, 3400 Vine St., 281-4700
The Museum Center (Museum of Natural
   History, Cincinnati Historical Society &
   Omnimax Theatre), 1301 Western Ave.,
   287-7000
Paramount's Kings Island (theme park),
   Kings Island, 241-5600
Showboat Majestic (live stage and musical
   shows), foot of Broadway, 241-6550
   (Jan.–Oct.)
Tower Place at the Carew Tower
   (shopping), 4th & Race sts., 241-7700
William Howard Taft National Historic
   Site, 2038 Auburn Ave., 684-3262

**CINCINNATI/NORTHERN KENTUCKY
INTERNATIONAL AIRPORT**

©1995-1983 Rand McNally & Co.

INFORMATION SOURCES:

Greater Cincinnati Convention and
Visitors Bureau
   300 W. 6th St.
   Cincinnati, Ohio 45202
   (513) 621-2142; Visitor Info.
   (800) 344-3445
Greater Cincinnati Chamber of Commerce
   300 Carew Tower
   441 Vine St.
   Cincinnati, Ohio 45202
   (513) 579-3100

Cleveland, Ohio

Population: (*1,831,122) 505,616 (1990C)
Altitude: 570 to 1,050 feet
Average Temp.: Jan., 29°F.; July, 74°F.
Telephone Area Code: 216
Time: 881-0880 **Weather:** 265-2370
Time Zone: Eastern

AIRPORT TRANSPORTATION:

Twelve miles from Hopkins to downtown
   Cleveland.
Taxicab, train, and limousine bus service.

SELECTED HOTELS:

Cleveland Airport Marriott, 4277 W. 150th
   St., 252-5333; FAX 251-1508
Cleveland Hilton South, 6200 Quarry Ln.
   at I-77 & Rockside Rd., 447-1300;
   FAX 642-9334
Cleveland Marriott East, Park East Dr.,
   I-271 & Chagrin Blvd., Beachwood,
   464-5950; FAX 464-6539
Harley Hotel East, 6051 SOM Center Rd.,
   Willoughby, 944-4300; FAX 944-5344
Harley Hotel West, 17000 Bagley Rd.,
   243-5200; FAX 243-5340
Holiday Inn Lakeside City Center, 1111
   Lakeside Ave., 241-5100; FAX 241-7437
Ritz Carlton, 1515 W. 3rd St., 623-1300;
   FAX 623-1492

Sheraton Cleveland City Center, 777 St.
   Clair Ave., 771-7600; FAX 566-0736
✈ Sheraton Hopkins Airport Hotel, 5300
   Riverside Dr. at the airport, 267-1500;
   FAX 267-1500
Stouffer's Tower City Plaza Hotel, 24
   Public Sq., 696-5600; FAX 696-0432

SELECTED RESTAURANTS:

Signature Food: kielbasa; pierogi
Brasserie, in Stouffer's Tower City Plaza
   Hotel, 696-5600
Cafe Sausilito, in the Galleria Mall, E. 9th
   St., 696-2233
Getty's at the Hanna, 1422 Euclid Ave.,
   771-1818
Sammy's, 1400 W. 10th St., 523-5560
Samurai Japanese Steak House, 23611
   Chagrin Blvd., Beachwood, 464-7575
That Place on Bellflower, 11401
   Bellflower Rd., 231-4469

SELECTED ATTRACTIONS:

Cleveland Children's Museum, 10730
   Euclid Ave., 791-7114
Cleveland Health Education Museum,
   8911 Euclid Ave., 231-5010
Cleveland Museum of Art, 11150 E.
   Boulevard, on University Circle,
   421-7340
Cleveland Museum of Natural History,
   University Circle, 231-4600
The Galleria (shopping), East 9th at St.
   Clair, 621-9999
Geauga Lake Amusement Park, 1060
   Aurora Rd., Aurora, 562-7131
Metroparks Zoo, 3900 Brookside Park,
   661-6500
NASA Visitor Center, 21000 Brookpark
   Rd., 433-4000
Sea World of Ohio, 1100 Sea World Dr.,
   Aurora, 562-8101
Trolley Tours of Cleveland, 771-4484

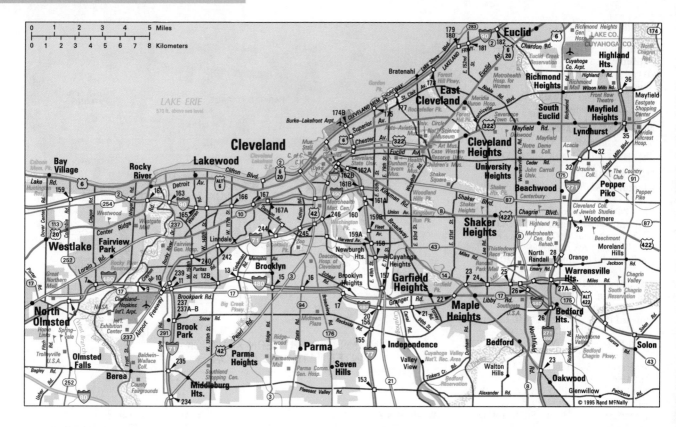

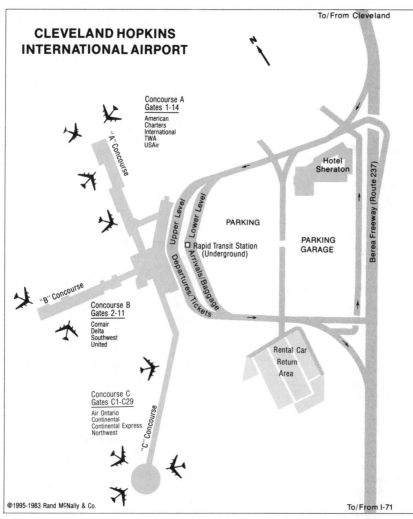

**CLEVELAND HOPKINS
INTERNATIONAL AIRPORT**

To/From Cleveland

Concourse A
Gates 1-14
American
Charters
International
TWA
USAir

"A" Concourse

Hotel
Sheraton

Upper Level
Lower Level

PARKING

PARKING
GARAGE

Berea Freeway (Route 237)

Arrivals/Baggage
Departures/Tickets

Rapid Transit Station
(Underground)

"B" Concourse

Concourse B
Gates 2-11
Comair
Delta
Southwest
United

Rental Car
Return
Area

Concourse C
Gates C1-C29
Air Ontario
Continental
Continental Express
Northwest

"C" Concourse

©1995-1983 Rand McNally & Co.

To/From I-71

INFORMATION SOURCES:

Convention and Visitors Bureau of Greater
Cleveland
    3100 Terminal Tower
    Cleveland, Ohio 44113
    (216) 621-4110; (800) 321-1001
Greater Cleveland Growth Association
    200 Tower City Center
    50 Public Square
    Cleveland, Ohio 44113-2291
    (216) 621-3300

Columbus, Ohio

Population: (*1,377,419) 632,910 (1990C)
Altitude: 685 to 893 feet
Average Temp.: Jan., 31°F.; July, 76°F.
Telephone Area Code: 614
Time and Weather: 469-1010
Time Zone: Eastern

AIRPORT TRANSPORTATION:

Eight miles to downtown Columbus.
Taxicab and limousine bus service.

SELECTED HOTELS:

Best Western North, 888 E. Dublin-
    Granville Rd., 888-8230; FAX 888-8223
Columbus Marriott North, 6500
    Doubletree Ave., 885-1885;
    FAX 885-7222
Columbus Sheraton, 2124 S. Hamilton
    Rd., 861-7220; FAX 866-9067
✈ Concourse Hotel & Conference Center,
    4300 International Gateway, 237-2515;
    FAX 237-6134
Harley Hotel of Columbus, 1000 E.
    Dublin-Granville Rd., 888-4300;
    FAX 888-3477
Hilton Inn North, 7007 N. High St.,
    436-0700; FAX 436-1208

✈ Holiday Inn Columbus Airport, 750
Stelzer/James Rd., 237-6360;
FAX 237-2978

Holiday Inn East, 4560 Hilton Corporate
Dr., 868-1380; FAX 863-1320

Holiday Inn on the Lane, 328 W. Lane
Ave., 294-4848; FAX 294-4848

Holiday Inn West, 2350 Westbelt Dr.
771-8999; FAX 771-8857

Hyatt Regency Columbus, 350 N. High St.,
463-1234; FAX 463-9161

✈ Radisson Airport Hotel, 1375 N.
Cassady Ave., 475-7551; FAX 476-1476

Ramada University Hotel & Conference
Center, 3110 Olentangy River Rd.,
267-7461; FAX 263-5299

SELECTED RESTAURANTS:

Signature Food: the hamburger

The Clarmont, 684 S. High St., 443-1125

Fifty-five on the Boulevard, 55
Nationwide Blvd., 228-5555

Jai Lai, 1421 Olentangy River Rd.,
421-7337

Kahiki, 3583 E. Broad St., 237-5425

One Nation, One Nationwide Plaza, 38th
Floor, 221-0151

River Club, 679 W. Long St., 469-0000

SELECTED ATTRACTIONS:

Columbus Museum of Art, 480 E. Broad
St., 221-6801

Columbus Zoo, 9990 Riverside Dr.,
Powell, 645-3400

COSI (science & industry museum), 280 E.
Broad St., 228-COSI

Fabulous Palace Theatre, 34 W. Broad St.,
469-1331

Franklin Park Conservatory, 1777 E. Broad
St., 645-3000

German Village Society, 588 S. 3rd St.,
221-8888

Ohio Historical Center/Ohio Village, 17th
Ave. & I-71, 297-2300

Wexner Center for the Arts, N. High St. at
15th Ave., 292-0330

Wyandot Lake (water & amusement park),
10101 Riverside Dr., 889-9283

INFORMATION SOURCES:

Greater Columbus Convention & Visitors
Bureau
10 W. Broad St., Suite 1300
Columbus, Ohio 43215
(614) 221-6623; (800) 354-4FUN

Columbus Visitor Center
Columbus Visitor Center, Third
Level (walk-in tourist info.)

Greater Columbus Chamber
37 N. High St.
Columbus, Ohio 43215
(614) 221-1321

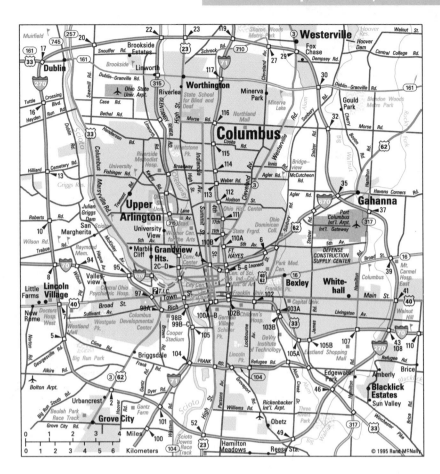

© 1995 Rand McNally

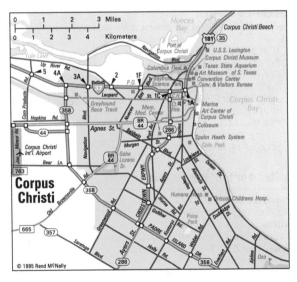

© 1995 Rand McNally

Corpus Christi, Texas

Population: (*349,894) 257,453 (1990C)
Altitude: 35 feet
Average Temp.: Jan., 56°F.; July, 85°F.
Telephone Area Code: 512
Time: 884-8463 **Weather:** 289-1861
Time Zone: Central

AIRPORT TRANSPORTATION:

Nine miles to downtown Corpus Christi.
Taxicab, airport shuttle, and hotel shuttle
service.

SELECTED HOTELS:

Bayfront Inn, 601 N. Shoreline, 883-7271;
FAX 883-2052

Best Western—Sandy Shores, 3200
Surfside, 883-7456; FAX 883-1437

Corpus Christi Marriott, 900 N. Shoreline,
887-1600; FAX 887-6715

Corpus Christi Sheraton, 707 N. Shoreline
Blvd., 882-1700; FAX 882-3113

Holiday Inn—Airport, 5549 Leopard,
289-5100; FAX 289-6209

Holiday Inn—Emerald Beach, 1102 S.
Shoreline, 883-5731; FAX 883-9079

Quality Hotel Bayfront, 601 N. Water,
882-8100; FAX 888-6540

Radisson Marina, 300 N. Shoreline,
883-5111; FAX 883-7702

SELECTED RESTAURANTS:

Signature Food: seafood

Catfish Charlie's, in Crossroads Shopping
Village, Airline & McArdle, 993-0363

Edelweiss, No. 10 Crossroads Shopping
Village, Airline & McArdle, 993-1901

Elmo's Staples St. Grille and Oyster Bar,
5253 S. Staples, 992-3474

Horizons, atop the Radisson Marina
Hotel, 883-5111

La Parisienne, in Lamar Park Center,
Alameda at Doddridge, 857-2736

The Lighthouse Restaurant and Oyster
Bar, Lawrence St., T-Head, 883-3982

Water Street Seafood Company, 309 N.
Water, 882-8683

SELECTED ATTRACTIONS:

Art Museum of South Texas, 1902 N.
Shoreline, 884-3844

Centennial House, 411 Upper N.
Broadway, 992-6003

The Columbus Fleet (Niná, Pinta, Santa
Maria replicas), Cargo Dock 1, 1900 N.
Chaparral, 883-2862

Corpus Christi Botanical Gardens, on S.
Staples past Oso Creek, 852-2100

Corpus Christi Museum of Science &
History, 1900 N. Chaparral, 883-2862

Hans A. Suter Wildlife Refuge, on Ennis
Joslin overlooking Oso Bay

Heritage Park (restored houses), bordered
by N. Chaparral & Mesquite sts.,
883-0639

International Kite Museum, in the Best
Western Sandy Shores, 3200 Surfside
Blvd., 883-7456

Museum of Oriental Cultures, 418 Peoples
St., Furman Plaza, 2nd Fl., 883-1303

Texas State Aquarium, underneath Harbor
Bridge on Corpus Christi Beach,
881-1200

U.S.S. *Lexington* Museum on the Bay, off
Corpus Christi Beach, 888-4873

INFORMATION SOURCES:

Corpus Christi Area Convention & Visitors
Bureau
1201 N. Shoreline
P.O. Box 2664
Corpus Christi, Texas 78403-2664
(512) 882-5603; (800) 678-OCEAN

Corpus Christi Chamber of Commerce
1201 N. Shoreline
P.O. Box 640
Corpus Christi, Texas 78403
(512) 882-6161

Dallas-Fort Worth, Texas

Population: (*3,885,415) (Dallas)
1,006,877; (Fort Worth) 447,619 (1990C)

Altitude: 463 to 750 feet

Average Temp.: Jan., 44°F.; July, 86°F.

Telephone Area Code: (Dallas) 214
(Fort Worth) 817

Time: (Dallas) 349-8367; (Fort Worth)
844-6611 **Weather:** (214) 787-1111

Time Zone: Central

AIRPORT TRANSPORTATION:

About 17 miles to downtown Dallas or
Fort Worth.

Taxicab and limousine bus service.

SELECTED HOTELS: DALLAS

Adolphus Hotel, 1321 Commerce St.,
742-8200; FAX 651-3588

Dallas Grand Hotel, 1914 Commerce St.,
747-7000; FAX 749-0231

Delux Inn, 3111 Stemmons Frwy.,
637-0060; FAX 637-3441

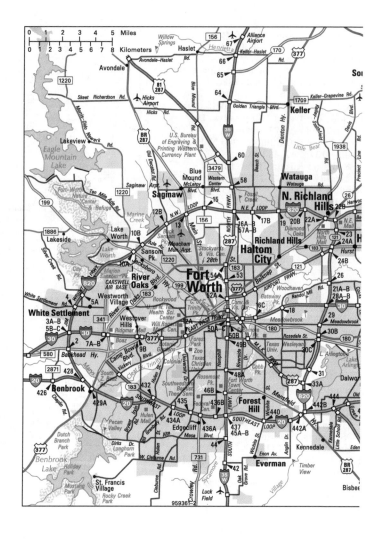

Fairmont Hotel, 1717 N. Akard St.,
720-2020; FAX 720-4015

Hyatt Regency Dallas, 300 Reunion Blvd.,
651-1234; FAX 742-8126

✈ Hyatt Regency—Dallas-Fort Worth
Airport, International Pkwy., 453-8400;
FAX 456-8668

Loews Anatole Hotel, 2201 Stemmons
Frwy., 748-1200; FAX 761-7520

Mansion on Turtle Creek, 2821 Turtle
Creek Blvd., 559-2100; FAX 528-4187

The Omni Mandalay, 221 E. Las Colinas
Blvd., Irving, 556-0800; FAX 556-0729

Plaza of the Americas Hotel, 650 N. Pearl
St., 979-9000; FAX 953-1931

Ramada Inn—Market Center, 1055 Regal
Row, 634-8550; FAX 634-8418

Sheraton Park Central Hotel & Towers,
12720 Merit Dr., 385-3000;
FAX 991-4557

Stouffer Dallas Hotel, 2222 N. Stemmons
Frwy., 631-2222; FAX 905-3814

The Westin Hotel, 13340 Dallas Pkwy.,
934-9494; FAX 851-2869

SELECTED RESTAURANTS: DALLAS

Signature Food: Dallas barbeque;
Mexican; home cooking

Baby Routh, 2708 Routh St., 871-2345

Beau Nash, Hotel Crescent Court, 400
Crescent Ct., 871-3200

Il Sorrento, 8616 Turtle Creek Blvd.,
352-8759

L'Entrecote, in the Loew's Anatole Hotel,
748-1200

Old Warsaw, 2610 Maple Ave., 528-0032

The Pyramid Room, Fairmont Hotel,
720-2020

650 North, Plaza of the Americas Hotel,
979-9000

SELECTED ATTRACTIONS: DALLAS

Biblical Arts Center, 7500 Park Lane,
691-4661

Dallas Arboretum & Botanical Gardens,
8525 Garland Rd., 327-8263

Dallas Museum of Art, 1717 N. Harwood
St., 922-1200

Dallas Zoo, in Marsalis Park, 621 E.
Clarendon, 946-5154

Fair Park (museums), 670-8400

The Meadows Museum (14th-20th
century Spanish art), Meadows School
of the Arts, Bishop Blvd., Southern
Methodist University campus, 768-2516

Old City Park (museum village), 1717
Gano, 421-5141

The Sixth Floor (President John F.
Kennedy educational exhibit), Houston
& Elm St., 653-6666

West End Historical District, on the west
end of Dallas' central business district

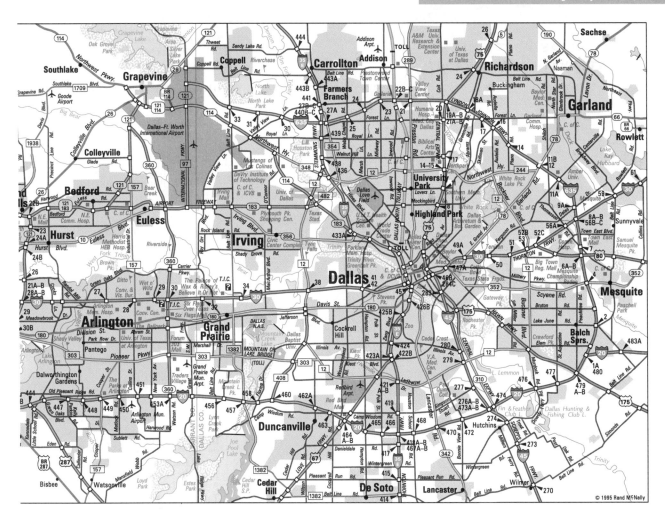

© 1995 Rand McNally

INFORMATION SOURCES: DALLAS

Dallas Convention & Visitors Bureau
1201 Elm St., Suite 2000
Dallas, Texas 75270
(214) 746-6677
Greater Dallas Chamber of Commerce
Information Department
1201 Elm St., Suite 2000
Dallas, Texas 75270
(214) 746-6700

SELECTED HOTELS: FORT WORTH

Best Western Inn, Jct. TX 183 & I-820, 284-9461; FAX 284-2126
Clarion Hotel Fort Worth, 2000 Beach St., 534-4801; FAX 534-3761
Days Inn, 1010 Houston St., 336-2011; FAX 336-0623
Green Oaks Inn, 6901 W. Frwy., 738-7311; FAX 377-1308
HoJo Inn, I-35 W. South at Seminary Exit, 923-8281; FAX 926-8756
La Quinta—Fort Worth West, 7888 I-30 W. at Cherry Ln., 246-5511; FAX 246-8870
Radisson Plaza, 815 Main St., 870-2100; FAX 882-1300
Ramada Hotel, 1701 Commerce St., 335-7000; FAX 335-3333
Remington Hotel, 600 Commerce St., 332-6900; FAX 877-5440
The Worthington, 200 Main St., 870-1000; FAX 338-9176

SELECTED RESTAURANTS: FORT WORTH

Signature Food: beef barbeque; Mexican; home cooking
The Balcony, 6100 Camp Bowie Blvd., 731-3719
Cactus Bar & Grill, in the Radisson Plaza, 870-2100
The Cattle Drive, 1900 Ben Ave., 534-4908
Juanita's, 115 W. 2nd St., 335-1777
The Keg, 1309 Calhoun, 332-1288
7th St. Cafe, 3500 W. 7th, 870-1672

SELECTED ATTRACTIONS: FORT WORTH

Botanic Gardens, 3220 Botanic Garden Blvd., 871-7686
Cattleman's Museum, 1301 W. Seventh St., 332-7064
Fort Worth Museum of Science and History, 1501 Montgomery St., 732-1631
Fort Worth Zoo, 1989 Colonial Pkwy., 871-7050 or -7051
Kimbell Art Museum, 3333 Camp Bowie Blvd., 332-8451
Log Cabin Village, on University across from the zoo, 926-5881
Modern Art Museum of Fort Worth, 1309 Montgomery St., 738-9215
Sid Richardson Collection of Western Art, 309 Main St., 332-6554
Stockyard Station Market, 130 E. Exchange, 625-9715

Tarantula Railroad, 2318 8th Ave., 625-RAIL

INFORMATION SOURCES: FORT WORTH

Fort Worth Convention & Visitors Burea
415 Throckmorton St.
Fort Worth, Texas 76102
(817) 336-8791; (800) 433-5747
Fort Worth Chamber of Commerce
777 Taylor, Suite 900
Fort Worth, Texas 76102-4997
(817) 336-2491

Dayton, Ohio

Population: (*951,270) 182,044 (1990C)
Altitude: 757 feet
Average Temp.: Jan., 26°F.; July, 74°F.
Telephone Area Code: 513
Time and Weather: 499-1212
Time Zone: Eastern

AIRPORT TRANSPORTATION:

Seven miles to downtown Dayton.
Taxicab and limousine service.

SELECTED HOTELS:

Days Inn Downtown, 330 W. 1st St., 223-7131; FAX 223-7131
Dayton Marriott Hotel, 1414 S. Patterson Blvd., 223-1000; FAX 223-7853

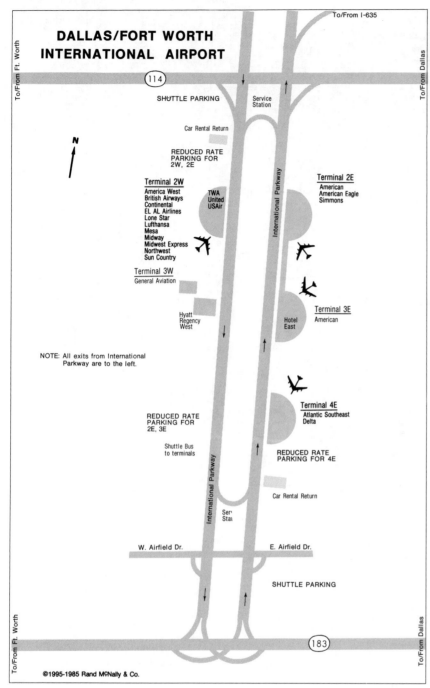

DALLAS/FORT WORTH INTERNATIONAL AIRPORT

To/From Ft. Worth

To/From I-635

To/From Dallas

114

SHUTTLE PARKING

Service Station

Car Rental Return

REDUCED RATE PARKING FOR 2W, 2E

N

Terminal 2W
America West
British Airways
Continental
EL AL Airlines
Lone Star
Lufthansa
Mesa
Midway
Midwest Express
Northwest
Sun Country

TWA
United
USAir

International Parkway

Terminal 2E
American
American Eagle
Simmons

Terminal 3W
General Aviation

Hyatt Regency West

Terminal 3E
American

Hotel East

NOTE: All exits from International Parkway are to the left.

Terminal 4E
Atlantic Southeast
Delta

REDUCED RATE PARKING FOR 2E, 3E

Shuttle Bus to terminals

International Parkway

REDUCED RATE PARKING FOR 4E

Car Rental Return

Serv Sta

W. Airfield Dr.

E. Airfield Dr.

SHUTTLE PARKING

To/From Ft. Worth

To/From Dallas

183

©1995-1985 Rand McNally & Co.

Carriage Hill Farm (1880s living history farm), 7860 Shull Rd., 879-0461

The Citizens Motorcar Company, A Packard Museum, 420 S. Ludlow St., 226-1917

Dayton Art Institute, Forest & Riverview, 223-5277

Dayton Museum of Natural History, 2600 DeWeese Pkwy., 275-7431

International Women's Air and Space Museum, 26 N. Main St., Centerville, 433-6766

SunWatch (archaeological Indian site), 2301 W. River Rd., 268-8199

United States Air Force Museum/IMAX Theatre, Springfield & Harshman, 255-3284

Wright Brothers Bicycle Shop (open weekends, or by appointment), 22 S. Williams St., 443-0793

INFORMATION SOURCES:

Dayton/Montgomery County Convention and Visitors Bureau
  1 Chamber Plaza
  Dayton, Ohio 45402-2400
  (513) 226-8211
  800-221-8234 (Ohio)
  800-221-8235 (Elsewhere)
Dayton Chamber of Commerce
  1 Chamber Plaza
  Dayton, Ohio 45402-2400
  (513) 226-1444

Denver, Colorado

Population: (*1,622,980) 467,610 (1990C)
Altitude: 5,280 feet
Average Temp.: Jan., 31°F.; July, 74°F.
Telephone Area Code: 303
Time: 850-7050 **Weather:** 337-2500
Time Zone: Mountain

AIRPORT TRANSPORTATION:

Twenty-three miles to downtown Denver. Bus, taxicab, and limousine bus service.

SELECTED HOTELS:

Brown Palace Hotel, 321 17th St., 297-3111; FAX 293-9204
Burnsley All Suite Hotel, 1000 Grant St., 830-1000; FAX 830-7676
Denver Marriott City Center, 1701 California, 297-1300; FAX 298-7474
Denver Marriott Southeast, 6363 E. Hampden Ave., 758-7000; FAX 782-3260
Holiday Inn Denver Southeast, I-225 & South Parker Rd., 695-1700; FAX 745-6958
Hotel Denver Downtown, 1450 Glenarm Pl., 573-1450; FAX 572-1113
Hyatt Regency Denver, 1750 Welton St., 295-1200; FAX 292-2472
Oxford Hotel, 1600 17th St., 628-5400; FAX 628-5413
Radisson Hotel Denver, 1550 Court Pl., 893-3333; FAX 892-0521
Red Lion Hotel, 3203 Quebec St., 321-3333; FAX 329-5233
Stapleton Plaza Hotel and Fitness Center, 3333 Quebec St., 321-3500; FAX 322-7343

Holiday Inn—Englewood, 10 Rockridge Rd., Englewood, 832-1234; FAX 832-3548
Radisson Hotel & Suites, Ludlow at 3rd, 461-4700; FAX 224-9160
Radisson Inn Dayton North, 2401 Needmore Rd., 278-5711; FAX 278-6048
Ramada Inn, 800 N. Broad St., Fairborn, 879-3920; FAX 879-3896
Ramada Inn North, 4079 Little York Rd., 890-9500; FAX 890-8525
Stouffer Center Plaza, 5th & Jefferson sts., 224-0800; FAX 224-3913

SELECTED RESTAURANTS:

Signature Food: American
Jay's Seafood, 225 E. 6th St., 222-2892
King Cole Restaurant, in the Kettering Tower, 2nd & Main sts., 222-6771
L'Auberge. 4120 Far Hills Ave., 299-5536
Peerless Mill Inn, 319 S. 2nd St., Miamisburg, 866-5968
The Pine Club, 1926 Brown St., 228-7463
Windows, in the Stouffer Center Plaza, 224-0800

SELECTED ATTRACTIONS:

Aviation Trail (self-guided driving tour), 443-0793 (weekends)

SELECTED RESTAURANTS:

Signature Food: buffalo

Ellyngton's, at the Brown Palace Hotel, 297-3111

European Café, 1515 Market, 825-6555

The Fort, CO Hwy. 8 near Morrison, 697-4771

Marlowe's, Glenarm & 16th St., 595-3700

Normandy French Restaurant, 1515 Madison St., 321-3311

Palace Arms, at the Brown Palace Hotel, 297-3111

Racines, 850 Bannock, 595-0418

Strings, 1700 Humboldt, 831-7310

Tante Louise, 4900 E. Colfax Ave., 355-4488

SELECTED ATTRACTIONS:

Buffalo Bill's Grave and Museum, top of Lookout Mountain, 526-0747

Children's Museum of Denver, 2121 Children's Museum Dr., 433-7444

Colorado Railroad Museum, 17155 W. 44th Ave., Golden, 279-4591

Coors Brewing Company (tours), 13th & Ford sts., Golden, 277-BEER

Denver Center for the Performing Arts, 14th & Curtis sts., 893-4100

Denver Museum of Natural History, 20th & Colorado Blvd. in City Park, 322-7009

Denver's Zoo, 23rd and Steele St., in City Park, 331-4110

Larimer Square (restored Victorian-era section of Denver), 1400 block of Larimer

Molly Brown House Museum, 1340 Pennsylvania, 832-4092

United States Mint, 320 W. Colfax Ave., 844-3582

INFORMATION SOURCES:

Denver Metro Convention and Visitors Bureau
225 W. Colfax Ave.
Denver, Colorado 80202
(303) 892-1112; (800) 888-1990

Denver Metro Chamber of Commerce
1445 Market St.
Denver, Colorado 80202
(303) 534-8500

Des Moines, Iowa

Population: (*392,928) 193,187 (1990C)
Altitude: 803 feet
Average Temp.: Jan., 21°F.; July, 76°F.
Telephone Area Code: 515
Time: 244-5611 **Weather:** 270-2614
Time Zone: Central

AIRPORT TRANSPORTATION:

Six miles to downtown Des Moines.
Taxicab and limousine service.

SELECTED HOTELS:

Best Western Bavarian Inn, 5220 NE 14th St., 265-5611; FAX 265-1669

✈ Best Western—Des Moines International, 1810 Army Post Rd., 287-6464; FAX 287-5818

Best Western—Des Moines West, 11040 Hickman Rd., 278-5575; FAX 278-4078

Best Western Starlite Village, 133 SE Delaware, Ankeny, 964-1717; FAX 964-8781

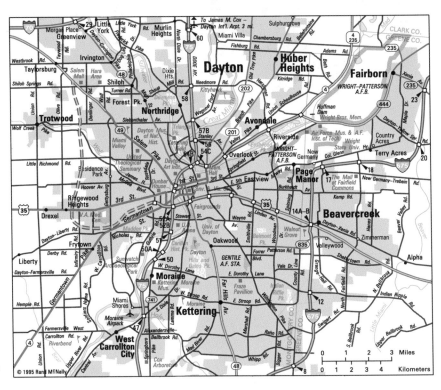

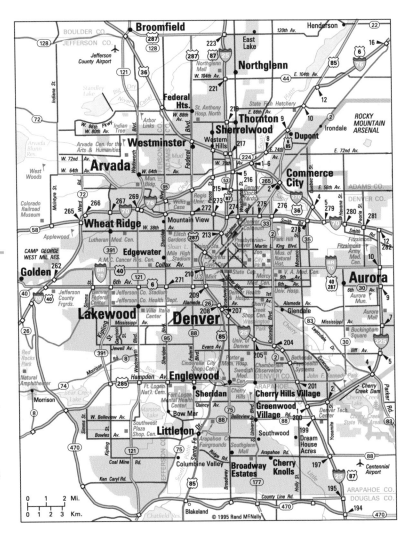

© 1995 Rand McNally

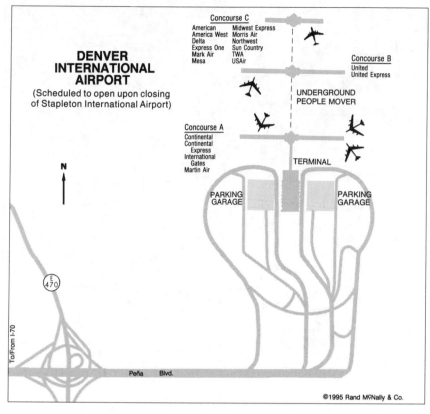

DENVER
INTERNATIONAL
AIRPORT
(Scheduled to open upon closing
of Stapleton International Airport)

Concourse C
American   Midwest Express
America West   Morris Air
Delta   Northwest
Express One   Sun Country
Mark Air   TWA
Mesa   USAir

Concourse B
United
United Express

UNDERGROUND
PEOPLE MOVER

Concourse A
Continental
Continental
Express
International
Gates
Martin Air

TERMINAL

N

PARKING
GARAGE

PARKING
GARAGE

E 470

To/From I-70

Peña   Blvd.

©1995 Rand McNally & Co.

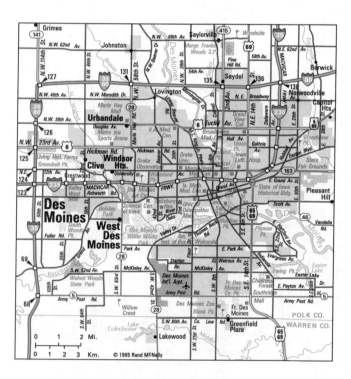

Best Western Starlite Village, 929 3rd St.,
   282-5251; FAX 282-5251
Comfort Suites, 11167 Hickman Rd.,
   276-1126; FAX 276-1126
✈ Crystal Inn & Conference Center, 6111
   Fleur Dr., 287-2400; FAX 287-4811
Des Moines Marriott, 700 Grand
   245-5500; FAX 245-5567
Executive Inn, 3530 Westown Pkwy., West
   Des Moines, 225-1144; FAX 225-6463
Holiday Inn—Downtown, 1050 6th Ave.,
   283-0151; FAX 283-0151, ext. 7055
Holiday Inn Merle Hay, 5000 Merle Hay
   Rd., 278-0271; FAX 276-8172
University Park Holiday Inn, 1800 50th
   St., West Des Moines, 223-1800;
   FAX 223-0894

SELECTED RESTAURANTS:

Signature Food: Iowa pork chop;
   corn-on-the-cob
Bavarian Haus, in the Best Western
   Bavarian Inn, 266-1173
The Gotham Club, in the Hotel Fort Des
   Moines, 10th & Walnut, 243-1161
The Mandarin Chinese Restaurant, 3520
   Beaver, 277-6263
Metz Continental Cuisine, 303 Locust,
   Homestead Bldg., 246-1656
The New Wall Street, 35th & University,
   Indian Hills Shopping Center 224-1184
Quenelles, in the Marriott Hotel, 700
   Grand, 245-5500
Shogun Steak House of Japan, 2900
   University, in Clocktower Square, West
   Des Moines, 225-3325
Waterfront Seafood Market, 2900
   University Ave., in Clocktower Square,
   West Des Moines, 223-5106

SELECTED ATTRACTIONS:

Adventureland Theme Park, Jct. I-80 &
   U.S. 65, Altoona, 266-2121
Blank Park Zoo, 7401 SW 9th St.,
   285-4722
Des Moines Art Center, 4700 Grand Ave.,
   277-4405
Historic Valley Junction, 5th St. and
   surrounding areas, West Des Moines,
   223-3286
Living History Farms, 2600 NW 111th St.,
   278-5286
Prairie Meadows (horse track), 1 Prairie
   Meadows Dr., Altoona, 967-1000
Salisbury House (museum: weekday tours
   by appointment), 4025 Tonawanda Dr.,
   279-9711
Science Center of Iowa, 4500 Grand Ave.,
   274-4138
Terrace Hill (mansion), 2300 Grand Ave.,
   281-3604
White Water University (water park), 5401
   University, 265-4904

INFORMATION SOURCES:

Greater Des Moines Convention & Visitors
Bureau
   Ruan II, Suite 222
   601 Locust
   Des Moines, Iowa 50309
   (515) 286-4960; (800) 451-2625
Greater Des Moines Chamber of
Commerce Federation
   601 Locust, Suite 100
   Des Moines, Iowa 50309
   (515) 286-4950

Detroit, Michigan

Population: (*4,382,299) 1,027,974
(1990C)
Altitude: 573 to 672 feet
Average Temp.: Jan., 26°F.; July, 73°F.
Telephone Area Code: 313
Time: 472-1212 **Weather:** 941-7192
Time Zone: Eastern

AIRPORT TRANSPORTATION:

Nineteen miles from Metropolitan Airport
to downtown Detroit.
Taxicab and limousine bus service.

SELECTED HOTELS:

✈ Clarion Hotel Detroit Metro Airport,
31200 Industrial Expwy., 728-2800;
FAX 728-2260
Hotel St. Regis, 3071 W. Grand Blvd.,
873-3000; FAX 873-2574
Hyatt Regency Dearborn, Fairlane Town
Center Dr., Dearborn, 593-1234;
FAX 593-3366
Northfield Hilton, 5500 Crooks Rd., Troy,
(810) 879-2100; FAX (810) 879-6054
Omni International, 333 E. Jefferson,
222-7700; FAX 222-6509
The Plaza, 16400 J.L. Hudson Dr.,
Southfield, (810) 559-6500;
FAX 559-3625
Radisson Hotel Pontchartrain, 2
Washington Blvd., 965-0200;
FAX 965-9464
Royce Hotel, 31500 Wick Rd., Romulus,
467-8000; FAX 721-8870
The Westin Hotel, Renaissance Center,
568-8000; FAX 568-8146

SELECTED RESTAURANTS:

Signature Food: Vernor ginger ale;
Coney Islands
Carl's Chop House, 3020 Grand River
Ave., 833-0700
Caucus Club, 150 W. Congress St.,
965-4970
Charley's Crab, 5498 Crooks Rd., Troy,
(810) 879-2060
The Golden Mushroom, 18100 W. 10 Mile
Rd., Southfield, (810) 559-4230
Joe Muer's Sea Food, 2000 Gratiot Ave.,
567-1088
Mario's Restaurant, 4222 2nd Ave.,
832-1616
Opus One, 565 E. Larned, 961-7766
St. Regis Restaurant, in the St. Regis
Hotel, 3071 W. Grand Blvd., 873-3000
The Summit, in the Westin Hotel,
568-8000
Van Dyke Place, 649 Van Dyke Ave.,
821-2620

SELECTED ATTRACTIONS:

Cranbrook Academy of Art Museum, 1221
N. Woodward, Bloomfield Hills,
(810) 645-3323
Detroit Institute of Arts, 5200 Woodward
Ave., 833-7900
Detroit Zoological Park, 8450 W. Ten Mile
Rd., Royal Oak, (810) 398-0903
Edsel & Eleanor Ford House, 1100
Lakeshore Rd., Grosse Pointe Shores,
884-3400 or 884-4222
Graystone International Jazz Museum,
1521 Broadway, 963-3813
Henry Ford Estate-Fair Lane, University of

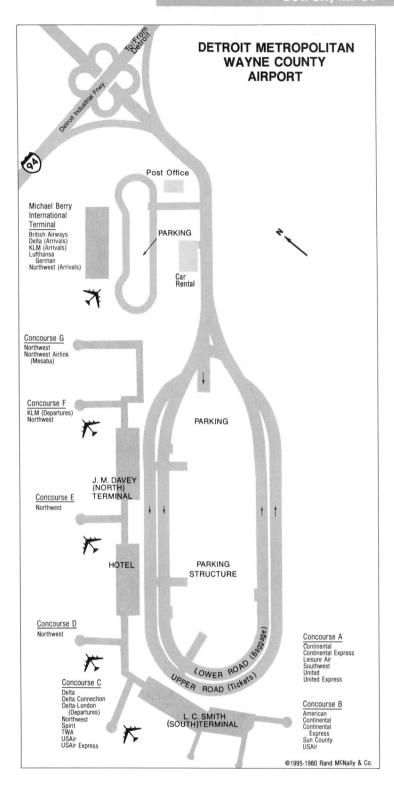

DETROIT METROPOLITAN
WAYNE COUNTY
AIRPORT

©1995-1980 Rand M°Nally & Co.

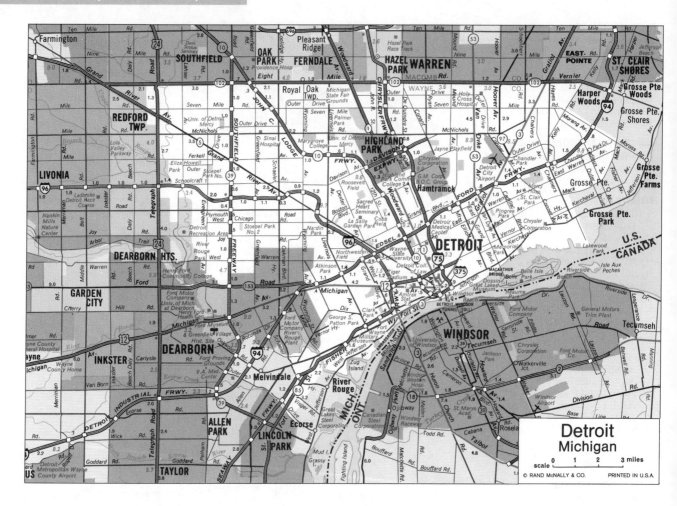

Detroit
Michigan

scale 0 1 2 3 miles

© RAND McNALLY & CO.   PRINTED IN U.S.A.

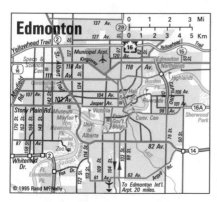

Detroit, Michigan 48243
(313) 259-4333; (800) DETROIT
Metropolitan Detroit Visitor Information
Center
  Suite 1900, 100 Renaissance Center
  Detroit, Michigan 48243
  (313) 567-1170; (800) 338-7648
Greater Detroit Chamber of Commerce
  600 W. Lafayette
  Detroit, Michigan 48226
  (313) 964-4000

Edmonton, Alberta, Canada

Population: (*785,465) 573,982 (1986C)
Altitude: 2,192 feet
Average Temp.: Jan., 6°F.; July, 62°F.
Telephone Area Code: 403
Time: 449-4444   **Weather:** 468-4940
Time Zone: Mountain

AIRPORT TRANSPORTATION:

Eighteen miles from Edmonton
International to downtown Edmonton;
three miles from Edmonton Municipal
to downtown Edmonton.
Taxicab; Airporter bus to and from
  Edmonton International; shuttle service
  between Edmonton International and
  Edmonton Municipal.

SELECTED HOTELS:

Delta Centre Suite Hotel, 103rd Ave. &
  102nd St., 429-3900; FAX 426-0562
Convention Inn, 4404 Calgary Trail
  Northbound, 434-6415; FAX 436-9247
Edmonton Hilton, 10235 101st St.,
  428-7111; FAX 441-3098
✈ Edmonton Inn, 11830 Kingsway Ave.,
  454-9521; FAX 453-7360
Holiday Inn Crowne Plaza, 10111 Bellamy
  Hill, 428-6611; FAX 425-6564
Howard Johnson, 10010 104th St.,
  423-2450; FAX 426-6090
✈ Nisku Inn, 4th St. & 11th Ave., across
  from International Airport, 955-7744;
  FAX 955-7743
Ramada Renaissance, 10155 105th St.,
  423-4811; FAX 423-3204
The Westin Hotel—Edmonton, 10135
  100th St., 426-3636; FAX 428-1454
Westwood Inn Best Western, 18035 Stony
  Plain Rd., 483-7770; FAX 486-1769

SELECTED RESTAURANTS:

Signature Food: Alberta beef
Boulevard Café, in the Ramada
  Renaissance Hotel, 423-4811
The Carvery, in the Westin Hotel,
  426-3636
Claude's, 9797 Jasper Ave., in the
  Convention Center, 429-2900
Japanese Village, 10126 100th St.,
  422-6083

Michigan—Dearborn campus, west side
  of Evergreen between Ford Rd. &
  Michigan Ave., Dearborn, 593-5590
Henry Ford Museum & Greenfield Village,
  20900 Oakwood Blvd., Dearborn,
  271-1620
Motown Museum, 2648 W. Grand Blvd.,
  875-2264

INFORMATION SOURCES:

Metropolitan Detroit Convention &
Visitors Bureau
  Suite 1900, 100 Renaissance Center

La Ronde, in the Holiday Inn Crowne
Plaza, 428-6611
The Mill Restaurant, 8109 101st St.,
432-1838
Pacific Fish Company, 1638 Bourbon St.,
in West Edmonton Mall, 444-1905
The Tin Palace, 11830 Jasper Ave.,
488-6582
Top of the Inn, in the Convention Inn
Hotel, 434-6415

SELECTED ATTRACTIONS:

Alberta Railway Museum (not open in
winter), 24215 34th St. NE, 472-6229
Edmonton Space & Science Centre, 11211
142nd St., 451-7722
Elk Island National Park, 35 mi. east on
Hwy. 16, 992-6380
Fort Edmonton Park, Whitemud & Fox
drs., 428-2992
Muttart Conservatory, 98th Ave. & 96A St.,
496-8755
Provincial Museum of Alberta, 128th St. &
102nd Ave., 427-1786
Ukrainian Cultural Heritage Museum, 30
mi. east on Hwy. 16, 662-3640
West Edmonton Mall/Canada
Fantasyland, 8770 170th St., 444-5200

INFORMATION SOURCE:

Edmonton Tourism
9797 Jasper Ave.
Edmonton, Alberta, Canada T5J 1N9
(403) 426-4715; (800) 463-4667
(Cont'l USA)

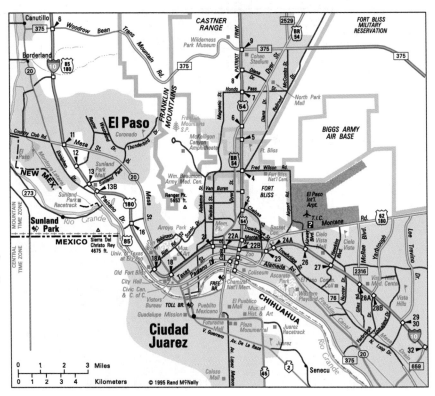

El Paso, Texas

Population: (*591,610) 515,342 (1990C)
Altitude: 3,762 feet
Average Temp.: Jan., 44°F.; July, 82°F.
Telephone Area Code: 915
Time: 532-9911   **Weather:** 562-4040
Time Zone: Mountain

AIRPORT TRANSPORTATION:

Eight miles to downtown El Paso.
Taxicab, airporter shuttle, and hotel
shuttle service.

SELECTED HOTELS:

Best Western Airport Inn, 7144 Gateway
East, 779-7700; FAX 772-1920
Camino Real Paso del Norte, 101 S. El
Paso St., 534-3000; FAX 534-3024
✈ El Paso Airport Hilton, 2027 Airway
Blvd., 778-4241; FAX 772-6871
✈ El Paso Marriott, 1600 Airway Blvd.,
779-3300; FAX 772-0915
✈ Embassy Suites, 6100 Gateway East,
779-6222; FAX 779-8846
✈ Holiday Inn—Airport, I-10 West &
Airway Blvd., 778-6411; FAX 778-6517
Holiday Inn—Park Place, 325 N. Kansas,
533-8241; FAX 544-9979
International Hotel, 113 W. Missouri,
544-3300; FAX (none)

SELECTED RESTAURANTS:

Signature Food: Tex-Mex
Bella Napoli Ristorante, 6331 N. Mesa,
584-3321
Bombay Bicycle Club, 6080 Gateway East,
778-4251
Chatfield's, in the El Paso Marriott,
779-3300

The Dome Grill, in the Westin Paso del
Norte, 534-3000
Forti's Mexican Elder, 321 Chelsea,
772-0066
Seafarer Seafood, 1711 Lee Trevino,
593-8388
The Starr Room Restaurant, in the
International Hotel, 544-3300

SELECTED ATTRACTIONS:

Americana Museum, 5 Civic Center Plaza,
542-0394
El Paso Museum of History, 12901
Gateway W., 858-1928
El Paso Zoo, 4001 E. Paisano, 544-1928
Fort Bliss Museums, Pleasanton Rd.,
Building 5000 & Buildings 5051-5054,
568-7345
Historic missions: San Elizario Presidio,
Socorro Mission, Ysleta Mission,
534-0630
Insights-El Paso Science Museum, 505 N.
Santa Fe St., 534-0000
Juarez trolley tour, El Paso-Juarez Trolley
Co., One Civic Center Plaza, 544-0061
Magoffin Home, 1120 Magoffin, 533-5147
Tigua Indian Reservation, 119 S. Old
Pueblo Rd., 859-7913
Wet 'n' Wild Waterworld, I-10 Anthony
Exit "0", 886-2222

TRADE EXHIBITION FACILITY:

El Paso Convention & Performing Arts
Center, One Civic Center Plaza,
534-0609

INFORMATION SOURCES:

El Paso Convention & Visitors Bureau
One Civic Center Plaza
El Paso, Texas 79901
(915) 534-0696; (800) 351-6024

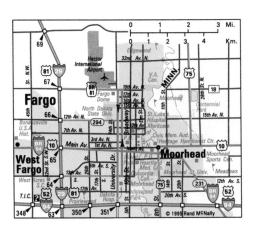

El Paso Chamber of Commerce
10 Civic Center Plaza
El Paso, Texas 79901
(915) 534-0500

Fargo, North Dakota

Population: (*153,296) 74,111 (1990C)
Altitude: 900 feet
Average Temp.: Jan., 6°F.; July, 71°F.
Telephone Area Code: 701
Time: (none)   **Weather:** 235-2600
Time Zone: Central

AIRPORT TRANSPORTATION:

Two miles to downtown Fargo.
Taxicab and hotel shuttle service.

SELECTED HOTELS:

AmericInn, 1423 35th St. SW, 234-9946;
FAX 234-9946

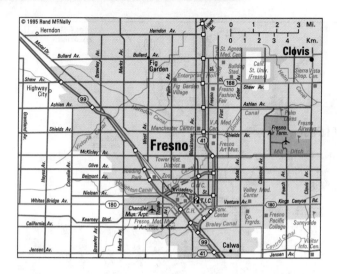

Best Western Doublewood Inn, 3333 13th
Ave. S., 235-3333; FAX 280-9482
Hampton Inn, 3431 14th Ave. SW,
235-5566; FAX 235-5566, ext. 302
Holiday Inn, 3803 13th Ave. S., 282-2700;
FAX 281-1240
The Kelly Inn, I-29 & Main Ave.,
282-2143; FAX 281-0243
Madison Hotel, 600 30th Ave. S.,
Moorhead, (218) 233-6171;
FAX (218) 233-0945
Radisson Hotel, 201 5th St. N., 232-7363;
FAX 232-7363, ext. 354
Town House Inn, 301 3rd Ave. N.,
232-8851; FAX 235-8701

SELECTED RESTAURANTS:

Signature Food: chippers (chocolate
covered potato chips)
Doublewood Inn, in the Best Western
Doublewood Inn, 235-3333
Fargo Cork, 3301 S. University Dr.,
237-6790
Gallery Terrace & Cafe, in the Holiday Inn,
282-2700
The Grainery, in West Acres Shopping
Center, 282-6262
Old Broadway, 22 Broadway, 237-6161
Passages, in the Radisson Hotel, 232-7363
Tree Top, 403 Center Ave., Moorhead,
(218) 233-1393

SELECTED ATTRACTIONS:

Black Swan Dinner Theatre, 613 1st Ave.
N., 293-6373
Bonanzaville (pioneer village & museum),
Main Ave., West Fargo, 282-2822
Children's Museum at Yunker Farm,
University Dr. & 28th Ave. N., 232-6102
Comstock House, 506 8th St. S.,
Moorhead, MN, (218) 233-0848
Fargo Theatre, 314 Broadway, 235-4152
Hjemkomst Center/Clay County Historical
Society, 202 First Ave. N., Moorhead,
MN, (218) 233-5604
Planetarium, Moorhead State University-
Bridges Hall, 11th St. & 8th Ave. S.,
Moorhead, MN, (218) 236-3982
Plains Art Museum, 521 Main Ave.,
Moorhead, MN, (218) 236-7383
Roger Maris Museum, West Acres
Shopping Mall, I-29 & 13th Ave. S.,
282-2222

Trollwood Park, Trollwood Dr. between N.
Broadway & Elm St., 241-8160

INFORMATION SOURCES:

Fargo—Moorhead Convention & Visitors
Bureau
2001 44th St. SW
P.O. Box 2164
Fargo, North Dakota 58107
(701) 282-3653; (800) 235-7654
Fargo Chamber of Commerce
321 N. 4th St.
P.O. Box 2443
Fargo, North Dakota 58108
(701) 237-5678

Fresno, California

Population: (*667,490) 354,202 (1990C)
Altitude: 296 feet
Average Temp.: Jan., 45°F.; July, 81°F.
Telephone Area Code: 209
Time: 767-8900 **Weather:** 291-1068
Time Zone: Pacific

AIRPORT TRANSPORTATION:

Seven miles to downtown Fresno.
Taxicab, city bus, and hotel shuttle
service.

SELECTED HOTELS:

Best Western Tradewinds, 2141 N.
Parkway Dr., 237-1881; FAX 237-9719
Centre Plaza Holiday Inn, 2233 Ventura
St., 268-1000; FAX 486-6625
Courtyard By Marriott, 140 E. Shaw,
221-6000; FAX 221-0368
Fresno Hilton, 1055 Van Ness St.,
485-9000; FAX 485-7666
✈ Holiday Inn—Airport, 5090 E. Clinton,
252-3611; FAX 456-8243
Piccadilly Inn—University, 4961 N. Cedar,
224-4200; FAX 227-2382
Ramada Inn, 324 E. Shaw Ave., 224-4040;
FAX 222-4017
San Joaquin Suites Hotel, 1309 W. Shaw,
225-1309; FAX 225-6021
Sheraton Smuggler's Inn, 3737 N.
Blackstone, 226-2200; FAX 222-7147

SELECTED RESTAURANTS:

Signature Food: raisins
Daily Planet, 1211 N. Wishon, 266-4259

Gaslight Steak House, in the Best Western
Tradewinds, 237-1881
Harland's, 722 W. Shaw, 225-7100
John Q's, in the Centre Plaza Holiday Inn,
268-1000
Nicola's, 3075 N. Maroa, 224-1660
The Old Spaghetti Factory, 2721 Ventura,
442-1066
The Remington, 927 S. Clovis Ave.,
251-8228
Richard's, 1609 E. Belmont, 266-4077
Stuart Anderson's Black Angus, 1737 E.
Shaw, 224-2205

SELECTED ATTRACTIONS:

Blackbeard's Family Entertainment
Center, 4055 N. Chestnut, 292-4554
The Blossom Trail (Feb.-March), 233-0836
Chaffee Zoological Gardens, in Roeding
Park, 894 W. Belmont, 498-2671
Forestiere Underground Gardens, 5021
W. Shaw Ave., 271-0734
Fresno Art Museum, 2233 N. First St.,
485-4810
Fresno Metropolitan Museum of Art,
History & Science, 1515 Van Ness Ave.,
441-1444
Kearney Mansion Museum, in Kearney
Park, 441-0862
Meux Home, Tulare & "R" sts., 233-8007
Simonian Farms, 2629 S. Clovis Ave.,
237-2294
Wild Water Adventures Water Park, 11413
E. Shaw, Clovis, 297-6500
Yosemite Mountain Sugar Pine Railroad,
56001 Hwy. 41, Fish Camp, 683-7273

INFORMATION SOURCES:

Fresno Convention & Visitors Bureau
808 "M" St.
Fresno, California 93721
(209) 233-0836; (800) 788-0836
Fresno Chamber of Commerce
2331 Fresno St.
Fresno, California 93721
(209) 233-4651

Grand Rapids, Michigan

Population: (*688,399) 189,126 (1990C)
Altitude: 657 feet
Average Temp.: Jan., 23°F.; July, 72°F.
Telephone Area Code: 616
Time: (none) **Weather:** 776-1234
Time Zone: Eastern

AIRPORT TRANSPORTATION:

Sixteen miles to downtown Grand Rapids.
Taxicab service.

SELECTED HOTELS:

✈ Airport Hilton Inn, 4747 28th St. SE,
957-0100; FAX 957-2977
Amway Grand Plaza Hotel, Pearl at
Monroe, 774-2000; FAX 776-6489
Best Western Midway Hotel, 4101 28th St.
SE, 942-2550; FAX 942-2446
Days Inn, 5500 28th St. SE, 949-8400;
FAX 949-8400
Days Inn—Downtown, 310 Pearl St. NW,
235-7611; 235-1995
✈ Hampton Inn, 4981 28th St. SE,
956-9304; FAX 956-6617
Harley Hotel, 4041 Cascade Rd.,
949-8800; FAX 949-4303
✈ Holiday Inn Crowne Plaza, 5700 28th
St. SE, 957-1770; FAX 957-0629

Quality Inn Terrace Club, 4495 28th St.
SE, 956-8080; FAX 956-0619

Radisson Hotel, 11 Monroe NW, 242-6000;
FAX 242-9580

SELECTED RESTAURANTS:

Signature Food: apples; peaches

Cascade Roadhouse, 6817 Cascade SE,
949-1540

Charley's Crab, 63 Market St. SW,
459-2500

Cygnus, in the Amway Grand Plaza Hotel,
774-2000

Duba's, 420 E. Beltline NE, 949-1011

Ducks, 740 Michigan NE, 451-2767

Gibson's, 1033 Lake Drive SE, 774-8535

The Hoffman House, in the Best Western
Midway Hotel, 942-2550

K. J. Cashmere's, 29 Pearl NW, 454-0299

The Legend, 420 Stocking NW, 459-6655

The 1913 Room, in the Amway Grand
Plaza Hotel, 774-2000

Pietro's Ristorante, 2780 Birchcrest SE,
452-3228

Sayfee's, 3555 Lake Eastbrook Blvd. SE,
949-5750

Yen Ching, 57 Monroe Center, 235-6969

SELECTED ATTRACTIONS:

Blandford Nature Center, 1715 Hillburn
NW, 453-6192

First (Park) Congregational Church
(Tiffany windows), 10 E. Park Pl. NE,
459-3203

Frederik Meijer Gardens—Michigan
Botanic Garden/Meijer Sculpture Park,
3411 Bradford NE, 957-1580 (due to
open in Jan., 1995)

Gerald R. Ford Museum, 303 Pearl St.
NW, 451-9290

Heritage Hill Historic District, Crescent to
Pleasant-Union to Lafayette, 459-8950

John Ball Zoo, Fulton & Valley, 336-4301

Roger B. Chaffee Planetarium, Pearl St. at
Front Ave., 456-3985

Van Andel Museum Center, Pearl St. at
Front Ave., 456-3977

Voigt House, 115 College SE, 456-4600

INFORMATION SOURCES:

Grand Rapids/Kent County Convention &
Visitors Bureau
    245 Monroe Ave. NW
    Grand Rapids, Michigan 49503-2832
    (616) 459-8287; (800) 678-9859

Grand Rapids Area Chamber of Commerce
    111 Pearl St. NW
    Grand Rapids, Michigan 49503
    (616) 771-0300

Hartford, Connecticut

Population: (*767,841) 139,739 (1990C)
Altitude: 10 to 290 feet
Average Temp.: Jan., 28°F.; July, 74°F.
Telephone Area Code: 203
Time: 524-8123 **Weather:** 1-936-1212
Time Zone: Eastern

AIRPORT TRANSPORTATION:

Thirteen miles to downtown Hartford.
Taxicab and limousine bus service.

SELECTED HOTELS:

Farmington Marriott Hotel, 15 Farm
    Springs Rd., Farmington, 678-1000;
    FAX 677-8849

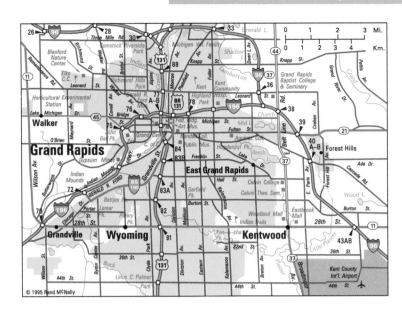

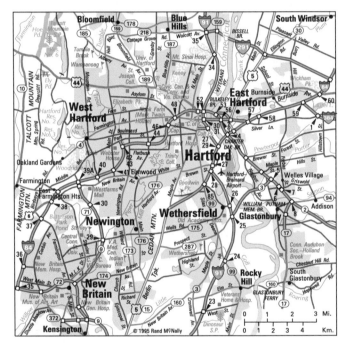

The Goodwin Hotel, 1 Haynes St.,
    246-7500; FAX 247-4576

Holiday Inn—Downtown, 50 Morgan St.,
    549-2400; FAX 527-2746

Ramada Hotel, 100 E. River Dr., E.
    Hartford, 528-9703; FAX 289-4728

Sheraton Hartford Hotel, 315 Trumbull
    St., 728-5151; FAX 240-7247

SELECTED RESTAURANTS:

Signature Food: oysters

Avon Old Farms Inn, U.S. 44 & CT 10,
    677-2818

The Blacksmith's Tavern, 2300 Main St.,
    Glastonbury, 659-0366

Carbone's, 588 Franklin Ave., 296-9646

Frank's, 185 Asylum, 527-9291

Gaetano's, One Civic Center Plaza,
    249-1629

Hearthstone, 678 Maple Ave., 246-8814

Hot Tomato's, One Union Place, 249-5100

Parson's Daughter, 2 Hopewell Rd., S.
    Glastonbury, 633-8698

SELECTED ATTRACTIONS:

Bushnell Park Carousel Society (hand-
    carved carousel), 246-7739

Butler-McCook Homestead (1782), 396
    Main St., 522-1806

Center Church & The Ancient Burying
    Ground, Main & Gold sts., 249-5631

Elizabeth Park (rose garden), Prospect
    Ave. & Asylum Ave.

Harriet Beecher Stowe House, Visitor
    Center, 71 Forest St., 525-9317

Historical Museum of Medicine &
    Dentistry, in the Hartford Medical
    Society building, 230 Scarborough St.,
    236-5613

Mark Twain House, 351 Farmington Ave.,
493-6411
Old State House, 800 Main St., 522-6766
Wadsworth Atheneum (oldest
continuously operating American
public art museum), 600 Main St.,
278-2670

INFORMATION SOURCES:

Greater Hartford Convention & Visitors
Bureau, Inc. (conventions)
One Civic Center Plaza
Hartford, Connecticut 06103
(203) 728-6789; (800) 446-7811
Visitor Information: (203) 275-6456
Greater Hartford Tourism District
(tourism)
One Civic Center Plaza
Hartford, Connecticut 06103
(203) 520-4480; (800) 793-4480;
Visitor Information: (203) 275-6456
Old Statehouse Information Center
800 Main St.
(brochures only; self-service)
Greater Hartford Chamber of Commerce
250 Constitution Plaza
Hartford, Connecticut 06103-1882
(203) 525-4451

Honolulu, Hawaii

Population: (*836,231) 365,272 (1990C)
Altitude: 18 feet
Average Temp.: Jan., 72°F.; July, 80°F.
Telephone Area Code: 808
Time: 983-3211   **Weather:** 833-2849
Time Zone: Hawaiian (Two hours earlier
than Pacific standard time)

AIRPORT TRANSPORTATION:

Nine miles to Waikiki.
Taxicab and limousine bus service.

SELECTED HOTELS:

Halekulani, 2199 Kalia Rd., 923-2311;
FAX 926-8004
Hawaiian Regent, 2552 Kalakaua Ave.,
922-6611; FAX 921-5255
Hilton Hawaiian Village, 2005 Kalia Rd.,
949-4321; FAX 955-3027
Hyatt Regency Waikiki, 2424 Kalakaua
Ave., 923-1234; FAX 926-3415
The Ilikai, 1777 Ala Moana Blvd.,
949-3811; FAX 947-4523
Kahala Hilton, 5000 Kahala Ave.,
734-2211; FAX 737-2478
Outrigger Waikiki Hotel, 2335 Kalakaua
Ave., 923-0711; FAX 921-9749
Queen Kapiolani, 150 Kapahulu Ave.,
922-1941; FAX 922-2694
Royal Hawaiian, 2259 Kalakaua Ave.,
923-7311; FAX 924-7098
Sheraton Moana Surfrider, 2365 Kalakaua
Ave., 922-3111; FAX 923-0308
Sheraton Waikiki, 2255 Kalakaua Ave.,
922-4422; FAX 923-8785

SELECTED RESTAURANTS:

Signature Food: poi; roast pork
Furusato, 2500 Kalakaua Ave., 922-5502
Golden Dragon Room (Chinese), in the
Hilton Hawaiian Village, 949-4321
The Hanohano Room, in the Sheraton
Waikiki, 922-4422
Maile Room, in the Kahala Hilton,
734-2211

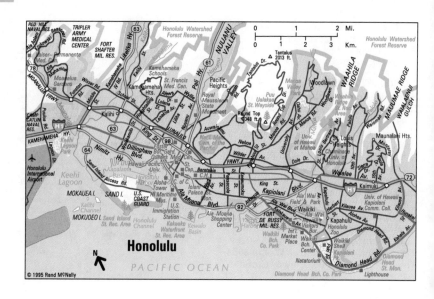

Honolulu

PACIFIC OCEAN

© 1995 Rand McNally

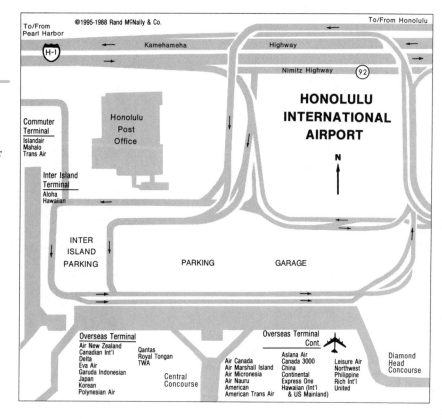

©1995-1988 Rand McNally & Co.

HONOLULU INTERNATIONAL AIRPORT

Michel's, in the Colony Surf Hotel, 2895
Kalakaua Ave., 923-6552
The Plantation Cafe, in the Ala Moana
Hotel, 410 Atkinson Dr., 955-4811
The Secret, in the Hawaiian Regent Hotel,
922-6611

SELECTED ATTRACTIONS:

Bernice P. Bishop Museum & Planetarium,
1525 Bernice St., 847-3511
Honolulu Zoo, Kapahulu & Kalakaua
aves., 971-7171
National Memorial Cemetery of the
Pacific, overlooking Honolulu,
541-1430

Pearl Harbor/The U.S.S. *Arizona*
Memorial, 422-0561
Polynesian Cultural Center, 55-370 Kam
Hwy., Laie, 293-3333
Sea Life Park, Makapuu Point, 259-7933
Waikiki Aquarium, 2777 Kalakaua Ave.,
923-9741
Waimea Falls Park, 59-864 Kam Hwy. on
the north shore of Oahu, 638-8511

INFORMATION SOURCES:

Hawaii Visitors Bureau
2270 Kalakaua Ave., 8th Floor
Honolulu, Hawaii 96815
(808) 923-1811

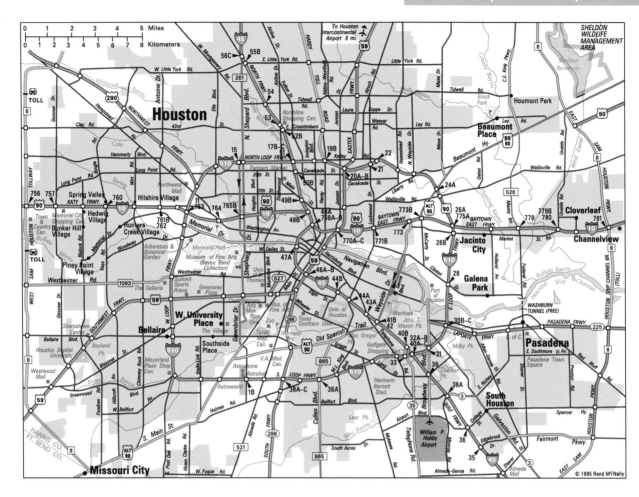

The Chamber of Commerce of Hawaii
1132 Bishop St., Suite 200
Honolulu, Hawaii 96813
(808) 522-8800

Houston, Texas

Population: (*3,301,937) 1,630,553
(1990C)
Altitude: Sea level to 50 feet
Average Temp.: Jan., 59°F.; July, 82°F.
Telephone Area Code: 713
Time: 222-8463   **Weather:** 529-4444
Time Zone: Central

AIRPORT TRANSPORTATION:

Twenty miles from Intercontinental
Airport to downtown Houston; 10 miles
from Hobby Airport to downtown.
Taxicab, limousine, coach bus, and
helicopter service.

SELECTED HOTELS:

Adam's Mark, 2900 Briarpark at
Westheimer, 978-7400; FAX 735-2726
Doubletree at Allen Center, 400 Dallas St.,
759-0202; FAX 752-2734
✈ Doubletree at Houston Intercontinental,
15747 JFK Blvd., 442-8000;
FAX 590-8461
Four Seasons Hotel, Houston Center, 1300
Lamar, 650-1300; FAX 650-8169
Guest Quarters—Galleria West, 5353
Westheimer, 961-9000; FAX 877-8835

Hyatt Regency Houston, 1200 Louisiana
St., 654-1234; FAX 951-0934
Omni Houston, 4 Riverway Dr., 871-8181;
FAX 871-0719
Ramada Astrodome, 2100 S. Braeswood,
797-9000; FAX 799-8362
The Sheraton Grand, 2525 W. Loop South,
961-3000; FAX 961-1490
✈ Sheraton Crown Hotel & Conference
Center, 15700 JFK Blvd., 442-5100;
FAX 987-1930
Stouffer Presidente Hotel, Southwest
Frwy. at Edloe, 629-1200;
FAX 629-4702
The Westin Galleria, 5060 W. Alabama St.,
960-8100; FAX 960-6553
The Westin Oaks Hotel, 5011 Westheimer
Rd., 623-4300; FAX 960-6554
The Wyndham Warwick Hotel, 5701 Main
St., 526-1991; FAX 639-4545

SELECTED RESTAURANTS:

Signature Food: barbeque; fajitas;
southwestern cuisine
Brennan's, 3300 Smith St., 522-9711
De Ville, in the Four Seasons Hotel,
Houston Center, 650-1300
The Great Caruso, 10001 Westheimer Rd.,
780-4900
La Tour d'Argent, 2011 Ella Blvd.,
864-9864
Maxim's, 3755 Richmond, 877-8899
The Rivoli, 5636 Richmond Ave., ½ block
west of Chimney Rock, 789-1900

Tony's, 1801 Post Oak Blvd., 622-6778
Vargo's, 2401 Fondren Rd., 782-3888

SELECTED ATTRACTIONS:

Battleship *Texas*, in San Jacinto
Battleground State Historic Park, 22
miles east of Houston off Texas
Highway 225 East, 479-2411
Bayou Bend (residence/museum of
American decorative arts/gardens), 1
Westcott, (reservations recommended
for tours of residence) 520-2600
Houston Zoo, in Hermann Park, off the
6300 block of Fannin St., 525-3300
Museum of Fine Arts, 1001 Bissonnet,
639-7375
San Jacinto Monument & Museum of
History, in San Jacinto Battleground
State Historic Park, 22 miles east of
Houston off Texas Highway 225 East,
479-2421
Space Center Houston, 1601 NASA Rd. 1,
244-2100
The Galleria (shopping, dining,
entertainment), Loop 610 at
Westheimer, 621-1907

INFORMATION SOURCES:

Greater Houston Convention & Visitors
Bureau
801 Congress St.
Houston, Texas 77002
(713) 227-3100; (800) 231-7799

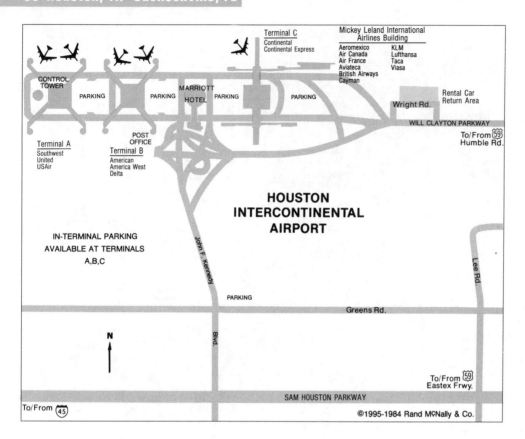

Terminal C
Continental
Continental Express

Mickey Leland International
Airlines Building

Aeromexico   KLM
Air Canada   Lufthansa
Air France   Taca
Aviateca   Viasa
British Airways
Cayman

CONTROL
TOWER

PARKING   PARKING

MARRIOTT
HOTEL   PARKING

PARKING

Rental Car
Return Area

Wright Rd.

WILL CLAYTON PARKWAY

To/From 59
Humble Rd.

Terminal A
Southwest
United
USAir

POST
OFFICE
Terminal B
American
America West
Delta

**HOUSTON
INTERCONTINENTAL
AIRPORT**

IN-TERMINAL PARKING
AVAILABLE AT TERMINALS
A,B,C

John F. Kennedy

Lee Rd.

PARKING

Blvd.

Greens Rd.

N

To/From 59
Eastex Frwy.

To/From 45

SAM HOUSTON PARKWAY

©1995-1984 Rand McNally & Co.

Greater Houston Partnership
1200 Smith, 7th Floor
Houston, Texas 77002-4309
(713) 651-2100

Indianapolis, Indiana

Population: (*1,249,822) 731,327 (1990C)
Altitude: 717 feet
Average Temp.: Jan., 29°F.; July, 76°F.
Telephone Area Code: 317
Time and Weather: 635-5959
Time Zone: Eastern standard all year

AIRPORT TRANSPORTATION:

Eight miles to downtown Indianapolis.
Taxicab and limousine bus service.

SELECTED HOTELS:

✈ Adam's Mark Hotel, 2544 Executive
Drive, 248-2481; FAX 381-6159
Courtyard by Mariott, 501 W. Washington
St., 635-4443; FAX 687-0029
✈ Holiday Inn Airport, 2501 S. High
School Rd., 244-6861; FAX 243-1059
Holiday Inn Crowne Plaza Union Station,
123 W. Louisiana, 631-2221;
FAX 236-7474
Hyatt Regency Indianapolis, 1 S. Capitol
Ave., 632-1234; FAX 236-6841
✈ Indianapolis Airport Ramada, 2500 S.
High School Rd., 244-3361;
FAX 241-9202
Indianapolis Marriott Hotel, 7202 E. 21st
St., 352-1231; FAX 352-1231
Indianapolis Motor Speedway Motel,
4400 W. 16th St., 241-2500;
FAX 241-2133
Omni Severin Hotel, 40 W. Jackson Pl.,
634-6664; FAX 687-3612

Ramada Plaza, 31 W. Ohio St., 635-2000;
FAX 638-0782
Westin Hotel, 50 S. Capitol Ave.,
262-8100; FAX 231-3928

SELECTED RESTAURANTS:

Signature Food: fried chicken with fried
biscuits and apple butter
Chanteclair Sur Le Toit, Holiday Inn
Airport, 244-6861
Chez Jean Restaurant Francais & Inn, 9027
S. State Rd. 67, Camby, 831-0870
The City Taproom & Grille, 28 S.
Pennsylvania St., 637-1334
The Eagle's Nest Restaurant, in the Hyatt
Regency Indianapolis, 632-1234
King Cole Restaurant, 7 N. Meridian St.,
638-5588
St. Elmo Steak House, 127 S. Illinois St.,
635-0636

SELECTED ATTRACTIONS:

Children's Museum of Indianapolis, 3000
N. Meridian St., 924-5431
Conner Prairie Pioneer Settlement, 13400
Allisonville Rd., Fishers, 776-6000
Eiteljorg Museum of American Indian and
Western Art, 500 W. Washington St.,
636-9378
Indiana State Museum & Historic Sites,
202 N. Alabama St., 232-1637
Indianapolis "500" Motor Speedway and
Hall of Fame Museum, 4790 W. 16th St.,
248-6746
Indianapolis Museum of Arts, 1200 W.
38th St., 923-1331
Indianapolis Zoo, 1200 W. Washington
St., 630-2030
Madame Walker Urban Life Center, 617
Indiana Ave., 236-2099

Union Station Festival Marketplace,
Meridian between South St. & W.
Jackson Place, 267-0701

INFORMATION SOURCES:

Indianapolis Convention and Visitors
Association
1 Hoosier Dome, Suite 100
Indianapolis, Indiana 46225
(317) 639-4282;
Visitor Info. (800) 323-INDY;
Convention Info. (800) 642-INDY
Indianapolis Chamber of Commerce
320 N. Meridian St., Suite 200
Indianapolis, Indiana 46204
(317) 464-2200
Indianapolis City Center
201 S. Capitol
Indianapolis, Indiana 46225
(317) 237-5200

Jacksonville, Florida

Population: (*906,727) 635,230 (1990C)
Altitude: 19 feet
Average Temp.: Jan., 55°F.; July, 80°F.
Telephone Area Code: 904
Time: 358-1212   **Weather:** 741-4311
Time Zone: Eastern

AIRPORT TRANSPORTATION:

Fifteen miles to downtown Jacksonville.
Taxicab and limousine service.

SELECTED HOTELS:

Best Western Spindrifter, 300 Park Ave.,
Orange Park, 264-1211; FAX 269-6756
Comfort Inn, 3233 Emerson St., 398-3331;
FAX 398-3331

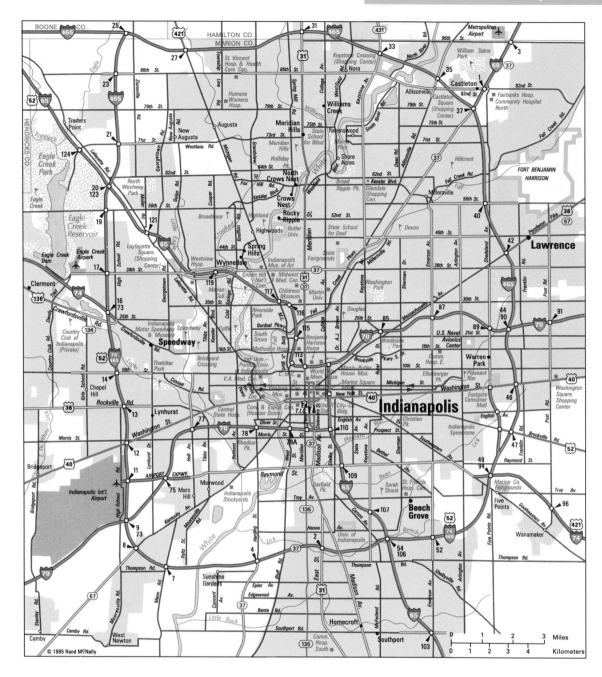

Holiday Inn and Conference Center, 5865
Arlington Expwy., 724-3410;
FAX 727-7606
✈ Holiday Inn—Jacksonville Airport, I-95
at Airport Rd., 741-4404; FAX 741-4907
Holiday Inn—Orange Park, 150 Park Ave.,
Orange Park, 264-9513; FAX 264-9513
Marina St. John's Place, 1515 Prudential
Dr., 396-5100; FAX 396-7154
Omni Jacksonville, 245 Water St.,
355-6664; FAX 354-2970
Paradise Inn, 6545 Ramona Blvd.,
781-1940; FAX 781-1940
Ramada Inn—East, 6237 Arlington
Expwy., 725-5093; FAX 720-0378
Ramada West, 510 S. Lane Ave., 786-0500;
FAX 786-3447
✈ Skycenter Inn, 2101 Dixie Clipper Dr.,
741-4747; FAX 741-0002

SELECTED RESTAURANTS:
Signature Food: barbeque; shrimp
Alhambra Dinner Theatre, 12000 Beach
Blvd., 641-1212
Cafe St. John's, in the Marina St. John's
Place, 396-5100
Crawdaddy's, 1643 Prudential Dr.,
396-3546
L & N Seafood Grill, 2 Independent Dr.,
358-7737
Patti's Italian & American Restaurant,
7300 Beach Blvd., 725-1662
The Wine Cellar, 1314 Prudential Dr.,
398-8989

SELECTED ATTRACTIONS:
Alhambra Dinner Theatre, 12000 Beach
Blvd., 641-1212

Anheuser-Busch Brewery (tours), 111
Busch Dr., 751-8117 or -8118
Cummer Gallery of Art, 829 Riverside
Ave., 356-6857
Jacksonville Art Museum, 4160 Boulevard
Center Dr., 398-8336
The Jacksonville Landing, 2 Independent
Dr., 353-1188
Jacksonville Zoo, Exit 124A (Heckscher
Dr.) off I-95, 757-4463
Kathryn Abbel Hanna Park, 500
Wonderwood Dr., 249-2317
Mayport Naval Station (weekend tours),
Mayport, 270-NAVY
Museum of Science & History, 1025
Museum Circle, 396-7062
The Riverwalk, 400 Wharfside Way,
396-4900

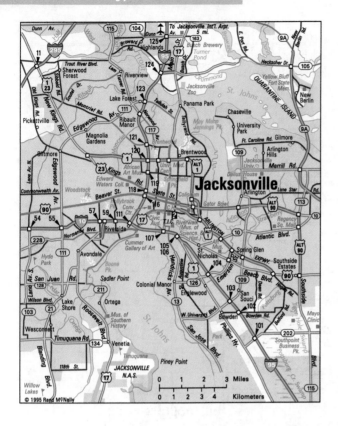

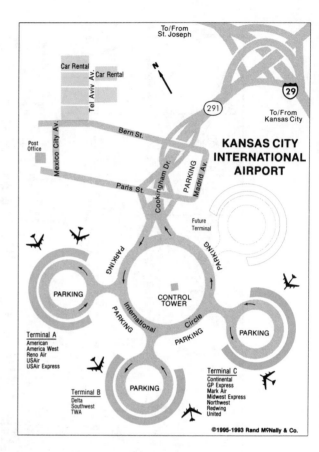

INFORMATION SOURCES:

Jacksonville & the Beaches Convention &
Visitors Bureau
   3 Independent Dr.
   Jacksonville, Florida 32202-5092
   (904) 798-9148; (800) 733-2668
Jacksonville Chamber of Commerce
   3 Independent Dr.
   Jacksonville, Florida 32202
   (904) 366-6600

Kansas City, Missouri

Population: (*1,566,280) 435,146 (1990C)
Altitude: 800 feet
Average Temp.: Jan., 30°F.; July, 81°F.
Telephone Area Code: 816
Time: 844-1212 **Weather:** 471-4840
Time Zone: Central

AIRPORT TRANSPORTATION:

Eighteen miles to downtown Kansas City.
Taxicab and limousine bus service.

SELECTED HOTELS:

Adam's Mark of Kansas City, 9103 E. 39th
  St., 737-0200; FAX 737-4712
✈ Airport Doubletree, I-29 & 112th St.
  NW., 891-8900; FAX 891-8030
Americana Hotel, 1301 Wyandotte St.,
  221-8800; FAX 472-7964
Holiday Inn Crowne Plaza, 4445 Main St.,
  531-3000; FAX 531-3007
Hyatt Regency Crown Center, 2345
  McGee, 421-1234; FAX 435-4190
✈ Kansas City Airport Marriott, 775
  Brasilia, 464-2200; FAX 464-5915

Marriott Downtown, 200 W. 12th St.,
  421-6800; FAX 855-4418
Park Place Hotel, 1601 N. Universal Ave.,
  483-9900; FAX 231-1418
Plaza Inn, 45th & Main sts., 753-7400;
  FAX 753-4777
Ritz Carlton Hotel, Kansas City, 401 Ward
  Pkwy., 756-1500; FAX 756-1635
Westin Crown Center, 1 Pershing Rd., 474-
  4400; FAX 391-4438

SELECTED RESTAURANTS:

Signature Food: barbeque; steak
American Restaurant, 2500 Grand
  426-1133
The Golden Ox, 1600 Genessee, 842-2866
The Hereford House, 20th & Main,
  842-1080
Jasper's Restaurant, 405 W. 75th St.,
  363-3003
La Mediterranee, Glenwood Shopping
  Center, 9058B Metcalf, Overland Park,
  561-2916
Peppercorn Duck Club, in the Hyatt
  Regency Crown Center, 421-1234
Plaza III, 4749 Pennsylvania Ave.,
  753-0000
Starker's, 200 Nichols Rd., 753-3565
Stephenson's Apple Farm, U.S. 40 & Lee's
  Summit Rd., 373-5400
Trader Vic's, in the Westin Crown Center,
  391-4444

SELECTED ATTRACTIONS:

Harry S Truman Library & Museum, U.S.
  24 & Delaware St., Independence,
  833-1400
Truman Home National Historic Site,
  Independence, 254-7199

Kansas City Museum, 3218 Gladstone
  Blvd., 483-8300
Kansas City Zoo, 6700 Zoo Dr. in Swope
  Park, 333-7408
Liberty Memorial, 100 W. 26th St.,
  221-1918
Nelson-Atkins Museum of Art, 4525 Oak
  St., 561-4000
Starlight Theater, 4600 Starlight Rd. in
  Swope Park, 363-7827
Watkins Woolen Mill State Historic Site,
  6½ mi. north of Excelsior Springs on
  U.S. 69, 296-3357
Worlds of Fun/Oceans of Fun
  (theme/water parks), 4545 Worlds of
  Fun Ave., 454-4545

INFORMATION SOURCES:

Convention and Visitors Bureau of Greater
Kansas City
  City Center Square, 1100 Main St.
  Suite 2550
  Kansas City, Missouri 64105
  (816) 221-5242; (800) 767-7700
  Visitor Info. Recording:
  (816) 691-3800
Kansas City Visitor Information Center
  The Historic River Market
  20 E. 5th St.
  Kansas City, Missouri 64106
  (816) 842-4386
Chamber of Commerce of Greater Kansas
City
  911 Main St., Suite 2600
  Kansas City, Missouri 64105
  (816) 221-2424

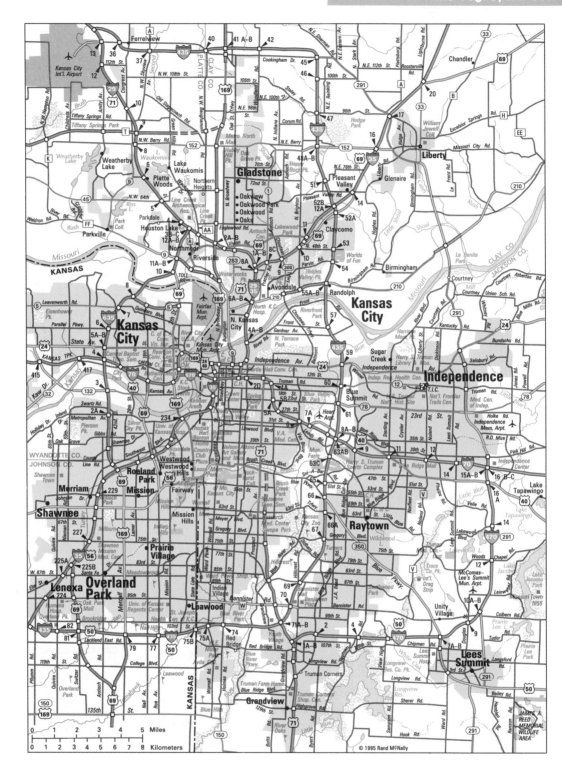

© 1995 Rand McNally

Las Vegas, Nevada

Population: (*741,459) 258,295 (1990C)
Altitude: 2,020 feet
Average Temp.: Jan., 44°F.; July, 90°F.
Telephone Area Code: 702
Time: 118   **Weather:** 734-2010
Time Zone: Pacific

AIRPORT TRANSPORTATION:

Eight miles to downtown Las Vegas.
Taxicab and limousine service.

SELECTED HOTELS:
Caesars Palace, 3570 Las Vegas Blvd. S.,
   731-7110; FAX 731-6636
Desert Inn Hotel and Country Club, 3145
   Las Vegas Blvd. S., 733-4444;
   FAX 733-4774
Excalibur, 3850 Las Vegas Blvd., S.,
   597-7777; FAX 597-7009
Golden Nugget Hotel & Casino, 129 E.
   Fremont St., 385-7111; FAX 386-8362
Hacienda Resort Hotel & Casino, 3950 Las
   Vegas Blvd. S., 739-8911; FAX 798-8289

Howard Johnson, 1322 E. Fremont St.,
   385-1150; FAX 385-4940
Imperial Palace Hotel & Casino, 3535 Las
   Vegas Blvd. S., 731-3311; FAX 735-8328
Las Vegas Hilton, 3000 Paradise Rd.,
   732-5111; FAX 794-3611
Luxor Hotel/Casino, 3900 Las Vegas Blvd.
   S., 262-4000; FAX 597-7455
MGM Grand Hotel, 3799 Las Vegas Blvd.
   S., 891-1111; FAX 891-3036
The Mirage, 3400 Las Vegas Blvd. S.,
   791-7111; FAX 791-7446

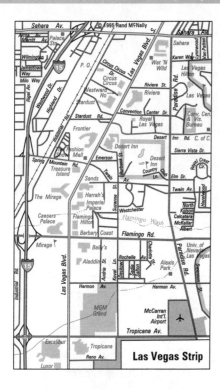

Las Vegas Strip

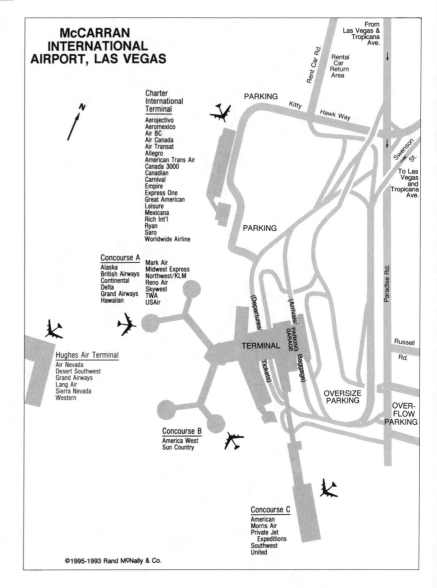

McCARRAN INTERNATIONAL AIRPORT, LAS VEGAS

Charter International Terminal
Aerojectivo
Aeromexico
Air BC
Air Canada
Air Transat
Allegro
American Trans Air
Canada 3000
Canadian
Carnival
Empire
Express One
Great American
Leisure
Mexicana
Rich Int'l
Ryan
Saro
Worldwide Airline

Concourse A
Alaska
British Airways
Continental
Delta
Grand Airways
Hawaiian
Mark Air
Midwest Express
Northwest/KLM
Reno Air
Skywest
TWA
USAir

Hughes Air Terminal
Air Nevada
Desert Southwest
Grand Airways
Lang Air
Sierra Nevada
Western

Concourse B
America West
Sun Country

Concourse C
American
Morris Air
Private Jet Expeditions
Southwest
United

©1995-1993 Rand McNally & Co.

Sahara Hotel & Casino, 2535 Las Vegas
 Blvd. S., 737-2111; FAX 737-1017
Sands Resort Hotel & Casino, 3355 Las
 Vegas Blvd. S., 733-5000; FAX 733-5624
Treasure Island at the Mirage, 3300 Las
 Vegas Blvd. S., 894-7111; FAX 894-7446
✈ Tropicana, 3801 Las Vegas Blvd. S.,
 739-2222; FAX 736-1883
Union Plaza Hotel/Casino, 1 Main St.,
 386-2110; FAX 382-8281

SELECTED RESTAURANTS:

Signature Food: buffet
Bacchanal, in Caesars Palace Hotel,
 734-7110
Burgundy Room, in the Sands Resort
 Hotel and Casino, 733-5000
Georges La Forge's Pamplemousse
 Restaurant, 400 E. Sahara Ave.,
 733-2066
Golden Steer Steak House, 308 W. Sahara
 Ave., 384-4470
The House of Lords, in the Sahara Hotel,
 737-2111
Nero's, in Caesars Palace Hotel, 731-7110
Palace Court, in Caesars Palace Hotel,
 734-7110
Portofino Room, in the Desert Inn Hotel
 and Country Club, 733-4444
The Rikshaw, in the Riviera Hotel, 2901
 Las Vegas Blvd. S., 734-5110

SELECTED ATTRACTIONS:

Belz Factory Outlet World Mall, Las Vegas
 Blvd. S. & Warm Springs Rd., 896-5599
The Forum Shops at Caesars, Las Vegas
 Blvd. S., adjacent to Caesars Palace
 Hotel, 893-4800
Gambling and Entertainment, at casinos
 and hotels throughout Las Vegas
Grand Slam Canyon (amusement park),
 2880 Las Vegas Blvd. S., 794-3912
Hoover Dam, 30 miles east of Las Vegas on
 US 93, 293-1081

Las Vegas Factory Outlet Stores of
 America, 9115 Las Vegas Blvd. S.,
 471-7755
Liberace Museum, 1775 E. Tropicana
 Ave., 798-5595
Valley of Fire, 55 miles northeast of Las
 Vegas off I-15, 397-2088

INFORMATION SOURCES:

Las Vegas Convention and Visitors
Authority
 Convention Center
 3150 Paradise Rd.
 Las Vegas, Nevada 89109
 (702) 892-0711
Las Vegas Chamber of Commerce
 711 E. Desert Inn
 Las Vegas, Nevada 89109
 (702) 735-1616

Lexington, Kentucky

Population: (*348,428) 225,366 (1990C)
Altitude: 983 feet
Average Temp.: Jan., 33°F.; July, 76°F.
Telephone Area Code: 606

Time: 259-2333 **Weather:** 293-9999
Time Zone: Eastern

AIRPORT TRANSPORTATION:

Eight miles to downtown Lexington.
Taxicab and limousine service.

SELECTED HOTELS:

Campbell House Inn, 1375 Harrodsburg
 Rd., 255-4281; FAX 254-4368
Continental Inn, 801 New Circle Rd. NE,
 299-5281; FAX 293-5905
Econo Lodge, 925 Newtown Pike,
 231-6300; FAX 231-6300
Harley Hotel of Lexington, 2143 N.
 Broadway, 299-1261; FAX 293-0048
Holiday Inn—North, 1950 Newtown Pike,
 233-0512; FAX 231-9285
Hyatt Regency Lexington, 400 W. Vine,
 253-1234; FAX 233-7974
Marriott Resort at Griffin Gate, 1800
 Newtown Pike, 231-5100;
 FAX 255-9944
Radisson Plaza Hotel, Broadway & Vine,
 231-9000; FAX 281-3737
Ramada Inn, 1938 Stanton Way, 259-1311;
 FAX 259-1311

SELECTED RESTAURANTS:

Signature Food: hot brown; burgoos; bluegrass pie

Cafe by the Park, in the Radisson Plaza Hotel, 231-9000

The Coach House, 855 S. Broadway, 252-7777

Fifth Quarter, 2305 Nicholasville Rd., 276-5223

The Glass Garden, in the Hyatt Regency Lexington, 253-1234

J. W. Steak House, in the Marriott Resort at Griffin Gate, 231-5100

The Mansion at Griffin Gate, 1720 Newtown Pike, 231-5152

Nagasaki Inn, 2013 Regency Rd., 278-8782

The Springs Inn, 2020 Harrodsburg Rd., 277-5751

SELECTED ATTRACTIONS:

Ashland (Henry Clay's home), Richmond Rd. at Sycamore, 266-8581

The Hunt-Morgan home, 201 N. Mill St. in Gratz Park, 233-3290

Kentucky Horse Park/International Museum of the Horse, 4089 Iron Works Pike, 233-4303

Lexington Cemetery, W. Main St., 255-5522

Mary Todd Lincoln House, 578 W. Main St., 233-9999

Old Fort Harrod State Park, U.S. 27 & U.S. 68, Harrodsburg, 734-3314

Shaker Village at Pleasant Hill, 7 miles northeast of Harrodsburg off U.S. 68, 734-5411

Star of Lexington Cruises, 263-STAR

Waveland State Historic Site, 225 Higbee Mill Rd. off U.S. 27 south, 272-3611

INFORMATION SOURCES:

Greater Lexington Convention and Visitors Bureau
430 W. Vine St., Suite 363
Lexington, Kentucky 40507
(606) 233-1221; (800) 848-1224

Lexington Chamber of Commerce
330 E. Main St.
Lexington, Kentucky 40507
(606) 254-4447

Little Rock, Arkansas

Population: (*513,117) 175,795 (1990C)
Altitude: 330 feet
Average Temp.: Jan., 40,°F.; July, 82°F.
Telephone Area Code: 501
Time and Weather: 376-4400
Time Zone: Central

AIRPORT TRANSPORTATION:

Four miles to downtown Little Rock.

Taxicab, bus and hotel shuttle service.

SELECTED HOTELS:

Best Western Governors Inn, 1501 Merrill, 224-8051; FAX 224-8051

Camelot Hotel, Markham & Broadway sts., 372-4371; FAX 372-0518

Capital Hotel, 111 W. Markham, 374-7474; FAX 370-7091

Excelsior, 3 Statehouse Plaza, 375-5000; FAX 375-4721

Holiday Inn—City Center, 617 S. Broadway, 376-2071; FAX 376-7733

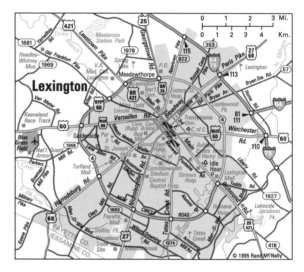

Holiday Inn North, 111 W. Pershing Blvd., North Little Rock, 758-1440; FAX 758-2094

Masters Economy Inn, 707 I-30, 372-4392; FAX 372-1732

Ramada Inn—North, 120 W. Pershing, North Little Rock, 758-1851; FAX 758-5616

Riverfront Hilton Inn, 2 Riverfront Place, 371-9000; FAX 371-9000

SELECTED RESTAURANTS:

Signature Food: catfish; chicken; watermelon

Alouette's, 11401 Rodney Parham Rd., 225-4152

Anderson's Cajun's Wharf, 2600 Cantrell Rd., 375-5351

Cafe Saint Moritz, 225 E. Markham, 372-0411

Josephine's, in the Excelsior Hotel, 375-5000

SELECTED ATTRACTIONS:

Arkansas Arts Center, MacArthur Park, 9th and Commerce, 372-4000

Arkansas State Capitol, Woodlawn and Capitol, for guided tours: 682-5080

Little Rock Zoo, 1 Jonesboro Dr., 666-2406

Old State House, 300 W. Markham, 324-9685

Quapaw Quarter Historic District, (tours) 371-0075

INFORMATION SOURCES:

Little Rock Convention & Visitors Bureau
P.O. Box 3232
Little Rock, Arkansas 72203
(501) 376-4781; (800) 844-4781

Greater Little Rock Chamber of Commerce
1 Spring St.
Little Rock, Arkansas 72201
(501) 374-4871

Los Angeles, California

Population: (*8,863,164) 3,485,398 (1990C)
Altitude: Sea level to 330 feet
Average Temp.: Jan., 55°F.; July, 73°F.
Telephone Area Code: 213
Time: 853-1212   **Weather:** 554-1212
Time Zone: Pacific

AIRPORT TRANSPORTATION:

Seventeen miles to downtown Los Angeles.

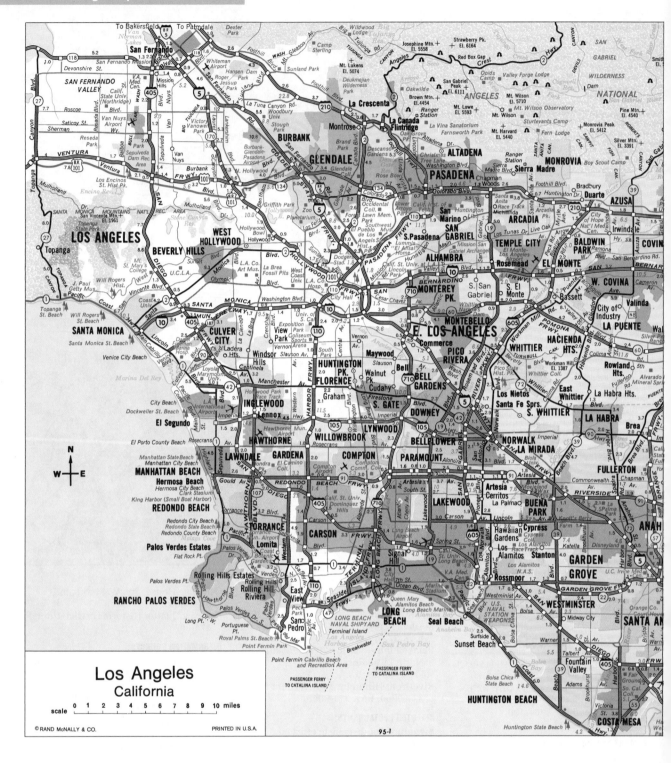

Los Angeles
California

scale 0 1 2 3 4 5 6 7 8 9 10 miles

© RAND McNALLY & CO.   PRINTED IN U.S.A.

95-1

Taxicab, limousine bus and city bus service.

SELECTED HOTELS:

Biltmore Hotel, 506 S. Grand Ave., 624-1011; FAX 612-1545

Century Plaza, 2025 Avenue of the Stars, (310) 277-2000; FAX (310) 551-3355

Holiday Inn—Hollywood, 1755 N. Highland Ave., 462-7181; FAX 466-9072

Hotel Bel-Air, 701 Stone Canyon Rd., (310) 472-1211; FAX (310) 476-5890

Hyatt Los Angeles Airport, 6225 W. Century Blvd., (310) 670-9000; FAX (310) 216-9334

Hyatt Regency Los Angeles, 711 S. Hope St., 683-1234; FAX 629-3230

Kawada Hotel, 200 S. Hill St., 621-4455; FAX 687-4455

Le Parc Hotel De Luxe, 733 N. West Knoll Dr., (310) 855-8888; FAX (310) 659-7812

Los Angeles Airport Marriott, 5855 W. Century Blvd., (310) 641-5700; FAX (310) 337-5358

Los Angeles Hilton and Towers, 930 Wilshire Blvd., 629-4321; FAX 612-3989

The New Otani Hotel & Garden, 120 S. Los Angeles St., 629-1200; FAX 622-0980

Sheraton–Universal, 333 Universal Terrace Pkwy., Universal City, (818) 980-1212; FAX (818) 985-4980

University Hilton, 3540 S. Figueroa St., 748-4141; FAX 746-3255

The Westin Bonaventure Hotel, 404 S. Figueroa St., 624-1000; FAX 612-4800

SELECTED RESTAURANTS:

Signature Food: American
Bernard's, in the Biltmore Hotel, 612-1580
Epicentre Restaurant, in the Kawada
  Hotel, 621-4455
Lawry's Prime Rib, 100 N. La Cienega
  Blvd., (310) 652-2827
L'Escoffier, in the Beverly Hilton, 9876
  Wilshire Blvd., (310) 274-7777
L'Orangerie, 903 N. La Cienega Blvd.,
  (310) 652-9770
Madame Wu's Garden, 2201 Wilshire
  Blvd., (310) 828-5656
Pacific Dining Car, 1310 W. 6th St.,
  483-6000
Stepps, 330 S. Hope St., 626-0900

SELECTED ATTRACTIONS:

Disneyland, 1313 S. Harbor Blvd.,
  Anaheim, (714) 999-4000
El Pueblo de Los Angeles State Historic
  Monument, bounded by Macy &
  Alameda, Arcadia and Main St.,
  628-1274
Farmer's Market & Shopping Village, 6333
  W. 3rd St., 933-9211
Griffith Park Observatory, 2800 E.
  Observatory Rd., 664-1191
Knott's Berry Farm, 8039 Beach Blvd.,
  Buena Park, (714) 220-5200
Los Angeles Zoo, 5333 Zoo Dr., 664-1100
Mann's Chinese Theatre (movie stars'
  hand and foot prints), 6925 Hollywood
  Blvd., 461-3331
Queen Mary, 1126 Queens Hwy., Long
  Beach, (310) 435-3511

Rodeo Drive (shopping), Beverly Hills
Universal Studios, on Lankershim, off the
  Hollywood 101 Freeway, Universal City,
  (818) 777-3750

INFORMATION SOURCES:

Los Angeles Convention & Visitors Bureau
  633 W. 5th St., Suite 6000
  Los Angeles, California 90071
  (213) 624-7300; (800) 228-2452
Los Angeles Area Chamber of Commerce
  404 S. Bixel St.
  Los Angeles, California 90017
  (213) 629-0602

Louisville, Kentucky

Population: (*952,662) 269,063 (1990C)
Altitude: 462 feet
Average Temp.: Jan., 35°F.; July, 78°F.
Telephone Area Code: 502
Time: 585-5961   **Weather:** 363-9655
Time Zone: Eastern

AIRPORT TRANSPORTATION:

Five miles to downtown Louisville.
Taxicab, limousine, and bus service.

SELECTED HOTELS:

Breckinridge Inn, 2800 Breckinridge Lane,
  456-5050; FAX 451-1577
The Brown Hotel, 335 W. Broadway,
  583-1234; FAX 587-7006
✈ Executive Inn, 978 Phillips Ln.,
  367-6161; FAX 363-1880
✈ Executive West Hotel, 830 Phillips Ln.,
  367-2251; FAX 363-2087

The Galt House, 140 N. 4th Ave.,
  589-5200; FAX 589-3444
Holiday Inn Louisville/Downtown, 120 W.
  Broadway, 582-2241; FAX 584-8591
Holiday Inn—S.W., 4110 Dixie Hwy.,
  448-2020; FAX 448-0808
Hurstbourne Hotel & Conference Center,
  9700 Bluegrass Pkwy., 491-4830;
  FAX 499-2893
Hyatt Regency Louisville, 320 W. Jefferson
  St., 587-3434; FAX 581-0133
Master Host Hotel, 100 E. Jefferson St.,
  582-2481; 582-3511
The Seelbach Hotel, 500 Fourth Ave.,
  585-3200; FAX 587-6564

SELECTED RESTAURANTS:

Signature Food: hot brown; burgoos
The Atrium, 1028 Barret Ave., 456-6789
Bristol Bar & Grille, 1321 Bardstown Rd.,
  456-1702
Cafe Metro, 1700 Bardstown Rd.,
  458-4830
Dell Frisco's, 4107 Oechsli Ave., 897-7077
Mama Grisanti, 3938 DuPont Circle,
  893-0141
New Orleans East, 9424 Shelbyville Rd.,
  426-1577
Oak Room, in The Seelbach Hotel,
  585-3200
The Spire, in the Hyatt Regency
  Louisville, 587-3434

SELECTED ATTRACTIONS:

Actors Theatre of Louisville, 316 W. Main
  St., 584-1205

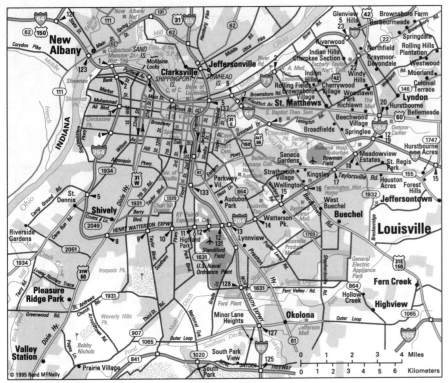

Churchill Downs, 700 Central Ave.,
636-4400

Farmington Historic Home, 3033
Bardstown Rd., 452-9920

Hillerich & Bradsby (Louisville Slugger
factory), Slugger Park, 1525
Charlestown-New Albany Rd.,
Jeffersonville, Indiana, 585-5226

J.B. Speed Art Museum, 2035 S. Third St.,
636-2920

Kentucky Center for the Arts, W. Main St.
between 5th & 6th, 584-7777

Kentucky Derby Museum, 704 Central
Ave., 637-1111

Locust Grove Historic Home, 561
Blankenbaker Ln., 897-9845

Louisville Science Center/IMAX Theater,
727 W. Main St., 561-6100

Louisville Zoo, 1100 Trevilian Way,
459-2181

INFORMATION SOURCES:

Louisville and Jefferson County
Convention & Visitors Bureau
400 S. 1st
Louisville, Kentucky 40202
(502) 584-2121; (800) 626-5646

Louisville Tourist Information
400 S. 1st
Louisville, Kentucky 40202
(502) 582-3732

Louisville Area Chamber of Commerce
600 W. Main St.
Louisville, Kentucky 40202
(502) 625-0000

Memphis, Tennessee

Population: (*981,747) 610,337 (1990C)
Altitude: 264 feet
Average Temp.: Jan., 43°F.; July, 80°F.
Telephone Area Code: 901
Time: 526-5261 **Weather:** 756-4141
Time Zone: Central

MEMPHIS INTERNATIONAL AIRPORT

To/From I-240

To/From U.S. 78

To/From I-55

Winchester Rd.

OVERFLOW PARKING

OVER-FLOW PARKING

LONG TERM PARKING

LONG TERM PARKING

N

PARKING GARAGE

TERMINAL

Concourse A
Delta
NW Airlink
TW Express

Concourse C
American
TW Express
United
USAir
Valujet

Concourse B
Northwest

©1995-1993 Rand McNally & Co.

AIRPORT TRANSPORTATION:
Ten miles to downtown Memphis.
Taxicab and limousine bus service.

SELECTED HOTELS:
Adams Mark Hotel, 939 Ridge Lake Blvd.,
684-6664; FAX 762-7411
Best Western, 2240 Democrat Rd.,
332-1130; FAX 398-5206
Brownestone Hotel, 300 N. Second St.,
525-2511; FAX 525-2511
Holiday Inn Crowne Plaza, 250 N. Main,
527-7300; FAX 526-1561
Holiday Inn Memphis Airport, 1441 E.
Brooks Rd., 398-9211; FAX 398-9211
Holiday Inn Midtown Medical Center,
1837 Union Ave., 278-4100;
FAX 272-3810
Memphis Marriott, 2625 Thousand Oaks
Blvd., 362-6200; FAX 360-8836
The Peabody Hotel, 149 Union Ave.,
529-4000; FAX 529-9600
✈ Sheraton Memphis Airport, at the
Airport, 332-2370; FAX 398-4085

SELECTED RESTAURANTS:
Signature Food: barbeque; pork ribs
Benihana of Tokyo, 912 Ridgelake,
683-7390
Grisanti's, 1489 Airways Blvd., 458-2648
Justine's, 919 Coward Pl., 527-3815
Paulette's, 2110 Madison Ave., 726-5128
The Pier, 100 Wagner Pl., 526-7381
The Rendezvous, 52 S. 2nd St., 523-2746

SELECTED ATTRACTIONS:
Beale Street (birthplace of the blues),
526-0110
Graceland, 3734 Elvis Presley Blvd.,
332-3322
Heritage Tours (African-American history
tours), 543-5333
Liberty Land (amusement park),
Mid-South Fairgrounds, 274-1776
Memphis Motor Sports Park, 5500 Taylor
Forge Rd., Millington, 358-7223
Memphis Pink Palace Museum,
Planetarium & IMAX Theater, 3050
Central Ave., 320-6320
Memphis Queen Line, 527-5694
Memphis Zoo & Aquarium, 2000
Gallaway, in Overton Park, 726-4787
National Civil Rights Museum, 450
Mulberry, 521-9699

INFORMATION SOURCES:
Memphis Convention & Visitors Bureau
47 Union Ave.
Memphis, Tennessee 38103
(901) 543-5300; (800) 873-6282
Visitor Information Center
340 Beale St.
Memphis, Tennessee 38103
(901) 543-5333
Memphis Area Chamber of Commerce
22 N. Front, Suite 200 Falls Building
Memphis, Tennessee 38103
(901) 575-3500

Mexico City (Ciudad de México), Mexico

Population: (*18,000,000) 9,377,300
(1991 estimate)
Altitude: 7,450 ft.
Average Temp.: Jan., 54°F.; July, 64°F.

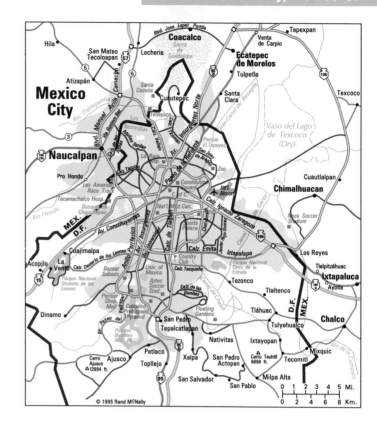

© 1995 Rand McNally

Telephone Code: 011-52-5
Time Zone: Central

AIRPORT TRANSPORTATION:
Four miles to downtown Mexico City.
Taxicab and limousine service.

SELECTED HOTELS:
Aristos, Paseo de la Reforma 276,
211-01-12; FAX 514-4473
Camino Real, Av. Mariano Escobedo 700,
203-2121; FAX 250-6723
Fiesta Americana Reforma, Paseo de la
Reforma 80, 705-15-15; FAX 705-0226
Four Seasons, Paseo de la Reforma 500,
230-1818; FAX 230-1817
Gran Hotel Howard Johnson, 16 de
Septiembre No. 82, 510-4040;
FAX 512-2085
Hotel Century Zona Rosa, Liverpool 152,
726-9911; FAX 525-7475
Hotel Maria Cristina, Rio Lerma 31,
566-9688; FAX 566-9194
Hotel Nikko, Campos Eliseos 204,
280-11-11; FAX 280-9191
Krystal Rosa, Liverpool 155, 228-99-28;
FAX 533-6442
Maria Isabel-Sheraton, Paseo de la
Reforma 325, 207-3933; FAX 207-0684
Marquis Reforma, Paseo de la Reforma
465, 211-36-00; FAX 211-5561
Radisson Paraiso, Cúspide 53, 606-42-11;
FAX 606-4006
Stouffer Presidente, Campos Eliseos 218,
327-77-00; FAX 327-7730
Westin Galeria Plaza, Hamburgo 159,
211-00-14; FAX 207-5867

SELECTED RESTAURANTS:
Signature Food: Mexican
Anderson's, Paseo de la Reforma 382

Antigua Hacienda de Tlalpan, Calz de
Tlalpan 4619
Azulejos, in the Camino Real Hotel
Bellinghausen, Londres 95
Chalet Suizo, Niza 37
Champs Elysses, Amberes 1
El Parador, Niza 17
Focolare, Hamburgo 87 Zona Rosa
Fonda Santa Anita, Humboldt 48
Fouquets de Paris, in the Camino Real
Hotel
La Hacienda de los Morales, Vázquez de
Mella 525
La Cava, Insurgentes Sur 2465
Les Moustaches, Rio Sena 88
Loredo, Hamburgo 29
Maxims, in the Stouffer Presidente Hotel
Mesón del Caballo Bayo, Av. Conscripcion
360
Rincón Argentina, Presidente Masaryk
177
Restaurant del Lago, Chapultepec Park
2nd Section
San Angel Inn, Palmas 50

SELECTED ATTRACTIONS:
Ballet Folklórico, at the Palace of Fine
Arts on Ave. Juárez
Chapultepec Castle, at the entrance to
Chapultepec Park
Floating Gardens of Xochimilco
Hipodromo de las Americas (horseracing)
National Museum of Anthropology, in
Chapultepec Park
National Palace
Polyforum Cultural Siqueiros,
Insurgentes, corner of Filadelfia
Pyramids of Teotihuacan (light and sound
show)

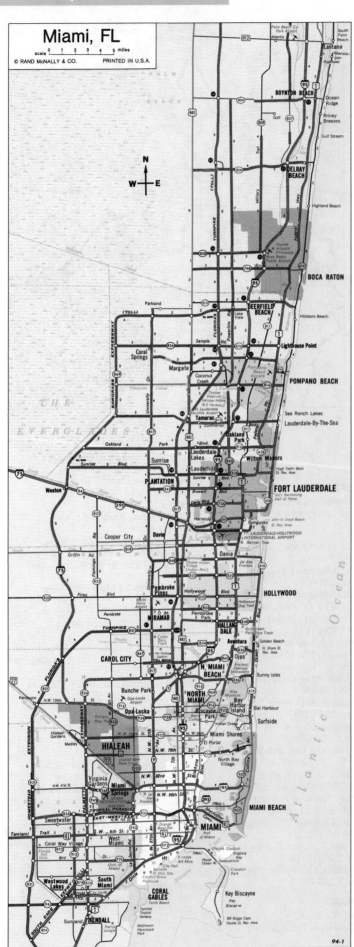

Miami, FL

scale 0 1 2 3 4 5 miles

© RAND McNALLY & CO.   PRINTED IN U.S.A.

Rufino Tamayo Museum (art), in
  Chapultepec Park on the west side of
  Reforma
Templo Mayor (Aztec temple remains),
  northeast corner of the Zocalo

INFORMATION SOURCES:

Mexico City Tourist Bureau
  Amberes No. 54 Esq. Londres
  Zona Rosa Col. Juarez
  06600 México, D.F.
  (011-52-5) 93-80
Federal District Chamber of Commerce
  Paseo de la Reforma No. 42
  Delegación Cuauhtémoc
  06048 México, D.F.
  (011-52-5) 592-26-77
Mexican Government Tourist Office
  405 Park Avenue, Suite 1401
  New York, New York 10022
  (212) 755-7261 or 421-6655
  (800) 446-3942

Miami, Florida

Population: (*1,937,094) 358,548 (1990C)
Altitude: Sea level to 30 feet
Average Temp.: Jan., 69°F.; July, 82°F.
Telephone Area Code: 305
Time: 324-8811 **Weather:** 661-5065
Time Zone: Eastern

AIRPORT TRANSPORTATION:

Five miles to downtown Miami.
Taxicab and limousine bus service.

SELECTED HOTELS:

Best Western Marina Park, 340 Biscayne
  Blvd., 371-4400; FAX 372-2862
Biscayne Bay Marriott, 1633 N. Bayshore
  Dr., 374-3900; FAX 375-0597
Don Shula's Hotel & Golf Club, Main St. &
  Bull Run Rd., Miami Lakes, 821-1150;
  FAX 819-8298; FAX 819-8298
Doubletree, 2649 S. Bayshore Dr.,
  858-2500; FAX 858-5776
Fontaine Bleau Hilton, 4441 Collins Ave.,
  Miami Beach, 538-2000; FAX 531-9274
Hyatt Regency Miami, 400 SE. 2nd Ave.,
  358-1234; FAX 358-0529
Inn at the Civic Center, 1170 NW. 11th St.,
  324-0800; FAX 547-1820
✈ Marriott Airport Hotel, 1201 NW.
  LeJeune Rd., 649-5000; FAX 642-3369
✈ Miami Airport Hilton, 5101 Blue
  Lagoon Dr., 262-1000; FAX 267-0038
The Omni Colonnade, 180 Aragon Ave.,
  Coral Gables, 441-2600; FAX 445-3929
Omni International Hotel, 1601 Biscayne
  Blvd., 374-0000; FAX 374-0020
✈ Ramada Hotel—Miami International
  Airport, 3941 NW. 22nd St., 871-1700;
  FAX 871-4830
Sheraton Brickell Point, 495 Brickell Ave.,
  373-6000; FAX 374-2279

SELECTED RESTAURANTS:

Signature Food: Cuban cuisine
Centro Vasco, 2235 SW. 8th St., 643-9606
The Chart House, 51 Charthouse Dr.,
  Coconut Grove, 856-9741
The Dining Galleries, in the Fontaine
  Bleau, 538-2000
Joe's Stone Crab, 227 Biscayne St., South
  Miami Beach, 673-0365
La Paloma, 10999 Biscayne Blvd.,
  891-0505

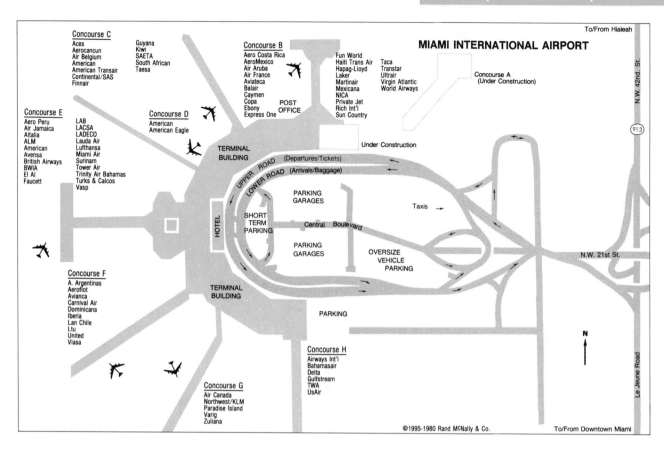

MIAMI INTERNATIONAL AIRPORT

Concourse C
Aces
Aerocancun
Air Belgium
American
American Transair
Continental/SAS
Finnair

Guyana
Kiwi
SAETA
South African
Taesa

Concourse B
Aero Costa Rica
AeroMexico
Air Aruba
Air France
Aviateca
Balair
Caymen
Copa
Ebony
Express One

POST OFFICE

Fun World
Haiti Trans Air
Hapag-Lloyd
Laker
Martinair
Mexicana
NICA
Private Jet
Rich Int'l
Sun Country

Taca
Transtar
Ultrair
Virgin Atlantic
World Airways

Concourse A
(Under Construction)

To/From Hialeah

N.W. 42nd. St.

913

Concourse E
Aero Peru
Air Jamaica
Altalia
ALM
American
Avensa
British Airways
BWIA
El Al
Faucett

LAB
LACSA
LADECO
Lauda Air
Lufthansa
Miami Air
Surinam
Tower Air
Trinity Air Bahamas
Turks & Calcos
Vasp

Concourse D
American
American Eagle

TERMINAL BUILDING

UPPER ROAD (Departures/Tickets)
LOWER ROAD (Arrivals/Baggage)

Under Construction

Taxis

PARKING GARAGES

SHORT TERM PARKING

Central Boulevard

PARKING GARAGES

OVERSIZE VEHICLE PARKING

HOTEL

N.W. 21st St.

N

Concourse F
A. Argentinas
Aeroflot
Avianca
Carnival Air
Dominicana
Iberia
Lan Chile
Ltu
United
Viasa

TERMINAL BUILDING

PARKING

Concourse H
Airways Int'l
Bahamasair
Delta
Gulfstream
TWA
UsAir

Concourse G
Air Canada
Northwest/KLM
Paradise Island
Varig
Zuliana

Le Jeune Road

©1995-1980 Rand McNally & Co.

To/From Downtown Miami

Renato's, 9561 E. Bay Harbor Dr., 866-8779

SELECTED ATTRACTIONS:
Ancient Spanish Monastery, 16711 W. Dixie Hwy., North Miami Beach, 945-1461
Biscayne National Park Tour Boats, Biscayne National Park Headquarters, 9 mi. east of Homestead, 247-2400
Fairchild Tropical Garden, 10901 Old Cutler Rd., Coral Gables, 667-1651
Miami Metrozoo, 12400 SW 152nd St., 251-0400
Miami Museum of Science & Space Transit Planetarium, 3280 S. Miami Ave., 854-4247
Miami Seaquarium, 4400 Rickenbacker Causeway, 361-5705
Miccosukee Indian Village & Airboat Tours, 30 miles west of Miami on U.S. 41, 223-8380 (weekdays)
Monkey Jungle, 14805 SW 216th St., 235-1611
Parrot Jungle & Gardens, 11000 SW 57th Ave., 666-7834
Vizcaya Museum and Gardens, 3251 S. Miami Ave., 579-2708

INFORMATION SOURCE:
Greater Miami Convention and Visitors Bureau
701 Brickell Ave., Suite 2700
Miami, Florida 33131
(305) 539-3000; (800) 933-8448

Milwaukee, Wisconsin

Population: (*1,432,149) 628,088 (1990C)
Altitude: 634 feet
Average Temp.: Jan., 21°F.; July, 71°F.
Telephone Area Code: 414
Time: 976-8463 **Weather:** 936-1212
Time Zone: Central

AIRPORT TRANSPORTATION:

Eight miles to downtown Milwaukee.
Taxicab, bus, and limousine bus service.

SELECTED HOTELS:

✈ The Grand Milwaukee Hotel, 4747 S. Howell Ave., 481-8000; FAX 481-8065
Hyatt Regency Milwaukee, 333 W. Kilbourn Ave., 276-1234; FAX 276-6338
Marc Plaza Hotel, 509 W. Wisconsin Ave., 271-7250; FAX 271-1039
Milwaukee River Hilton Inn, 4700 N. Port Washington Rd., 962-6040; FAX 962-6166
The Pfister Hotel, 424 E. Wisconsin Ave., 273-8222; FAX 273-8222
Sheraton Mayfair Inn, 2303 N. Mayfair, Wauwatosa, 257-3400; FAX 257-0900
Wyndham Hotel, 139 E. Kilbourn Ave., 276-8686; FAX 276-8007

SELECTED RESTAURANTS:

Signature Food: brats; frozen custard
Benson's, in the Marc Plaza Hotel, 271-7250
The English Room, in The Pfister Hotel, 273-8222
Grenadier's, 747 N. Broadway, 276-0747

John Ernst's, Ogden at Jackson, 273-1878
Karl Ratzsch's, 320 E. Mason St., 276-2720
Mader's German Restaurant, 1037 N. 3rd St., 271-3377
Pieces of Eight, 550 N. Harbor Dr., 271-0597
Whitney's, in the Milwaukee Marriott, 375 S. Moorland Rd., Brookfield, 786-1100

SELECTED ATTRACTIONS:

Miller Brewing Company, 4251 W. State St., 931-BEER
Milwaukee Art Museum, 750 N. Lincoln Memorial Dr., 224-3200
Milwaukee County Zoo, 10001 W. Bluemound Rd., 771-3040
Milwaukee Public Museum, 800 W. Wells St., 278-2702
Mitchell Park Horticultural Conservatory (The Domes), 524 S. Layton Blvd., 649-9830
Pabst Brewing Company, 915 W. Juneau Ave., 223-3709
Pabst Mansion, 2000 W. Wisconsin Ave., 931-0808
Potawatomi Bingo, 1721 W. Canal St., 645-6888

INFORMATION SOURCES:

Greater Milwaukee Convention & Visitors Bureau
510 W. Kilbourn Ave.
Milwaukee, Wisconsin 53203
(414) 273-3950; (800) 231-0903
Metropolitan Milwaukee Association of Commerce

756 N. Milwaukee St.
Milwaukee, Wisconsin 53202
(414) 287-4100

Minneapolis-St. Paul, Minnesota

Population: (*2,464,124) (Minneapolis)
368,383; (St. Paul) 272,235 (1990C)
Altitude: (Minneapolis) 840 feet;
(St. Paul) 874 feet
Average Temp.: Jan., 15°F.; July, 74°F.
Telephone Area Code: 612
Time: 546-8463   **Weather:** 725-6090
Time Zone: Central

AIRPORT TRANSPORTATION:

About 8 miles to downtown Minneapolis
or St. Paul.
Taxicab, bus, and limousine bus service to
Minneapolis and St. Paul.

SELECTED HOTELS: MINNEAPOLIS

Bloomington Marriott Hotel, 2020 E. 79th
St., 854-7441; FAX 854-7671
Holiday Inn—Airport #2, 5401 Green
Valley Dr., 831-8000; FAX 831-8426
Holiday Inn Crowne Plaza Northstar, 618
2nd Ave. S., 338-2288; FAX 338-2288
Hotel Sofitel, 5601 W. 78th St., 835-1900;
FAX 835-2696
Hyatt Regency Minneapolis, 1300 Nicollet
Mall, 370-1234; FAX 370-1463
Mall of America GrandHotel, 7901 24th
Ave. S., 854-2244; FAX 854-4421
The Marquette Hotel, 7th St. & Marquette
Ave., 332-2351; FAX 376-7419

Minneapolis Hilton, 1001 Marquette,
376-1000; FAX 397-4875
Minneapolis Marriott City Center, 30 S.
7th St., 349-4000; FAX 332-7165
Radisson Hotel South, 7800 Normandale
Blvd., 835-7800; FAX 893-8419
Sheraton Airport Inn, 2500 E. 79th St.,
854-1771; FAX 854-5898
Wyndham Garden Hotel, 4460 W. 78th St.
Circle, 831-3131; FAX 831-6372

SELECTED RESTAURANTS: MINNEAPOLIS

Signature Food: wild rice; walleye
The Anchorage Restaurant, 1330
Industrial Blvd., 379-4444
Goodfellow's, 800 Nicollet Mall, 332-4800
Gustino's, in the Minneapolis Marriott
City Center, 349-4000
Lord Fletcher's of the Lake, 3746 Sunset
Dr., Spring Park, 471-8513
Taxi Restaurant, in the Hyatt Regency
Minneapolis, 370-1234

SELECTED ATTRACTIONS: MINNEAPOLIS

American Swedish Institute, 2600 Park
Ave., 871-4907
Bell Museum (Minnesota flora and fauna
in natural habitat), University of
Minnesota, 17th Ave. & University Ave.
SE, 624-7083
Frederick R. Weisman Art Museum, 333
E. River Rd., Univ. of Minnesota
campus, 625-9494
Guthrie Theatre, 725 Vineland Pl.,
377-2224
The Mall of America, MN Hwy. 77 &
I-494, Bloomington, 851-3500

Minneapolis Institute of Arts, 2400 Third
Ave. S., 870-3046
Minneapolis Planetarium, inside the
Public Library, 300 Nicollet Mall,
372-6644
Nicollet Mall, downtown Minneapolis
Riverplace (European-theme festival
marketplace), 43 Main St. SE, 378-1969
Walker Art Center/Minneapolis Sculpture
Garden, 725 Vineland Pl., 375-7600

INFORMATION SOURCES: MINNEAPOLIS

Greater Minneapolis Convention &
Visitors Association
4000 Multifoods Tower
33 S. 6th St.
Minneapolis, Minnesota 55402
(612) 661-4700 (meetings &
conventions); (900) 860-0092 (tourist
information: $1.99 first minute; 99¢
each additional minute)

SELECTED HOTELS: ST. PAUL

Best Western Kelly Inn State Capitol, 161
St. Anthony Blvd., 227-8711;
FAX 227-1698
Country Inn by Carlson—Woodbury, 6003
Hudson Rd., Woodbury, 739-7300;
FAX 731-4007
Crown Sterling Suites, 175 E. 10th St.,
224-5400; FAX 224-0957
Holiday Inn Express, 1010 Bandana Blvd.,
647-1637; FAX 647-0244
Holiday Inn St. Paul East, I-94 at
McKnight, 731-2220; FAX 731-0243

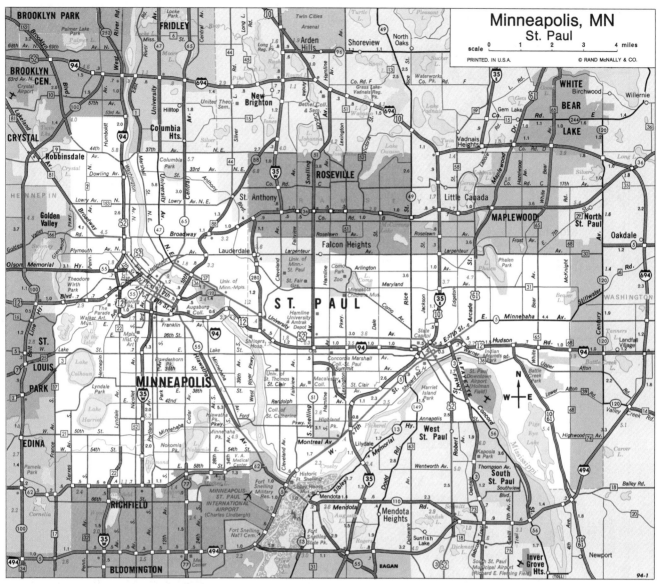

Minneapolis, MN
St. Paul

scale 0   1   2   3   4 miles

PRINTED IN U.S.A.   © RAND McNALLY & CO.

MINNEAPOLIS-ST. PAUL INTERNATIONAL AIRPORT

Red Concourse
Gates 21-38

Canadian Airlines-Time Air
KLM
Northwest
TWA

Gold Concourse
Gates 1–15

Mesaba
Northwest

PARKING
Bearskin Airlines

HUBERT H HUMPHREY TERMINAL

34th Ave.

Blue Concourse
Gates 41-55

American
Continental
Comair
Delta
Northwest
United
USAir

Green Concourse
Gates 62-79

America West
Express II
Mark Air
Northwest

Regional/Commuter
Gates 87-89

Airvantage
Express II
Great Lakes/Aviation
United Express

PARKING STRUCTURE

Hiawatha Ave.

To/From Minneapolis

55

To/From Bloomington

To/From St. Paul

5

©1995-1983 Rand McNally & Co.

The Radisson Hotel St. Paul, 11 E. Kellogg
  Blvd., 292-1900; FAX 224-8999
Ramada Hotel St. Paul, 1870 Old Hudson
  Rd., 735-2330; FAX 735-1953
The St. Paul Hotel, 350 Market St.,
  292-9292; FAX 228-9506
Sheraton Midway, 400 N. Hamline Ave.,
  642-1234; FAX 642-1126

SELECTED RESTAURANTS: ST. PAUL

Signature Food: wild rice; walleye
Bali Hai, 2305 White Bear Ave. and Hwy
  36, Maplewood, 777-5500
The Carrousel, in The Radisson Hotel St.
  Paul, 292-1900
Forepaugh's Restaurant, 276 S. Exchange
  St., 224-5606
Francesca's Bakery and Cafe, 33 W. 7th
  Pl., 292-0027
Gallivan's, 354 Wabasha St., 227-6688
McGovern's Pub, 225 W. 7th St., 224-5821
Venetian Inn, 2814 Rice St., 484-7215

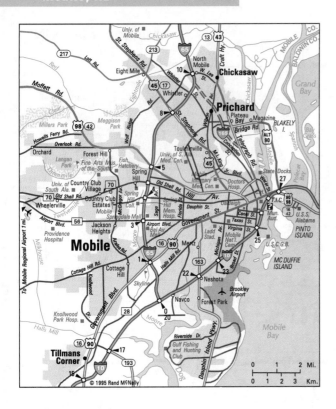

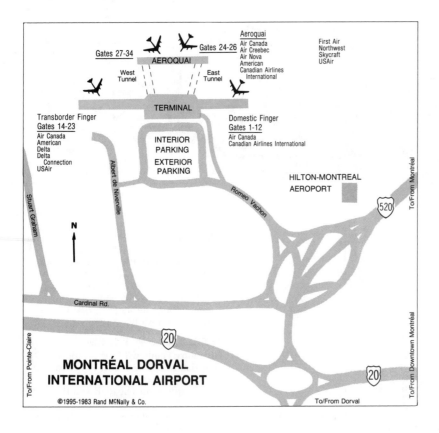

SELECTED ATTRACTIONS: ST. PAUL

Alexander Ramsey House, 265 S.
Exchange St., 296-8681
The Caves (cabaret theater), 215 S.
Wabasha, 224-8319
Children's Museum, 1217 Bandana Blvd.
N., 644-3818
Como Park Zoo and Conservatory,
Midway Pkwy. & Kaufman Dr., 488-5571
Historic Fort Snelling, Fort Rd. at MN
Hwys. 5 & 55, 725-2413
James J. Hill House, 240 Summit Ave.,
297-2555
Jonathan Padelford & Josiah Snelling
Riverboats, Harriet Island, 227-1100
Minnesota History Center, 345 Kellogg
Blvd. W., 296-6126
Minnesota Museum of American Art, in
the Landmark Center, 75 W. 5th St.,
292-4355
Minnesota State Capitol, Cedar & Aurora
sts., 297-3521 (recorded info.); 296-2881
(tour reservations)
Science Museum of Minnesota/William L.
McKnight Omnitheater, Exchange &
Wabasha sts., 221-9454

INFORMATION SOURCES: ST. PAUL

St. Paul Convention & Visitors Bureau
101 Norwest Center
55 E. 5th St.
St. Paul, Minnesota 55101
(612) 297-6985; (800) 627-6101;
FAX 297-6879
Minnesota Office of Tourism
101 Metro Square Building
121 7th Place E.
St. Paul, Minnesota 55101-2112
(612) 296-5029; (800) 657-3700
St. Paul Area Chamber of Commerce
101 Norwest Center
55 E. 5th St.
St. Paul, Minnesota 55101
(612) 223-5000

Mobile, Alabama

Population: (*476,923) 196,278 (1990C)
Altitude: 7 feet
Average Temp.: Jan., 51°F.; July, 82°F.
Telephone Area Code: 334
Time: 660-0044   **Weather:** 478-6666
Time Zone: Central

AIRPORT TRANSPORTATION:

Twelve miles to downtown Mobile.
Taxicab and limousine bus service.

SELECTED HOTELS:

Adams Mark Riverview Plaza Hotel, 64
Water St., 438-4000; FAX 438-3718
Clarion Hotel, 3101 Airport Blvd.,
476-6400; FAX 476-9360
Holiday Inn, 6527 Hwy. 90 West,
666-5600; FAX 666-2773
Holiday Inn I-65, 850 S. Beltline Hwy.,
342-3220; FAX 342-8919
Malaga Inn, 359 Church St., 438-4701;
FAX 438-4701, ext. 123
Radisson Admiral Semmes, 251
Government St., 432-8000;
FAX 432-8000
Ramada Inn, 600 S. Beltline Hwy.,
344-8030; FAX 344-8055

SELECTED RESTAURANTS:

Signature Food: seafood; shrimp; oysters

La Louisiana, 2400 Airport Blvd.,
476-8130

Nautilus, US Hwy. 98, Daphne, 626-0783

The Pillars, 1757 Government, 478-6341

Port City Brewery, 225 Dauphin St.,
438-2739

Ruth's Chris Steak House, 271 Glenwood
St., 476-0516

Stonewall Jackson's Restaurant, in the
Malaga Inn, 438-4701

SELECTED ATTRACTIONS:

Battleship USS *Alabama* Memorial Park,
Battleship Pkwy., 433-2703

Bellingrath Gardens and Home, 12401
Bellingrath Gardens Rd., Theodore,
973-2217

Bragg-Mitchell Mansion, 1906 Spring Hill
Ave., 471-6364

Museum of the City of Mobile, 355
Government St., 434-7569

Condé-Charlotte House, 104 Theater St.,
432-4722

The Exploreum, 1906 Spring Hill Ave.,
476-6873 or 471-5923

Fort Condé, 150 S. Royal St., 434-7658

Oakleigh Mansion, 350 Oakleigh Pl.,
432-1281

The Phoenix Fire Museum, 203 S.
Claiborne St., 434-7569

Richards DAR House, 256 N. Joachim St.,
434-7320

INFORMATION SOURCES:

Mobile Convention & Visitors Corporation
1 S. Water St.
Mobile, Alabama 36602
(334) 415-2000; (800) 5MOBILE

City of Mobile Department of Tourism
150 S. Royal St.
Mobile, Alabama 36602
(334) 434-7304; (800) 252-3862

Mobile Area Chamber of Commerce
451 Government
Mobile, Alabama 36602
(334) 433-6951

Montréal, Québec, Canada

Population: (*2,921,357) 1,015,420
(1986C)

Altitude: 50 ft.

Average Temp.: Jan., 16°F.; July, 71°F.

Telephone Area Code: 514

Time: (none)   **Weather:** 283-4006

Time Zone: Eastern

AIRPORT TRANSPORTATION:

Fourteen miles from Dorval Airport to
downtown Montréal, 34 miles from
Mirabel Airport.

Taxicab and bus service.

Transit service between Dorval and
Mirabel airports.

SELECTED HOTELS:

✈ Aeroport Hilton International, 12505
Cote de Liesse Rd., 631-2411;
FAX 631-0192

Bonaventure Hilton International,
Mansfield & Lagauchetiére, 878-2332;
FAX 878-1442

Centre Sheraton, 1201 René-Lévesque
Blvd. W., 878-2000; FAX 878-3958

Four Seasons Hotel, 1050 Sherbrooke St.
W., 284-1110; FAX 845-3025

Grand Hotel, University St. & St. James,
879-1370; FAX 879-1761

Hotel Meridien-Montreal, 4 Complexe
Desjardins, 285-1450; FAX 285-1243

Le Chateau Champlain, 1050 Ouest de
Lagauchetiére, 878-9000; FAX 878-6761

Queen Elizabeth, 900 Blvd.
René-Lévesque W., 861-3511;
FAX 954-2256

Ritz-Carlton, 1228 Sherbrooke St. W.,
842-4212; FAX 842-3383

Ruby Foo's Hotel, 7655 Decarie Blvd.,
731-7701; FAX 731-7158

SELECTED RESTAURANTS:

Signature Food: smoked meat
sandwiches; Montréal bagel

Beaver Club, in the Queen Elizabeth

Hotel, 861-3511

Café de Paris, in the Ritz-Carlton,
842-4212

Chez la Mére Michel, 1209 Guy, 934-0473

Desjardins, 1175 Mackay, 866-9741

Le Castillon, in the Bonaventure Hilton
International, 878-2332

Le Neufchatel, in Le Chateau Champlain,
878-9000

Les Halles, 1450 Crescent, 844-2328

SELECTED ATTRACTIONS:

Biodôme, 4777 Pierre-de-Coubertin Ave.,
868-3000

Botanical Gardens, 4101 Sherbrooke St.
E., 872-1427

Canadian Centre for Architecture, 1920
Baile St., 939-7000

Contemporary Art Museum, 185
Ste-Catherine St. W., 847-6212

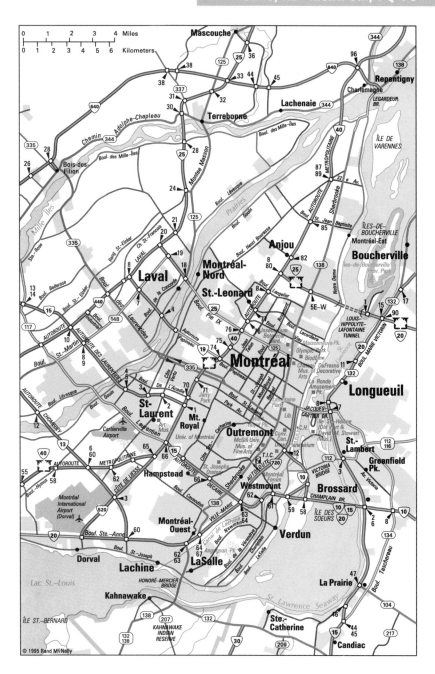

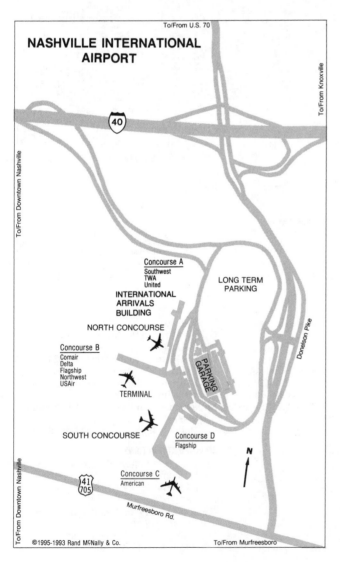

Insectarium, in the Botanical Gardens,
4101 Sherbrooke St. E., 872-0663
Jet boating on the Lachine Rapids, the Old
Port at the foot of Berri St. & De la
Commune St., 284-9607
Montréal Museum of Fine Arts, 1379
Sherbrooke St. W., 285-1600
Notre Dame Basilica, in front of Place
d'Armes, Old Montréal, 842-2925
Olympic Park and Tower, 3200 Viau,
252-8687
The Underground City, Ste-Catherine St.
Downtown

INFORMATION SOURCES:
Greater Montréal Convention & Tourism
Bureau,
1555 Peel St., Suite 600
Montréal, Québec H3A 1X6
(514) 844-5400; (800) 363-7777 (tourism
information)
Canadian Consulate General
1251 Avenue of the Americas
New York, New York 10020-1175
(212) 596-1600

Nashville, Tennessee

Population: (*985,026) 487,969 (1990C)
Altitude: 440 feet
Average Temp.: Jan., 38°F.; July, 80°F.
Telephone Area Code: 615
Time: 259-2222   **Weather:** 244-9393
Time Zone: Central

AIRPORT TRANSPORTATION:
Seven miles to downtown Nashville.
Taxicab, limousine and bus service.

SELECTED HOTELS:
Days Inn, 1 International Plaza, 361-7666;
FAX 399-0283
Doubletree Hotel, 315 4th Ave. N.,
244-8200; FAX 747-4894
The Hermitage, 231 6th Ave. N., 244-3121;
FAX 254-6909
✈ Holiday Inn—Briley Pkwy., 2200 Elm
Hill Pike, 883-9770; FAX 391-4521
Holiday Inn Express, 981 Murfreesboro
Rd., 367-9150; FAX 361-4865
Nashville Marriott Hotel, 600 Marriott Dr.,
889-9300; FAX 889-9315
Opryland Hotel, 2800 Opryland Dr.,
889-1000; FAX 871-7741
Quality Inn Hall of Fame, 1407 Division
St., 242-1631; FAX 244-9519
Ramada Inn—Southwest, 709 Spence Ln.,
361-0102; FAX 361-4765
Ramada South Inn & Convention Center,
737 Harding Pl., 834-5000;
FAX 834-5600
Regal Maxwell House, 2025 MetroCenter
Blvd., 259-4343; FAX 259-4343
Stouffer Nashville, 611 Commerce St.,
255-8400; FAX 255-8202

SELECTED RESTAURANTS:
Signature Food: fried catfish; fried
chicken; meat and three
Arthur's of Nashville, in the Union
Station Hotel, 1001 Broadway, 255-1494
Crown Court, in the Regal Maxwell
House, 259-4343
The Hermitage Hotel Dining Room, in The
Hermitage Hotel, 244-3121
Hunt Room, in the Doubletree Hotel,
244-8200

©1995-1988 Rand McNally & Co.

Mario's, 2005 Broadway, 327-3232
New Orleans Manor, 1400 Murfreesboro
Rd., 367-2777
Old Hickory, in the Opryland Hotel,
889-1000
Praline's, in the Regal Maxwell House,
259-4343
Stockyard, 901 2nd Ave. N., 255-6464

SELECTED ATTRACTIONS:
Belle Meade Plantation, 5025 Harding
Rd., 356-0501
Cheekwood, 1200 Forest Park Dr.,
356-8000
Country Music Hall of Fame, 4 Music
Square E., 256-1639
Cumberland Science Museum, 800 Fort
Negley Blvd., 862-5160
General Jackson Showboat, Exit 11 on
Briley Pkwy., 889-6611
Grand Ole Opry, in Opryland USA Park,
889-6611
The Hermitage, 4580 Rachel's Lane,
Hermitage, 889-2941
Opryland USA, 2802 Opryland Dr.,
889-6600
The Parthenon, West End & 25th aves., in
Centennial Park, 862-8431
Tennessee State Museum, Polk Cultural
Center, 505 Deaderick St., 741-2692

INFORMATION SOURCES:
Nashville Convention & Visitors Bureau
161 4th Ave. N.
Nashville, Tennessee 37219
(615) 259-4730
Nashville Area Chamber of Commerce
161 4th Ave. N.
Nashville, Tennessee 37219
(615) 259-4755

New Orleans, Louisiana

Population: (*1,238,816) 496,938 (1990C)
Altitude: −5 to 25 feet
Average Temp.: Jan., 55°F.; July, 82°F.
Telephone Area Code: 504
Time: 529-6111 **Weather:** 465-9212
Time Zone: Central

AIRPORT TRANSPORTATION:
Eleven miles to downtown New Orleans.
Taxicab, airport shuttle, and limousine
bus service.

SELECTED HOTELS:
Chateau Sonesta, 800 Iberville St.,
553-2339; FAX 553-2387 (due to open
in Jan., 1995)
Clarion Hotel, 1500 Canal St., 522-4500;
FAX 525-2644
Dauphine Orleans, 415 Dauphine St.,
586-1800; FAX 586-1409
Fairmont Hotel, 123 Baronne St.,
529-7111; FAX 522-2303
Hotel Intercontinental, 444 St. Charles
Ave., 525-5566; FAX 523-7310
Hyatt Regency New Orleans, 500 Poydras
Plaza, 561-1234; FAX 587-4141
The Monteleone, 214 Royal St., 523-3341;
FAX 528-1019
New Orleans Hilton Riverside and
Towers, Poydras St. at the Mississippi
River, 561-0500; FAX 568-1721
New Orleans Marriott, 555 Canal St.,
581-1000; FAX 523-6755
The Omni Royal Orleans Hotel, 621 St.
Louis St., 529-5333; FAX 523-5046
The Pontchartrain Hotel, 2031 St. Charles
Ave., 524-0581; FAX 529-1165
Royal Sonesta Hotel, 300 Bourbon St.,
586-0300; FAX 586-0335

SELECTED RESTAURANTS:
Signature Food: gumbo; jambalaya;
beignets; cafe au lait
Arnaud's, 813 Bienville, 523-5433
Brennan's, 417 Royal St., 525-9711
Broussard's, 819 Conti St., 581-3866
Caribbean Room, in The Pontchartrain
Hotel, 524-0581
Commander's Palace Restaurant, 1403
Washington Ave., 899-8221
Copeland's, 1001 S. Clearview Pkwy.,
733-7843
Galatoire's Restaurant, 209 Bourbon St.,
525-2021
Kabby's, in the New Orleans Hilton
Riverside and Towers, 561-0500

Louis XVI French Restaurant, in the St.
Louis Hotel, 730 Bienville, 581-7000
Praline Connection, 542 Frenchman St.,
943-3934
Rib Room, in The Omni Royal Orleans
Hotel, 529-5333
Sazerac Restaurant, in the Fairmont Hotel,
529-7111

SELECTED ATTRACTIONS:
Aquarium of the Americas, 1 Canal St.,
565-3033, ext. 549 (information)
Audubon Zoo, 6500 Magazine St.,
861-2537
French Quarter, 78-block area bounded by
the Mississippi River, Esplanade Ave.,
Rampart St., & Canal St.
Jax Brewery, 620 Decatur St., 586-8015
New Orleans Museum of Art, Lelong Ave.
in City Park, 488-2631
New Orleans Paddlewheels (cruises),
529-4567
New Orleans Steamboat Company
(cruises), 586-8777
Riverwalk (festival marketplace), 1
Poydras St., 522-1555
New Orleans Centre, 1400 Poydras St.,
568-0000

INFORMATION SOURCES:
Greater New Orleans Tourist &
Convention Commission
1520 Sugar Bowl Dr.
New Orleans, Louisiana 70112
(504) 566-5011
The Chamber/New Orleans and the River
Region
301 Camp St.
New Orleans, Louisiana 70130
(504) 527-6900

New York, New York

Population: (*8,546,846) 7,322,564
(1990C)
Altitude: Sea level to 30 feet
Average Temp.: Jan., 33°F.; July, 75°F.
Telephone Area Code: 212
Time: 976-8463 **Weather:** 315-2705
Time Zone: Eastern

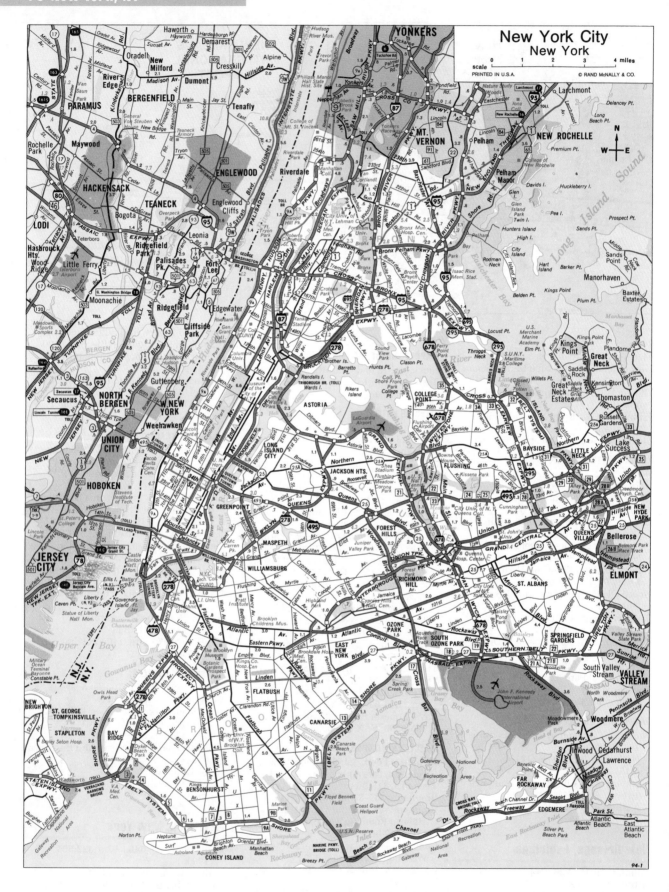

New York City
New York

scale
0 1 2 3 4 miles
PRINTED IN U.S.A.   © RAND McNALLY & CO.

94-1

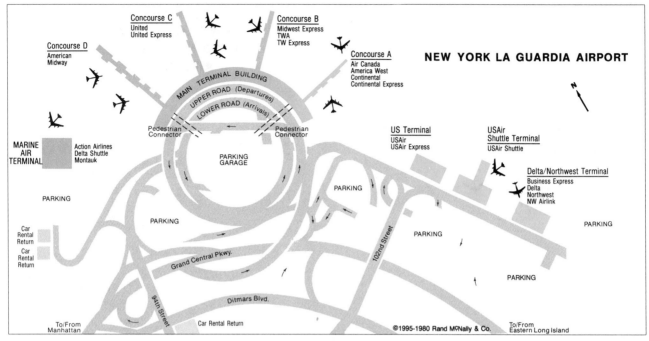

NEW YORK LA GUARDIA AIRPORT

Concourse C
United
United Express

Concourse D
American
Midway

Concourse B
Midwest Express
TWA
TW Express

Concourse A
Air Canada
America West
Continental
Continental Express

MAIN TERMINAL BUILDING
UPPER ROAD (Departures)
LOWER ROAD (Arrivals)

Pedestrian Connector

Pedestrian Connector

MARINE AIR TERMINAL

Action Airlines
Delta Shuttle
Montauk

PARKING GARAGE

US Terminal
USAir
USAir Express

USAir Shuttle Terminal
USAir Shuttle

Delta/Northwest Terminal
Business Express
Delta
Northwest
NW Airlink

PARKING

PARKING

PARKING

PARKING

PARKING

PARKING

Car Rental Return

Car Rental Return

Grand Central Pkwy.

102nd Street

94th Street

Ditmars Blvd.

Car Rental Return

To/From Manhattan

To/From Eastern Long Island

©1995-1980 Rand McNally & Co.

AIRPORT TRANSPORTATION:

Fifteen miles to Manhattan from JFK Airport; 8 miles from La Guardia Airport to Manhattan; 10 miles from Newark Airport to Manhattan.

Taxicab; limousine bus service to and from JFK, La Guardia, and Newark airports and East Side Airlines Terminal. Also JFK Express Subway/bus service from Manhattan to JFK Airport.

SELECTED HOTELS:

Carlyle, Madison Ave. at E. 76th St., 744-1600; FAX 717-4682

The Hotel Pierre, 2 E. 61st St. at 5th Ave., 838-8000; FAX 940-8109

The Lowell Hotel, 28 E. 63rd St., 838-1400; FAX 319-4230

Marriott Marquis, 1535 Broadway at 46th St., 398-1900; FAX 704-8930

The New York Hilton & Towers at Rockefeller Center, 1335 Avenue of the Americas, 586-7000; FAX 315-1374

The New York Palace, 455 Madison Ave., 888-7000; FAX 303-6000

The Plaza, 59th St. & 5th Ave., 759-3000; FAX 759-3167

Regency, 540 Park Ave., 759-4100; FAX 826-5674

Sheraton New York, 811 7th Ave. at 53rd St., 581-1000; FAX 262-4410

Sherry-Netherland, 781 5th Ave., 355-2800; FAX 319-4306

United Nations Plaza Park Hyatt, 1 U.N. Plaza, 355-3400; FAX 702-5051

Waldorf–Astoria, 301 Park Ave. at 50th St., 355-3000; FAX 872-7272

SELECTED RESTAURANTS:

Signature Food: hot pastrami sandwich; knishes; bagels with cream cheese

The Four Seasons, 99 E. 52nd St., 754-9494

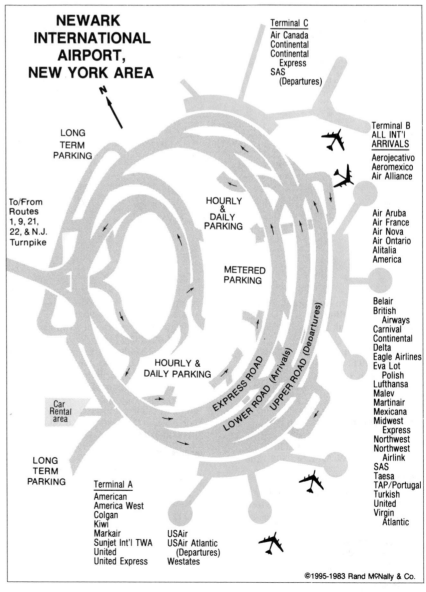

NEWARK INTERNATIONAL AIRPORT, NEW YORK AREA

Terminal C
Air Canada
Continental
Continental Express
SAS
(Departures)

Terminal B
ALL INT'L ARRIVALS

Aerojecativo
Aeromexico
Air Alliance

Air Aruba
Air France
Air Nova
Air Ontario
Alitalia
America

Belair
British Airways
Carnival
Continental
Delta
Eagle Airlines
Eva Lot
Polish
Lufthansa
Malev
Martinair
Mexicana
Midwest Express
Northwest
Northwest Airlink
SAS
Taesa
TAP/Portugal
Turkish
United
Virgin Atlantic

LONG TERM PARKING

To/From Routes 1, 9, 21, 22, & N.J. Turnpike

HOURLY & DAILY PARKING

METERED PARKING

HOURLY & DAILY PARKING

Car Rental area

LONG TERM PARKING

EXPRESS ROAD

LOWER ROAD (Arrivals)

UPPER ROAD (Departures)

Terminal A
American
America West
Colgan
Kiwi
Markair
Sunjet Int'l TWA
United
United Express

USAir
USAir Atlantic (Departures)
Westates

©1995-1983 Rand McNally & Co.

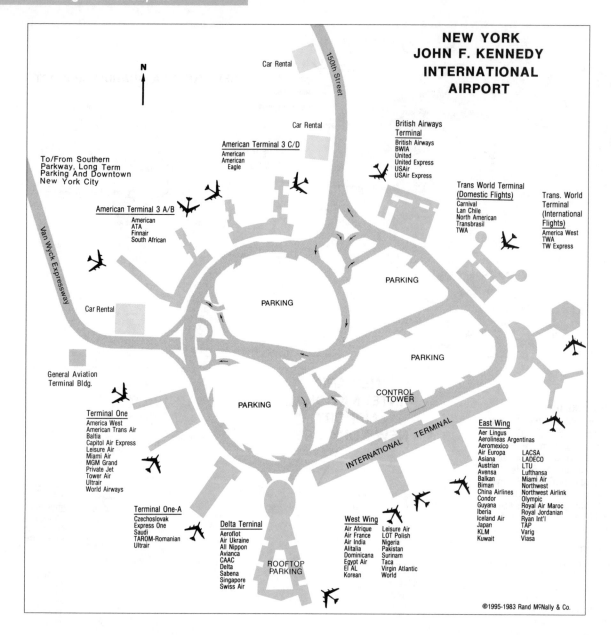

NEW YORK JOHN F. KENNEDY INTERNATIONAL AIRPORT

Car Rental

150th Street

Car Rental

American Terminal 3 C/D
American
American
Eagle

To/From Southern Parkway, Long Term Parking And Downtown New York City

British Airways Terminal
British Airways
BWIA
United
United Express
USAir
USAir Express

Trans World Terminal (Domestic Flights)
Carnival
Lan Chile
North American
Transbrasil
TWA

Trans. World Terminal (International Flights)
America West
TWA
TW Express

American Terminal 3 A/B
American
ATA
Finnair
South African

Van Wyck Expressway

Car Rental

PARKING

PARKING

PARKING

General Aviation Terminal Bldg.

CONTROL TOWER

Terminal One
America West
American Trans Air
Baltia
Capitol Air Express
Leisure Air
Miami Air
MGM Grand
Private Jet
Tower Air
Ultrair
World Airways

PARKING

INTERNATIONAL TERMINAL

East Wing
Aer Lingus
Aerolineas Argentinas
Aeromexico
Air Europa   LACSA
Asiana   LADECO
Austrian   LTU
Avensa   Lufthansa
Balkan   Miami Air
Biman   Northwest
China Airlines   Northwest Airlink
Condor   Olympic
Guyana   Royal Air Maroc
Iberia   Royal Jordanian
Iceland Air   Ryan Int'l
Japan   TAP
KLM   Varig
Kuwait   Viasa

Terminal One-A
Czechoslovak
Express One
Saudi
TAROM-Romanian
Ultrair

Delta Terminal
Aeroflot
Air Ukraine
All Nippon
Avianca
CAAC
Delta
Sabena
Singapore
Swiss Air

West Wing
Air Afrique   Leisure Air
Air France   LOT Polish
Air India   Nigeria
Alitalia   Pakistan
Dominicana   Surinam
Egypt Air   Taca
El AL   Virgin Atlantic
Korean   World

ROOFTOP PARKING

©1995-1983 Rand McNally & Co.

Fraunces Tavern Restaurant, Pearl & Broad sts., 269-0144

Gallagher's Steak House, 228 W. 52nd, 245-5336

La Cote Basque, 5 E. 55th St., 688-6525

Lutèce, 249 E. 50th St., 752-2225

Mitsukoshi, 461 Park Ave., 935-6444

The Russian Tea Room, 150 W. 57th, 265-0947

Sardi's, 234 W. 44th, 221-8440

Stage Delicatessen, 834 7th Ave., 245-7850

Toots Shore, 233 W. 33rd, 630-0333

"21" Club, 21 W. 52nd St., 582-7200

SELECTED ATTRACTIONS:

American Museum of Natural History, 79th & Central Park W., 769-5100

Empire State Building, Fifth Ave. & 34th St., 736-3100

Guggenheim Museum, Fifth Ave. at 89th St., 423-3600

Intrepid Sea-Air-Space Museum, W. 46th St. & 12th Ave., 245-0072

Metropolitan Museum of Art, Fifth Ave. & 82nd St., 535-7710

Museum of Modern Art, 53rd St., between 5th & 6th aves., 708-9480

NBC Tours, 30 Rockefeller Plaza, 664-7174

South Street Seaport, Water & Fulton sts., 732-7678

Statue of Liberty, Liberty Island, 363-3200

World Trade Center Observation Deck, 2 World Trade Center, 435-7397

INFORMATION SOURCE:

New York Convention and Visitors Bureau, Inc.
  Two Columbus Circle
  New York City, New York 10019
  (212) 397-8222

Norfolk-Virginia Beach, Virginia

Population: (*1,396,107 Norfolk-Virginia Beach-Newport News Metro Area)
Norfolk 261,229;
Virginia Beach 393,069 (1990C)

Altitude: Sea level to 12 feet

Average Temp.: Jan., 41°F.; July, 77°F.

Telephone Area Code: 804

Time: 622-9311   **Weather:** 853-3013

Time Zone: Eastern

AIRPORT TRANSPORTATION:

Ten miles to downtown Norfolk; twenty-two miles to downtown Virginia Beach.

Taxicab, limousine and bus service to Norfolk; taxicab and airport limousine service to Virginia Beach.

SELECTED HOTELS: NORFOLK

✈ Airport Hilton, 1500 N. Military Hwy., 466-8000; FAX 466-8000

Holiday Inn Executive Center, 5655
   Greenwich Rd., 499-4400;
   FAX 473-0517
Howard Johnson Downtown Norfolk, 700
   Monticello Ave., 627-5555;
FAX 533-9651
Omni International Hotel, 777 Waterside
   Dr., 622-6664; FAX 625-4930
Ramada Inn, 6360 Newtown Rd.,
   461-1081; FAX 461-1081
Ramada Norfolk Hotel, Granby &
   Freemason sts., 622-6682; FAX 623-5949
Sheraton Inn—Military Circle, 880 N.
   Military Hwy., 461-9192; FAX 461-8290
Waterside Marriott Hotel, Main & Atlantic
   sts. 627-4200; FAX 628-6466

SELECTED RESTAURANTS: NORFOLK

Signature Food: seafood
The Dumb Waiter, 117 Tazewell St.,
   623-3663

Elliot's, 1421 Colley Ave., 625-0259
Freemason Abbey, 209 W. Freemason St.,
   622-3966
La Galleria, 120 College Pl., 623-3939
Lockhart's Seafood Restaurant, 8440
   Tidewater Dr., 588-0405
Riverwalk Cafe, in the Omni International
   Hotel, 622-6664
San Antonio Sam's, 1501 Colley Ave.,
   623-0233
The Ship's Cabin Seafood Restaurant,
   4110 E. Ocean View Ave., 362-2526

SELECTED ATTRACTIONS: NORFOLK

American Rover Tall Sailing Ship Tours,
   627-SAIL
Carrie B Harbor Tours, 393-4735
Chrysler Museum of Art, Olney Rd. &
   Mowbray Arch, 622-1211
Douglas MacArthur Memorial Museum,
   City Hall Ave. & Bank St., 441-2965
Hermitage Foundation Museum, 7637 N.
   Shore Rd., 423-2052
Nauticus—The National Maritime Center,
   west end of Main St. at the Waterfront,
   623-9084 or (800) 664-1080
Norfolk Botanical Gardens, Azalea Garden
   Rd., 441-5830
Norfolk Naval Base (tour), 9079 Hampton
   Blvd., adjacent to Gate 5, 444-7955
Virginia Zoological Park, 3500 Granby St.,
   441-2706
The Waterside (festival marketplace), 333
   Waterside Dr., 627-3300

Waterside Live!, 333 Waterside Dr.,
   625-LIVE

INFORMATION SOURCES: NORFOLK

Norfolk Convention and Visitors Bureau
   236 E. Plume St.
   Norfolk, Virginia 23510
   (804) 441-5266; (800) 368-3097
Hampton Roads Chamber of Commerce
   420 Bank St.
   Norfolk, Virginia 23510
   (804) 622-2312

SELECTED HOTELS: VIRGINIA BEACH

Colonial Inn Motel, 29th & Oceanfront,
   428-5370; FAX 422-5902
Comfort Inn, 2800 Pacific Ave., 428-2203;
   FAX 422-6043
Courtyard by Marriott, 5700 Greenwich
   Rd., 490-2002; FAX 490-0169
✈ Days Inn, 5708 Northampton Blvd.,
   460-2205; FAX 363-8089
Founders Inn & Conference Center, I-64 &
   Indian River Rd., 424-5511;
   FAX 366-0613
Holiday Inn, US 13 & I-64, 464-9351;
   FAX 464-9351
Holiday Inn Executive Center, 5655
   Greenwich Rd. & Newtown Rd. Exit,
   499-4400; FAX 473-0517
Holiday Inn on the Ocean, 39th &
   Oceanfront, 428-1711; FAX 425-5742
Quality Inn Pavilion, Parks Ave. at 21st,
   422-3617; FAX 428-7434

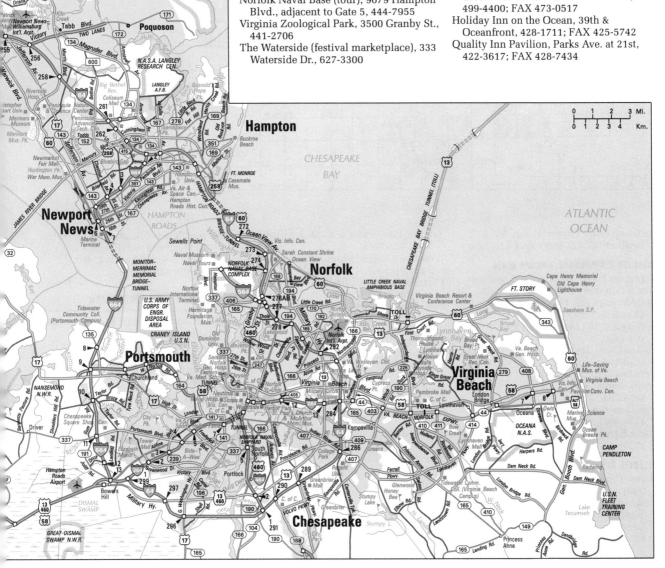

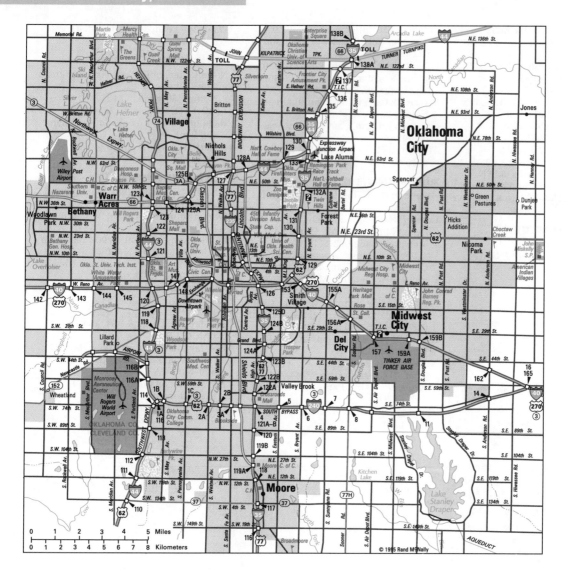

Radisson Hotel Virginia Beach, 1900
Pavilion Dr., 422-8900; FAX 425-8460

SELECTED RESTAURANTS: VIRGINIA BEACH

Signature Food: crabs; oysters; seafood

Beach Pub, 1001 Laskin Rd., 422-8817

Belle Monte Cafe, 134 Hilltop E.,
425-6290

Bennigan's, 757 Lynnhaven Pkwy.,
463-7100

Casual Clam, 3157 Virginia Beach Blvd.,
463-5106

Joe's Sea Grill, 981 Laskin Rd., 422-5637

Olive Garden, 6831 Lynnhaven Pkwy.,
486-8234

Szechuan Garden, 2720 N. Mall Dr.,
463-1680

Three Ships Inn, 3800 Shore Dr.,
460-0055

SELECTED ATTRACTIONS: VIRGINIA BEACH

Adam Thoroughgood House (17th-
century), 1636 Parish Rd., 460-0007

Association for Research & Enlightenment
(headquarters for the work of psychic
Edgar Cayce), 67th St. & Atlantic Ave.,
428-3588

Atlantic Fun Center, 25th St. & Atlantic
Ave., 422-1742

Back Bay National Wildlife Refuge, 4005
Sandpiper Rd., 721-2412

Christian Broadcasting Network, I-64 at
Exit 286B, 523-7123

First Landing Cross, Fort Story

Lynnhaven House (circa 1725), 4405
Wishart Rd., 460-1688

Ocean Breeze Fun Park, 849 General
Booth Blvd., 422-4444 or -0718

Seashore State Park, 2500 Shore Dr.,
481-4836 or -2131

Virginia Marine Science Museum, 717
General Booth Blvd., 437-4949 or
425-FISH (recording)

INFORMATION SOURCES: VIRGINIA BEACH

Virginia Beach Convention & Visitor
Development
2101 Parks Ave., Suite 500
Virginia Beach, Virginia 23451
(804) 437-4700

Virginia Beach Visitor Information Center
2100 Parks Ave.
P.O. Box 200
Virginia Beach, Virginia 23458
(804) 437-4888; (800) 446-8038

Hampton Roads Chamber of Commerce
4512 Virginia Beach Blvd.
Virginia Beach, Virginia 23462
(804) 490-1223

Oklahoma City, Oklahoma

Population: (*958,839) 444,719 (1990C)
Altitude: 1,243 feet
Average Temp.: Jan., 37°F.; July, 82°F.
Telephone Area Code: 405
Time: 681-1600   **Weather:** 478-3377
Time Zone: Central

AIRPORT TRANSPORTATION:

Ten miles to downtown Oklahoma City.
Taxicab, limousine bus service.

SELECTED HOTELS:

Century Center Hotel & Towers, One N.
Broadway Ave., 235-2780;
FAX 272-0369

Clarion Comfort Inn, 4445 N. Lincoln
Blvd., 528-2741; FAX 525-8185

Days Inn, 2616 S. I-35, 677-0521;
FAX 677-0521

Days Inn, 2801 NW. 39th St., 946-0741;
FAX 942-0181

Embassy Suites, 1815 S. Meridian,
682-6000; FAX 682-9835
Hilton Inn Northwest, 2945 NW. Expwy.,
848-4811; FAX 843-4829
Holiday Inn East, 5701 Tinker Diagonal,
Midwest City, 737-4481; FAX 732-5706
Oklahoma City Marriott, 3233 NW.
Expwy., 842-6633; FAX 842-3152
Radisson, 401 S. Meridian Ave., 947-7681;
FAX 947-4253
Waterford Hotel, 6300 Waterford Blvd.,
848-4782; FAX 843-9161

SELECTED RESTAURANTS:
Signature Food: barbeque
Applewoods, 4301 SW. 3rd, 947-8484
Eagle's Nest, 5900 Mosteller Dr., Top floor,
840-5655
Eddy's Steak House, 4227 N. Meridian
Ave., 787-2944
Harry's, 5705 Mosteller Dr., 840-9912
Harry's American Grill, 4540 NW. 23rd
St., 946-1421
Oklahoma County Line, 1226 NE. 63rd,
478-4955
Shorty Small's, Meridian & Reno,
947-0779
Sleepy Hollow, 1101 NE. 50th, 424-1614
Sunshine Express, in the Century Center
Hotel & Towers, 235-2780
Texanna Red's, 4600 W. Reno Ave.,
947-8665

SELECTED ATTRACTIONS:
Enterprise Square, 2501 E. Memorial Rd.,
425-5030
Frontier City (theme park), 11501 NE
Expressway, 478-2414
Horse Shows (year-round), at the State
Fairgrounds, 948-6700
Kirkpatrick Center Complex, 2100 NE
52nd, 427-5461
Myriad Gardens/Crystal Bridge, Reno &
Robinson, 297-3995
National Cowboy Hall of Fame & Western
Heritage Center, 1700 NE 63rd St.,
478-2250
Oklahoma City Zoo, NE 50th & Martin
Luther King Blvd., 424-3344
Remington Park (racetrack), One Remington
Pl., 424-9000

INFORMATION SOURCES:
Oklahoma City Convention & Visitors
Bureau
123 Park Ave.
Oklahoma City, Oklahoma 73102
(405) 297-8912; (800) 225-5652
Oklahoma City Chamber of Commerce
One Santa Fe Plaza
Oklahoma City, Oklahoma 73102
(405) 278-8900

Omaha, Nebraska

Population: (*618,262) 335,795 (1990C)
Altitude: 1,040 feet
Average Temp.: Jan., 23°F.; July, 77°F.
Telephone Area Code: 402
Time: 342-8463   **Weather:** 571-8111
Time Zone: Central

AIRPORT TRANSPORTATION:
Three miles to downtown Omaha.
Taxicab, bus and hotel limousine service.

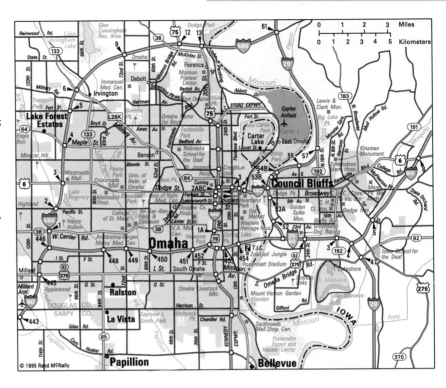
© 1995 Rand MNally

SELECTED HOTELS:
✈ Airport Inn, at Eppley Airfield,
348-0222; FAX (none)
Best Western Central, 3650 S. 72nd.,
397-3700; FAX 397-8632
Best Western Omaha Inn, 4706 S. 108th
St., 339-7400; FAX 339-5155
Best Western Regency West, 107th and
Pacific sts., 397-8000; FAX 397-8000
Embassy Suites, 7220 Cedar St., 397-5141;
FAX 397-3266
Holiday Inn Central, 3321 S. 72nd St.,
393-3950; FAX 393-8718
New Tower Inn, 7764 Dodge St.,
393-5500; FAX 393-5500, ext. 2240
Omaha Marriott, 10220 Regency Circle,
399-9000; FAX 399-0223
Ramada—Central, 7007 Grover St.,
397-7030; FAX 397-8449
✈ Ramada Inn Airport, Abbott Dr. &
Locust St., 342-5100; FAX 342-5100,
ext. 150
Red Lion Inn Omaha, 1616 Dodge St.,
346-7600; FAX 346-5722
Sheraton Inn Southwest, 4888 S. 118th
St., 895-1000; FAX 896-9247

SELECTED RESTAURANTS:
Signature Food: steak
Anthony's, 7220 F St., 331-7575
Cascio, 1620 S. 10th St., 345-8313
Chardonnay, in the Omaha Marriott,
399-9000
French Cafe, 1017 Howard St., 341-3547
Gallagher's, 10730 Pacific, 393-1421
Gorats Steak House, 4917 Center St.,
551-3733
Johnny's Cafe, 4702 S. 27th St., 731-4774
Maxine's, in the Red Lion Inn Omaha
Hotel, 346-7600
Neon Goose, 1012 S. 10th St., 341-2063
Ross' Steak House, 909 S. 72nd St.,
393-2030

V. Mertz, 1022 Howard, 345-8980

SELECTED ATTRACTIONS:
AKsarben Thoroughbred Racetrack, 63rd
& Shirley, 444-4000
Belle at Bellevue Riverboat Cruises,
Haworth Park, Bellevue, 292-BOAT
Boys Town, W. Dodge Rd. between 132nd
& 144th, 498-1140
Fun Plex, 70th & Q sts., 331-8436
Gene Leahy Mall, 14th & Farnam,
444-5900
Heartland of America Park & Fountain,
8th & Douglas, 444-PARK
Henry Doorly Zoo/Lied Jungle, 3701 S.
10th, 733-8400 or -8401
Joslyn Art Museum, 24th & Dodge,
342-3300
Old Market, 10th to 13th, Harney to
Jackson
Strategic Air Command (SAC) Museum,
2510 SAC Pl., Bellevue, 292-2001

INFORMATION SOURCES:
Greater Omaha Convention & Visitors
Bureau
AKsarben Field
6300 Shirley St.
Omaha, Nebraska 68106
(402) 444-4660; (800) 332-1819
Greater Omaha Chamber of Commerce
1301 Harney St.
Omaha, Nebraska 68102
(402) 346-5000

Orlando, Florida

Population: (*1,072,748) 164,693 (1990C)
Altitude: 106 feet
Average Temp.: Jan., 62°F.; July, 82°F.
Telephone Area Code: 407
Time: 976-1611   **Weather:** 851-7510
Time Zone: Eastern

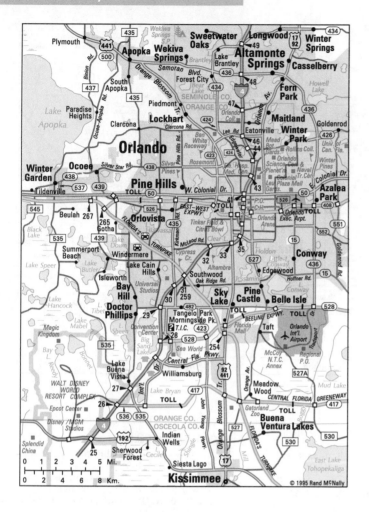

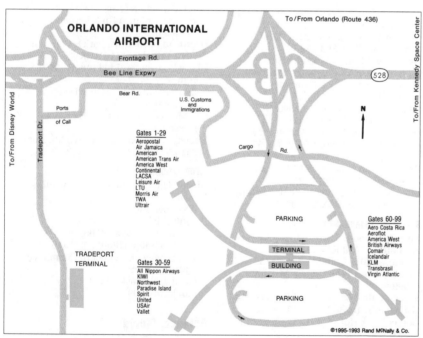

AIRPORT TRANSPORTATION:
Fifteen miles to downtown Orlando.
Taxicab, limousine and bus service.

SELECTED HOTELS:
Delta Orlando Resort, 5715 Major Blvd.,
351-3340; FAX 351-5117
Gold Key Inn, 7100 S. Orange Blossom
Trail, 855-0050; FAX 855-6230
Holiday Inn at the Orlando Arena, 304 W.
Colonial Dr. at jct. FL 50 & I-4,
843-8700; FAX 841-4978
Holiday Inn—Central Park, 7900 S.
Orange Blossom Trail, 859-7900;
FAX 859-7442
Holiday Inn—Centroplex, 929 W. Colonial
Dr., 843-1360; FAX 839-3333
Marriott's Orlando World Center, 8701
World Center Dr., 239-4200;
FAX 238-8777
Orlando Marriott, 8001 International Dr.,
351-2420; FAX 345-5611
Park Inn International—Orlando North,
736 Lee Rd., 647-1112; FAX 740-8964
Peabody Orlando, 9801 International Dr.,
352-4000; FAX 351-0073
Ramada Resort, 7400 International Dr.,
351-4600; FAX 363-0517
Ramada Orlando Central, 3200 W.
Colonial Dr., 295-5270; FAX 291-2092
Sheraton World Resort, 10100 International
Dr., 352-1100; FAX 352-3679
The Stouffer Orlando Resort, 6677 Sea
Harbor Dr., 351-5555; FAX 351-9991

SELECTED RESTAURANTS:
Signature Food: barbeque; alligator
Cafe on the Park, in the Harley Hotel of
Orlando, 151 E. Washington, 841-3220
Charlie's Lobster House, 8445 International
Dr., 352-6929
Christini's, 7600 Dr. Phillips Blvd.,
345-8770
Church Street Station, 129 W. Church St.,
422-2434
4th Fighter Group, 494 Rickenbacker Dr.,
898-4251
Maison et Jardin, 430 S. Wymore Rd.,
862-4410
Ming Court, 9188 International Dr.,
351-9988
Piccadilly, in the Gold Key Inn Motel,
855-0050

SELECTED ATTRACTIONS:
Church Street Station (dining, shopping,
entertainment), 129 W. Church St.,
422-2434
Cypress Gardens, off US 27 at FL 540W,
near Winter Haven, (813) 324-2111
Florida Citrus Tower, Clermont,
(904) 394-8585
Gatorland Zoo, 14501 S. Orange Blossom
Trail, 855-5496
Ripley's Believe It or Not, 8201
International Dr., 363-4418
Sea World of Florida, 7007 Sea World Dr.,
351-3600
Spaceport USA, Kennedy Space Center,
452-2121
Universal Studios Florida, off I-4 Exit
30B, 363-8000
Walt Disney World, Lake Buena Vista,
824-4321

INFORMATION SOURCES:

Orlando/Orange County Convention and
Visitors Bureau
  7208 Sand Lake Rd., Suite 300
  Orlando, Florida 32819
  (407) 363-5800
Greater Orlando Chamber of Commerce
  75 E. Ivanhoe Blvd.
  Orlando, Florida 32802
  (407) 425-1234

Ottawa, Ontario, Canada

Population: (*819,263) 300,763 (1986C)
Altitude: 374 ft.
Average Temp.: Jan., 12°F.; July, 70°F.
Telephone Area Code: 613
Time: 745-1576   **Weather:** 998-3439
Time Zone: Eastern

AIRPORT TRANSPORTATION:

Fifteen miles to downtown Ottawa.
Taxicab, limousine, and hotel shuttle bus
  service.

SELECTED HOTELS:

Best Western Macies, 1274 Carling,
  728-1951; FAX 728-1955
Chateau Laurier, 1 Rideau St., 241-1414;
  FAX 241-2958
Chimo Hotel, 1199 Joseph Cyr, 744-1060;
  FAX 744-7845
The Citadel Inn, 101 Lyon St., 237-3600;
  FAX 237-2351
Delta Ottawa, 361 Queen St., 238-6000;
  FAX 238-2291
Hilton Ottawa, 150 Albert St., 238-1500;
  FAX 235-2723
Holiday Inn Market Square, 350
  Dalhousie, 241-1000; FAX 241-4804
Howard Johnson, 140 Slater, 238-2888;
  FAX 235-8421
Parkway Motor Hotel, 475 Rideau,
  789-3781; FAX 789-0207
Radisson Ottawa Centre, 100 Kent,
  238-1122; FAX 783-4229
WelcomINNS, 1220 Michael St.,
  748-7800; FAX 748-0499
The Westin Hotel, 11 Colonel By Dr.,
  560-7000; FAX 234-5396

SELECTED RESTAURANTS:

Signature Food: beavan tarts
The Courtyard Restaurant, 21 George St.,
  241-1516
Friday's Roast Beef House, 150 Elgin St.,
  237-5353
Hy's, 170 Queen St., Metropolitan Life
  Building, 234-4545
Leone's Ristorante, 412 Preston St.,
  230-6021
Marble Works, 14 Waller St., 241-6764
Nick & Jerry's Simply Seafood, 253 Slater,
  232-4895
The Siam Bistro, 1268 Wellington St.,
  728-3111
Wilfrid's, in the Chateau Laurier Hotel,
  241-1414

SELECTED ATTRACTIONS:

Bytown Museum-Ottawa Locks, off Elgin
  St. beside the National Arts Centre,
  234-4570
Canadian Museum of Civilization, 100
  Laurier St., Hull, (819) 776-7002

Canadian Museum of Contemporary
  Photography, 1 Rideau Canal, 991-4896
Canadian Museum of Nature, Metcalfe St.
  at McLeod, 996-3102
Canadian War Museum, 330 Sussex Dr.,
  992-2774
Currency Museum, 245 Sparks St.,
  782-8914
Laurier House, 335 Laurier Ave. E.,
  992-8142
National Aviation Museum, 1 Aviation
  Pkwy., 993-2010
National Gallery of Canada, 380 Sussex
  Dr., 990-1985
National Museum of Science & Technology,
  1867 St. Laurent Blvd., 991-3044
Rideau Hall (Governor General's
  residence), 1 Sussex Dr., 998-7113
Royal Canadian Mint, 320 Sussex Dr.,
  993-3500

INFORMATION SOURCES:

Ottawa Tourism and Convention Authority
  111 Lisgar St., 2nd Floor
  Ottawa, Ontario K2P 2L7
  (613) 237-5150
Canadian Consulate General
  1251 Avenue of the Americas
  New York, New York 10020-1175
  (212) 596-1600

Philadelphia, Pennsylvania

Population: (*4,856,881) 1,585,577
  (1990C)
Altitude: 45 feet
Average Temp.: Jan., 35°F.; July, 78°F.
Telephone Area Code: 215
Time: 846-1212   **Weather:** 936-1212
Time Zone: Eastern

AIRPORT TRANSPORTATION:

Eight miles to downtown Philadelphia.
Taxicab, limousine bus, and rail line
  service.

SELECTED HOTELS:

Adam's Mark, City Ave. & Monument Rd.,
  581-5000; FAX 581-5089
The Barclay Hotel, 237 S. 18th St.,
  545-0300; FAX 545-2896
The Four Seasons, 18th St. & Benjamin
  Franklin Pkwy., 963-1500;
  FAX 963-9506
Hotel Atop the Bellevue, Broad & Walnut
  sts., 893-1776; FAX 732-8518
The Latham Hotel, 17th & Walnut sts.,
  563-7474; FAX 568-0110
Marriott Hotel, 12th & Filbert sts. (due to
  open in Jan., 1995)
Penn Tower Hotel, 34th & Civic Center
  Blvd., 387-8333; FAX 386-8306
✈ Philadelphia Airport Marriott Hotel,
  4509 Island Ave., 365-4150;
  FAX 365-3875
Ramada Inn, 2400 Old Lincoln Hwy.,
  Trevose, 638-8300; FAX 638-4377
Sheraton Society Hill, 1 Dock St.,
  238-6000; FAX 922-2709
The Warwick Hotel, 17th at Locust St.,
  735-6000; FAX 790-7766
Wyndham Franklin Plaza Hotel, 17th &
  Race, 448-2000; FAX 448-2864

SELECTED RESTAURANTS:

Signature Food: soft pretzels; cheesesteaks

Le Bec Fin, 1523 Walnut St., 567-1000
Deux Cheminees, 1221 Locust St.,
  790-0200
Di Lullo Centro, 1407 Locust, 546-2000
La Famiglia, 8 S. Front St., 922-2803
The Garden, 1617 Spruce St., 546-4455
Harry's Bar and Grill, 22 S. 18th St.,
  561-5757
The Monte Carlo Living Room, 2nd &
  South sts., 925-2220
Old Original Bookbinders, 125 Walnut St.,
  925-7027
The Palm Restaurant of Philadelphia, 200
  S. Broad St. at Walnut, 546-7256
Zanzibar Blue, 305 S. 11th St., 829-0300

SELECTED ATTRACTIONS:

Academy of Natural Sciences, 19th &
  Benjamin Franklin Pkwy., 299-1000
Afro-American Historical & Cultural
  Museum, NW corner of 7th & Arch sts.,
  574-0380
Betsy Ross House, 239 Arch St., 627-5343
Franklin Institute-Futures Center &
  Omniverse Theater, 20th & Benjamin
  Franklin Pkwy., 448-1208
Independence National Historical Park,
  3rd & Chestnut (Vistors Center),
  597-8974
The New Jersey State Aquarium at
  Camden, Riverside Dr., Camden, New
  Jersey, (609) 365-3300
Philadelphia Museum of Art, 26th &
  Benjamin Franklin Pkwy., 763-8100
Philadelphia Zoo, 34th & Girard Ave.,
  243-1100
Please Touch Museum for Children, 210
  N. 21st St., 963-0666
U.S. Mint, 5th & Arch sts., 597-7353

INFORMATION SOURCES:

Philadelphia Convention & Visitors
Bureau
  1515 Market St., Suite 2020
  Philadelphia, Pennsylvania 19102
  (215) 636-3300; (800) 225-5745
Philadelphia Visitors Center
  16th St. & JFK Blvd.
  Philadelphia, Pennsylvania 19102
  (215) 636-1666; (800) 537-7676
Greater Philadelphia Chamber of
Commerce

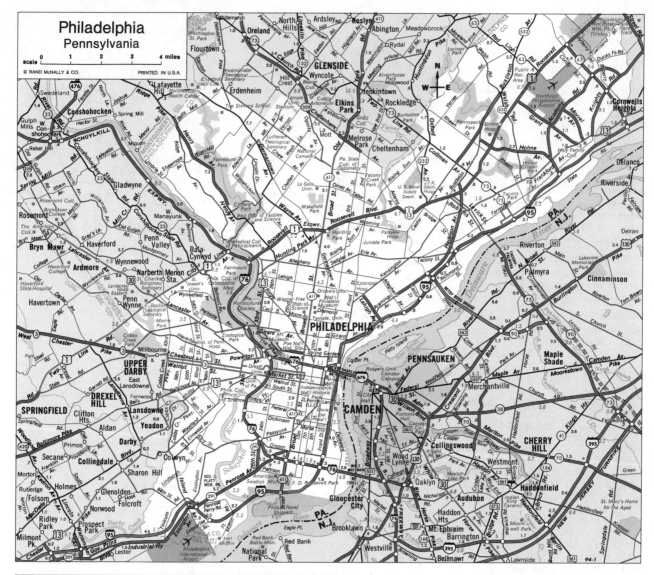

Philadelphia
Pennsylvania

scale 0   1   2   3   4 miles

© Rand McNally & Co.   PRINTED IN U.S.A.

PHILADELPHIA INTERNATIONAL AIRPORT

To/From Philadelphia

To/From Philadelphia

To/From Wilmington

PARKING

Recirculation Rd.

Access Rd.

Parking

SHORT TERM PARKING E

SHORT TERM PARKING A

PARKING GARAGE A-B

PARKING GARAGE C

PARKING GARAGE D

Bag Claim Rd.

Departures Rd.

Tunnel to Parking Garage D

DOMESTIC TERMINAL

Tunnel to Parking Garage C

Terminal A
Air Jamaica
Air Mobility
Command
  (Int'l Arrivals)
American
American Eagle
British Airways
Charter Airlines
Midway Airlines
Midwest Express
Swissair
USAir
  (Int'l Arrivals)

Terminal B
USAir
USAir Express

Terminal C
USAir
  (Int'l Departures)

Terminal D
Air Mobility
  Command
America West
Continental
Continental
  Express
United
United Express

Terminal E
Delta
Delta Connection
Northwest
Northwest
  Airlink
Spirit
TWA
TW Express
Valujet

©1995-1988 Rand McNally & Co.

1234 Market St., 18th Floor
Philadelphia, Pennsylvania 19107
(215) 545-1234

Phoenix, Arizona

Population: (*2,122,101) 983,403 (1990C)
Altitude: 1,090 feet
Average Temp.: Jan., 51°F.; July, 85.9°F.
Telephone Area Code: 602
Time and Weather: 976-7600
Time Zone: Mountain Standard all year

AIRPORT TRANSPORTATION:

Four miles to downtown Phoenix.
Taxicab, bus, and limousine bus service.

SELECTED HOTELS:

Arizona Biltmore, 24th St. & Missouri
  Ave., 955-6600; FAX 381-7600
Embassy Suites Hotel, 2333 E. Thomas
  Rd., 957-1910; FAX 955-2861
Holiday Inn Crowne Plaza, Central &
  Adams sts., 257-1525; FAX 495-1209
Hyatt Regency Phoenix, 122 N. 2nd St.,
  252-1234; FAX 254-9472

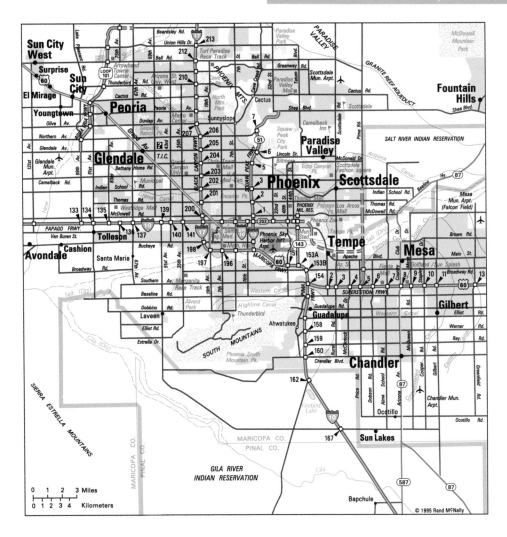

0 1 2 3 Miles
0 1 2 3 4 Kilometers

© 1995 Rand McNally & Co.

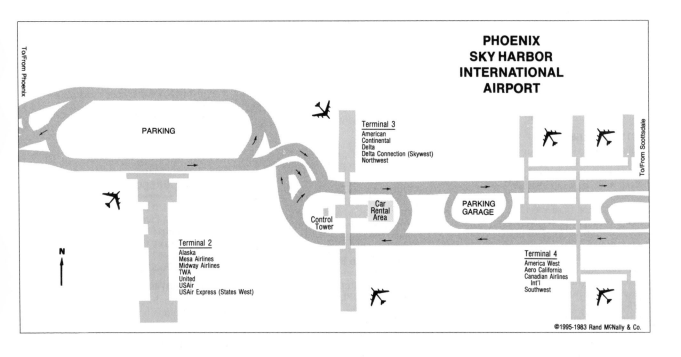

PHOENIX SKY HARBOR INTERNATIONAL AIRPORT

To/From Phoenix

To/From Scottsdale

PARKING

N

Terminal 3
American
Continental
Delta
Delta Connection (Skywest)
Northwest

Car Rental Area

PARKING GARAGE

Control Tower

Terminal 2
Alaska
Mesa Airlines
Midway Airlines
TWA
United
USAir
USAir Express (States West)

Terminal 4
America West
Aero California
Canadian Airlines Int'l
Southwest

©1995-1983 Rand McNally & Co.

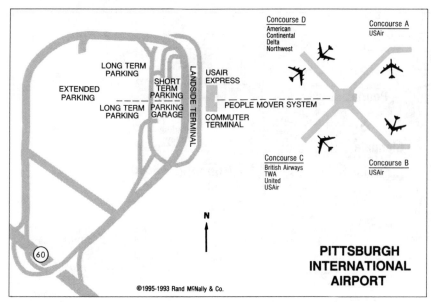

LONG TERM PARKING
EXTENDED PARKING
SHORT TERM PARKING
LONG TERM PARKING
PARKING GARAGE
LANDSIDE TERMINAL
USAIR EXPRESS
PEOPLE MOVER SYSTEM
COMMUTER TERMINAL
Concourse D
American
Continental
Delta
Northwest
Concourse A
USAir
Concourse C
British Airways
TWA
United
USAir
Concourse B
USAir
N
60
©1995-1993 Rand M°Nally & Co.

PITTSBURGH INTERNATIONAL AIRPORT

Orange Tree Golf & Conference Resort, 10601 N. 56th St., Scottsdale, 948-6100; FAX 483-6074

The Phoenix Hilton, 2435 S. 47th St., 894-1600; FAX 921-7844

The Pointe Hilton Resort, 7677 N. 16th St., 997-2626; FAX 997-2391

Quality Hotel Central, 3600 N. 2nd Ave., 248-0222; FAX 265-6331

Ritz-Carlton Phoenix, 24th St. & Camelback Rd., 468-0700; FAX 468-9883

SELECTED RESTAURANTS:

Signature Food: Mexican

Compass, in the Hyatt Regency Phoenix, 252-1234

Don & Charlie's, 7501 E. Camelback, 990-0900

Etienne's Different Pointe of View, 11111 N. 7th St., 863-0912

The Hungry Hunter, 3102 E. Camelback Rd., 957-7180

Orangerie, in the Arizona Biltmore Hotel, 954-2507

Ruth's Chris Steak House, 7701 N. Scottsdale Rd., Scottsdale, 991-5988

Trumps, in the Hotel Westcourt, 10220 N. Metro Pkwy. E., 997-5900

SELECTED ATTRACTIONS:

Arabian Productions & Tours, 18001 N. Tatum Blvd., 867-8275

Champlin Fighter Museum, inside Falcon Field Airfield, 4636 Fighter Aces Dr., Mesa, 830-4540

Desert Botanical Gardens, 1201 N. Galvin Pkwy., 941-1225

Dolly's Steamboat (Canyon Lake tours), 827-9144

Gila River Arts & Crafts Center, off I-10 Exit 175, 963-3981

Hall of Flame Museum, between Center Pkwy. & Van Buren St. on Project Dr., 275-3473

The Heard Museum, 22 E. Monte Vista, ___ ___40

___rande Museum, 4619 E.
___ngton, 495-0901

Rawhide (recreated 1880s Western town), Scottsdale Rd. 4 mi. north of Bell Rd., Scottsdale, 563-1880

Taliesin West, Scottsdale, 860-8810

INFORMATION SOURCE:

Phoenix & Valley of the Sun Convention & Visitors Bureau
One Arizona Center
400 E. Van Buren St., Suite 600
Phoenix, Arizona 85004-2290
(602) 254-6500
Visitor Info. Hotline (602) 252-5588;
Meeting Planners Line
(800) 535-8898

Pittsburgh, Pennsylvania

Population: (*2,056,705) 369,879 (1990C)
Altitude: 760 feet
Average Temp.: Jan., 33°F.; July, 75°F.
Telephone Area Code: 412
Time: 391-9500   **Weather:** 936-1212
Time Zone: Eastern

AIRPORT TRANSPORTATION:

Seventeen miles to downtown Pittsburgh. Taxicab and limousine bus service.

SELECTED HOTELS:

Holiday Inn—Pittsburgh, 1406 Beers School Rd., Coraopolis, 262-3600; FAX 262-6221

Hyatt Pittsburgh at Chatham Center, 112 Washington Pl., 471-1234; FAX 355-0315

Marriott—Greentree, 101 Marriott Dr., 922-8400; FAX 922-8981

Pittsburgh Best Western, Parkway West at the Montour Run Exit, 262-3800; FAX 695-1068

The Pittsburgh Hilton & Towers, 600 Commonwealth Pl., 391-4600; FAX 594-5161

Royce Hotel, 1160 Thorn Run Rd. Extension, Coraopolis, 262-2400; FAX 264-9373

Sheraton Hotel at Station Square, Carson & Smithfield sts., 261-2000; FAX 261-2932

Vista International Hotel, 1000 Penn Ave., 281-3700; FAX 227-4500

Westin William Penn, 530 William Penn Pl., 281-7100; FAX 553-5252

SELECTED RESTAURANTS:

Signature Food: chipped ham; pierogi; kielbasa; klondike bar

Christopher's, 1411 Grandview Ave., Mt. Washington, 381-4500

Colony, Greentree & Cochran rds., 561-2060

Common Plea, 310 Ross St., 281-5140

D'Imperio's, 3412 Wm. Penn Hwy., 823-4800

Grand Concourse, 1 Station Sq., 261-1717

Le Mont Restaurant, 1114 Grandview Ave., 431-3100

Tambellini, S. of town on PA 51, 481-1118

The Terrace Room, in the Westin William Penn, 281-7100

Top of the Triangle, 600 Grant St., 62nd Floor, 471-4100

SELECTED ATTRACTIONS:

Andy Warhol Museum, 117 Sandusky St., 237-8300

Benedum Center for the Performing Arts, 7th St. between Penn & Liberty aves., 456-2600

Carnegie Museum of Art & Natural History, 4400 Forbes Ave., 622-3172

Carnegie Science Center, 1 Allegheny Ave., 237-3400

Frank Lloyd Wright's Fallingwater, between Mill Run & Ohiopyle on PA 381, 329-8501

Heinz Hall (Pittsburgh Symphony), 600 Penn Ave., 392-4800

Monongahela Incline (cable car), 205 W. Carson St., The South Side, 442-2000

Phipps Conservatory, Schenley Park, Oakland, 622-6914

Pittsburgh Zoo, in Highland Park, 665-3639

Station Square, W. Carson St. & Smithfield St. bridge, The South Side, 261-9911

INFORMATION SOURCES:

Greater Pittsburgh Convention & Visitors Bureau
4 Gateway Center
Pittsburgh, Pennsylvania 15222
(412) 281-7711; (800) 366-0093

Greater Pittsburgh Chamber of Commerce
3 Gateway Center
Pittsburgh, Pennsylvania 15222
(412) 392-4500

Portland, Oregon

Population: (*1,239,842) 437,319 (1990C)
Altitude: Sea level to 1,073 feet
Average Temp.: Jan., 40°F.; July, 69°F.
Telephone Area Code: 503
Time: 976-8463   **Weather:** 281-1911
Time Zone: Pacific

AIRPORT TRANSPORTATION:

Nine miles to downtown Portland. Taxicab, limousine bus service and public mass transit system.

SELECTED HOTELS:

The Benson, 309 SW. Broadway, 228-2000; FAX 226-4603

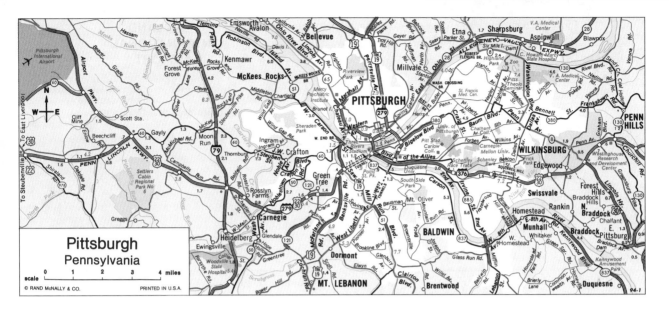

Pittsburgh
Pennsylvania

scale 0 1 2 3 4 miles
© RAND McNALLY & CO.   PRINTED IN U.S.A.

✈ Flamingo Motor Inn, 9727 NE. Sandy
  Blvd., 255-1400; FAX 256-3842
The Heathman Hotel, 1009 SW. Broadway,
  241-4100; FAX 790-7110
✈ Holiday Inn—Portland Airport, 8439
  NE. Columbia Blvd., 256-5000;
  FAX 257-4742
Portland Hilton, 921 SW. 6th Ave.,
  226-1611; FAX 220-2565
The Portland Inn, 1414 SW. 6th Ave.,
  221-1611; FAX 226-0447
Portland Marriott, 1401 SW. Front Ave.,
  226-7600; FAX 221-1789
✈ Ramada Inn, 6221 NE. 82nd Ave.,
  255-6511; FAX 255-8417
Red Lion Coliseum, 1225 N. Thunderbird
  Way, 235-8311; FAX 232-2670
Red Lion Columbia River, 1401 N. Hayden
  Island Dr., 283-2111; FAX 283-4718
Red Lion Inn Downtown, 310 SW. Lincoln
  Ave., 221-0450; FAX 226-6260
Red Lion Motor Inn Jantzen Beach, 909 N.
  Hayden Island Dr., 283-4466;
  FAX 283-4743
✈ Sheraton Inn—Portland Airport, 8235
  NE. Airport Way, 281-2500;
  FAX 249-7602

SELECTED RESTAURANTS:

Signature Food: salmon
Alexander's, atop Portland Hilton,
  226-1611
Brickstone's Restaurant, in the Red Lion
  Columbia River, 283-2111
Couch Street Fish House, 105 NW. 3rd St.,
  223-6173
Huber's, 411 SW. 3rd., 228-5686
Jake's Famous Crawfish Restaurant, 401
  SW. 12th Ave., 226-1419
The London Grill, in The Benson Hotel,
  228-2000
Ocean Palace, 336 NW. Davis, 223-7475
Polo Restaurant, 718 NE. 12th Ave.,
  232-1801
Ringside, 2165 W. Burnside St., 223-1513
River Queen, 1300 NW. Front Ave.,
  228-8633

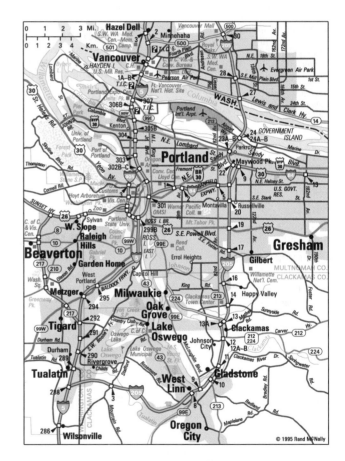

SELECTED ATTRACTIONS:

International Rose Test Gardens, in
  Washington Park, 823-2223
Japanese Gardens, in Washington Park,
  223-4070
Oregon Historical Society's Library &
  Museum, 1200 SW. Park Ave., 222-1741
Oregon Museum of Science & Industry,
  1945 SE. Water Ave., 797-4000
Pittock Mansion, 3229 NW. Pittock Dr.,
  823-3624

Portland Art Museum, 1219 SW. Park
  Ave., 226-2811
Sanctuary of Our Sorrowful Mother (the
  "Grotto"), NE. 85th & Sandy Blvd.,
  254-7371
Washington Park Zoo, 4001 SW. Canyon
  Ct., 226-1561
World Forestry Center, 4033 SW. Canyon
  Rd. in Washington Park, 228-1367

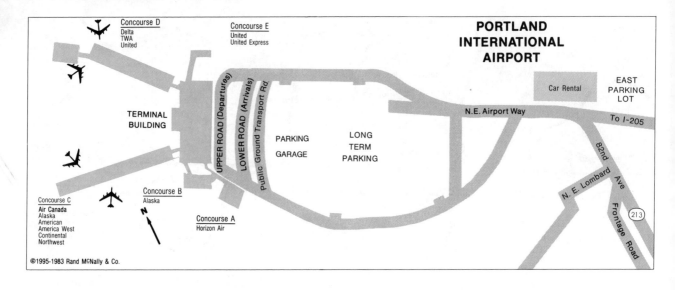

PORTLAND INTERNATIONAL AIRPORT

Concourse D — Delta, TWA, United
Concourse E — United, United Express
Concourse C — Air Canada, Alaska, American, America West, Continental, Northwest
Concourse B — Alaska
Concourse A — Horizon Air
TERMINAL BUILDING
UPPER ROAD (Departures)
LOWER ROAD (Arrivals)
Public Ground Transport Rd.
PARKING GARAGE
LONG TERM PARKING
Car Rental
EAST PARKING LOT
N.E. Airport Way
To I-205
82nd Ave.
N. E. Lombard
Frontage Road
213
©1995-1983 Rand McNally & Co.

INFORMATION SOURCES:
Portland/Oregon Visitors Association Marketing, Tourism & Conventions
26 SW. Salmon St.
Portland, Oregon 97204
(503) 275-9750; Visitor Infomation
(503) 222-2223
Portland Chamber of Commerce
221 NW. 2nd Ave.
Portland, Oregon 97209
(503) 228-9411

Providence, Rhode Island

Population: (*654,854) 160,728 (1990C)
Altitude: 24 feet
Average Temp.: Jan., 29°F.; July, 72°F.
Telephone Area Code: 401
Time and Weather: 738-1211
Time Zone: Eastern

AIRPORT TRANSPORTATION:
Nine miles to downtown Providence.
Taxicab and airport limousine service.

SELECTED HOTELS:
Days Hotel on the Harbor, 220 India Pt., 272-5577; FAX 272-5577, ext. 199
Holiday Inn Downtown, 21 Atwells Ave., 831-3900; FAX 751-0007
Inn at the Crossing—Holiday Inn, 801 Greenwich Ave., Warwick, 732-6000; FAX 732-4839
Omni Biltmore, Kennedy Plaza, 11 Dorrance St., 421-0700; FAX 421-0210
Providence Marriott, Charles & Orms sts., 272-2400; FAX 273-2686
✈ Sheraton Tara, 1850 Post Rd., Warwick, 738-4000; FAX 738-8206
Westin Hotel, 1 W. Exchange St., 598-8000 (due to open late 1994)

SELECTED RESTAURANTS:
Signature Food: coffee milk; frozen lemonade; cannoli; pasta
Camille's, 71 Bradford St., 751-4812
Ca___'o, Dyer and Pine sts., 421-1320
_ Mill Tavern, 390 Fall River Ave., _k, MA, (508) 336-8460
_s, 345 S. Water, 421-4646

Stacey's, in the Providence Marriott, 272-2400

SELECTED ATTRACTIONS:
The Arcade, 65 Weybosset St., 272-2340
Benefit Street ("Mile of History"), 831-7440
First Baptist Meeting House, 75 N. Main St., 454-3418
Governor Stephen Hopkins House, Benefit & Hopkins sts., 421-0694
John Brown House, 52 Power St., 331-8575
Lippitt House, 199 Hope St., 453-0688
Museum of Art, Rhode Island School of Design, 224 Benefit St., 454-6507
Roger Williams National Memorial, 282 N. Main St., 521-7266
State House (tours), 82 Smith St., 277-2357
WaterPlace (riverwalk area), Franacis St. & Memorial Blvd.

INFORMATION SOURCES:
Greater Providence Convention & Visitors Bureau
30 Exchange Terrace
Providence, Rhode Island 02903
(401) 274-1636; (800) 233-1636 (out of state)
Greater Providence Chamber of Commerce
30 Exchange Terrace
Providence, Rhode Island 02903
(401) 521-5000

Raleigh, North Carolina

Population: (*735,480) 207,951 (1990C)
Altitude: 363 feet
Average Temp.: Jan., 42°F.; July, 78°F.
Telephone Area Code: 919
Time: 976-2511 **Weather:** 860-1234
Time Zone: Eastern

AIRPORT TRANSPORTATION:
Fifteen miles to downtown Raleigh.
Taxicab and airport limousine service.

SELECTED HOTELS:
Brownstone Hotel, 1707 Hillsborough St., 828-0811; FAX 834-0904
Days Inn—North, 2805 Highwood Blvd., 872-3500; FAX 878-7817

Holiday Inn—North, 2815 Capital Blvd., 872-7666; FAX 872-3915
✈ Holiday Inn—Raleigh-Durham Airport, I-40 Exit 282 at Page Rd., Research Triangle Park, 941-6000; FAX 941-6030
Holiday Inn—Raleigh State Capital, 320 Hillsborough St., 832-0501; FAX 833-1631
North Raleigh Hilton Convention Center and Towers, 3415 Wake Forest Rd., 872-2323; FAX 876-0890
Plantation Inn, 6401 Capital Blvd., 876-1411; FAX 790-7093
Radisson Plaza Hotel Raleigh, 420 Fayetteville St. Mall, 834-9900; FAX 833-1217
Raleigh Marriott Hotel, 4500 Marriott Dr., 781-7000; FAX 781-3059
Ramada Inn—Apex, on US 1 at jct. NC 55, 362-8621; FAX 362-9383
Sheraton—Crabtree Inn, 4501 Creedmoor Rd., 787-7111; FAX 783-0024
✈ Sheraton Imperial Hotel & Convention Center, I-40 Exit 282 at Page Rd., Research Triangle Park, 941-5050; FAX 941-9109
Velvet Cloak Inn, 1505 Hillsborough St., 828-0333; FAX 828-2656

SELECTED RESTAURANTS:
Signature Food: barbeque
Allies, in the Raleigh Marriott Hotel, 781-7000
Angus Barn Ltd., 9401 Glenwood Ave. (US Hwy. 70), 787-3505
Big Ed's City Market Restaurant, 220 Wolfe St., 836-9909
42nd St. Oyster Bar, 508 W. Jones St., 831-2811
Fox & Hound, 107 Edinburgh S., MacGregor Village Shopping Center, Cary, 380-0080
Greenshields Brewery & Pub, 214 E. Martin St., 829-0214
Jacqueline's, in the Plantation Inn, 876-1411
Top of the Tower, in the Holiday Inn—Raleigh State Capital, 832-0501

SELECTED ATTRACTIONS:
Artspace, 201 E. Davie St., 821-2787

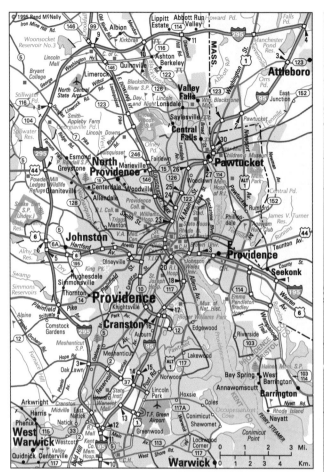

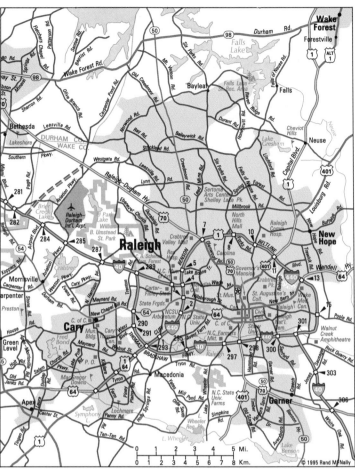

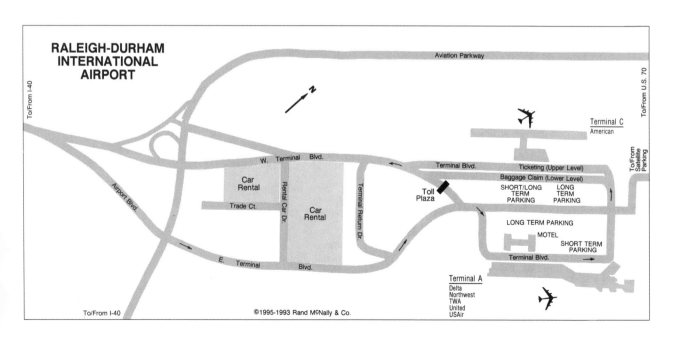

RALEIGH-DURHAM
INTERNATIONAL
AIRPORT

Aviation Parkway

To/From I-40

To/From U.S. 70

Terminal C
American

W. Terminal Blvd.

Terminal Blvd.

Ticketing (Upper Level)

Baggage Claim (Lower Level)

Car
Rental

Rental Car Dr.

Car
Rental

Terminal Return Dr.

Toll
Plaza

SHORT/LONG
TERM
PARKING

LONG
TERM
PARKING

Trade Ct.

To/From Satellite Parking

LONG TERM PARKING

MOTEL

SHORT TERM
PARKING

E. Terminal Blvd.

Terminal Blvd.

Terminal A
Delta
Northwest
TWA
United
USAir

To/From I-40

©1995-1993 Rand McNally & Co.

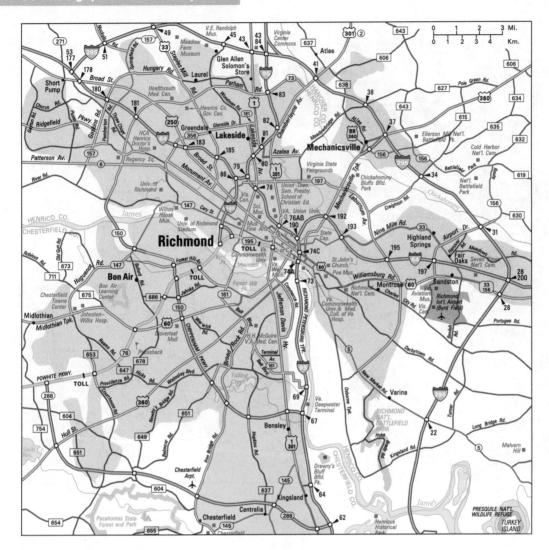

City Cemetery (1798), S. East & Hargett sts.

Executive Mansion, 200 N. Blount St., (Visitor Center, for tours) 733-3456

Hardee's Walnut Creek Amphitheatre, 3801 Rock Quarry Rd., 831-6400

Mordecai Historic Park, 1 Mimosa St., 834-4844

North Carolina Museum of Art, 2110 Blue Ridge Blvd., 833-1935

North Carolina Museum of History, 1 E. Edenton St., 715-0200

Oakwood Cemetery (1866), Oakwood Ave. at Watauga St.

State Capitol, bounded by Wilmington, Edenton, Salisbury, and Morgan sts., 733-4994

Wakefield (oldest dwelling in Raleigh), Hargett & St. Mary's sts., 733-3456

INFORMATION SOURCES:

Greater Raleigh Convention & Visitors Bureau
225 Hillsborough St., Suite 400
P.O. Box 1879
…igh, North Carolina 27602
…834-5900
…849-8499

Greater Raleigh Chamber of Commerce
800 S. Salisbury St.
P.O. Box 2978
Raleigh, North Carolina 27602
(919) 664-7000

Richmond, Virginia

Population: (*865,640) 203,056 (1990C)
Altitude: 150 feet
Average Temp.: Jan., 38°F.; July, 78°F.
Telephone Area Code: 804
Time: 844-3711   **Weather:** 222-7411
Time Zone: Eastern

AIRPORT TRANSPORTATION:

Ten miles to downtown Richmond.
Taxicab and limousine service.

SELECTED HOTELS:

Commonwealth Park Suites Hotel, 9th & Bank sts., 343-7300; FAX 343-1025

Courtyard by Marriott, 6400 W. Broad St., 282-1881; FAX 288-2934

Embassy Suites Hotel, 2925 Emerywood Pkwy., 672-8585; FAX 672-3749

Holiday Inn Historic District, 301 W. Franklin St., 644-9871; FAX 344-4380

Hyatt Richmond, 6624 W. Broad St., 285-1234; FAX 288-3961

Jefferson Hotel, Franklin & Adams sts., 788-8000; FAX 225-0334

Omni Richmond Hotel, 100 S. 12th St., 344-7000; FAX 648-6704

Radisson Hotel, 555 East Canal St., 788-0900; FAX 788-0791

Richmond Marriott, 500 E. Broad St., 643-3400; FAX 788-1230

✈ Sheraton Airport Inn, 4700 S. Laburnum Ave., 226-4300; FAX 226-6516

Sheraton Park South, 9901 Midlothian Turnpike, 323-1144; FAX 320-5255

SELECTED RESTAURANTS:

Signature Food: Virginia ham; peanut soup; spoonbread pecan pie; Brunswick stew; corn pudding

Byram's Lobster House, 3215 W. Broad St., 355-9193

Ellingtons, in the Embassy Suites Hotel, 672-8585

Gallego, in the Omni Richmond Hotel, 344-7000

Hugo's, in the Hyatt Richmond, 285-1234

Julian's Restaurant, 2617 W. Broad St., 359-0605

Kabuto—Japanese House of Steaks, 8052 W. Broad St., 747-9573

Sal Federico's, 1808 Staples Mill Rd., 358-9111

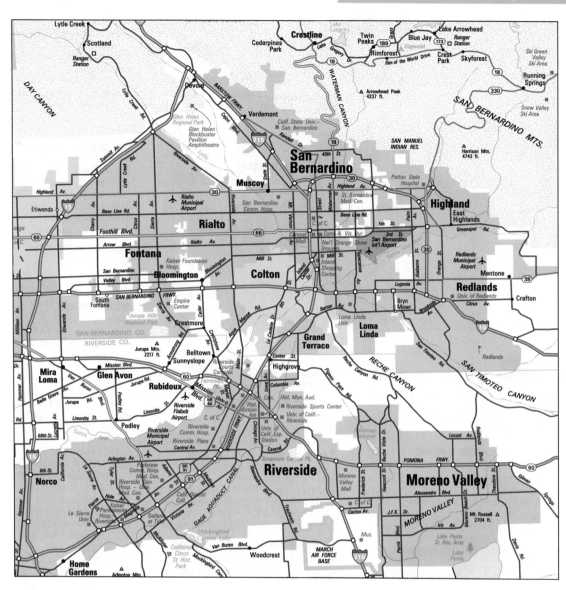

SELECTED ATTRACTIONS:

Berkeley Plantation, VA Hwy. 5, Charles City County, 829-6018

Edgar Allan Poe Museum, 1914 E. Main St., 648-5523

Hollywood Cemetery, Albemarle & Cherry sts., 648-8501

Maymont House and Park (turn-of-the-century estate), 1700 Hampton St., 358-7166

Museum and White House of the Confederacy, 12th & Clay sts., 649-1861

Richmond National Battlefield Park (Visitor Center), 3215 E. Broad St., 226-1981

St. John's Church ("Give me Liberty or give me death" speech site), 24th & E. Broad sts., 648-5015

Shockoe Slip (historic district), Cary St. between 12th & 14th sts., 232-3919

Virginia Aviation Museum, Huntsman Rd. next to Richmond International Airport, 236-3622

Virginia Museum of Fine Arts, 2800 Grove Ave., 367-0844

INFORMATION SOURCES:

Metropolitan Richmond Convention & Visitors Bureau
  550 E. Marshall St.
  Box C—250
  Richmond, Virginia 23219
  (804) 782-2777; (800) 365-7272

Metro Richmond Visitors Center
  1710 Robin Hood Rd.
  Richmond, Virginia 23220
  (804) 358-5511

Metro Richmond Chamber of Commerce
  201 E. Franklin
  Richmond, Virginia 23219
  (804) 648-1234

Riverside/San Bernardino, California

Population: (*2,588,793—San Bernardino-Riverside Metro Area) Riverside 226,505; San Bernardino 164,164 (1990C)

Altitude: Riverside 858 feet; San Bernardino 1,049 feet

Average Temp.: Jan., 46°F.; July, 74°F.

Telephone Area Code: 909

Time: 853-1212 **Weather:** (none)

Time Zone: Pacific

AIRPORT TRANSPORTATION:

Sixteen miles from Ontario International to downtown Riverside; twenty-four miles to downtown San Bernardino from Ontario International

Taxicab, bus and airport shuttle service to Riverside. Taxicab, limousine, and van service to San Bernardino.

SELECTED HOTELS: RIVERSIDE

Days Inn, 1510 University Ave., 787-1191; FAX 787-6783

Econolodge, 1971 University Ave., 684-6363; FAX 684-9228

The Hampton Inn, 1590 University Ave., 683-6000; FAX 782-8052

Holiday Inn—Riverside, 3400 Market St., 784-8000; FAX 369-7127

✈ Marriott Hotel, 2200 E. Holt Blvd., Ontario, 986-8811; FAX 391-6151

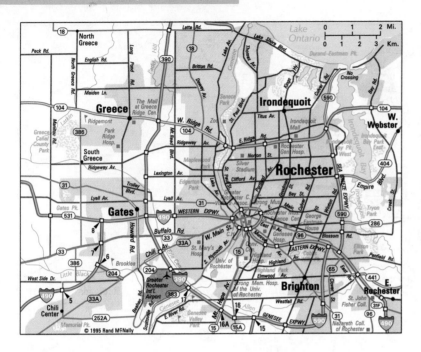

201 N. E St., Suite 103
San Bernardino, California 92401
(909) 889-3980; (800) T0-RTE-66
San Bernardino Area Chamber of
Commerce
546 W. Sixth St.
San Bernardino, California 92402
(909) 885-7515

Rochester, New York

Population: (*1,002,410) 231,636 (1990C)
Altitude: 515 feet
Average Temp.: Jan., 24°F.; July, 71°F.
Telephone Area Code: 716
Time and Weather: 475-1111
Time Zone: Eastern

AIRPORT TRANSPORTATION:

Six miles to downtown Rochester.
Taxicab, major hotel courtesy car and city
bus service.

SELECTED HOTELS:

Genesee Plaza Holiday Inn, 120 E. Main
St., 546-6400; FAX 546-3908
✈ Holiday Inn—Airport, 911 Brooks Ave.,
328-6000; FAX 328-1012
Hyatt Regency Rochester, 125 E. Main St.,
546-1234; FAX 546-6777
Radisson Hotel Rochester Plaza, 70 State
St., 546-3450; FAX 546-8712
Radisson Inn, 175 Jefferson Rd., 475-1910;
FAX 475-9633
✈ Ramada Inn, 1273 Chili Ave., 464-8800;
FAX 464-8800
✈ Rochester Airport—Howard Johnson,
1100 Brooks Ave., 235-6030;
FAX 235-7051
Rochester Marriott Airport, 1890 W. Ridge
Rd., 225-6880; FAX 225-8188
Rochester Marriott Thruway Hotel, 5257
W. Henrietta Rd., 359-1800;
FAX 359-1349
Strathallan Hotel, 550 East Ave., 461-5010;
FAX 461-3387

SELECTED RESTAURANTS:

Signature Food: white hots;
marshmallows
Chapels, 30 W. Broad St., 232-2300
Daisy Flour Mill, 1880 Blossom Rd.,
381-1880
Depot Restaurant, 41 N. Main St.,
Pittsford, 381-9991
Edward's, 13 S. Fitzhugh, 423-0140
Lloyd's, 289 Alexander St., 546-2211
Maplewood Inn, 3500 East Ave., 381-7700
Richardson's Canal House, 1474 Marsh
Rd., Pittsford, 248-5000
Sabrinas, in the Strathallan Hotel,
461-5010
Spring House, 3001 Monroe Ave.,
586-2300

SELECTED ATTRACTIONS:

Casa Larga Vineyards (tours), 2287 Turk
Hill Rd., Fairport, 223-4210
Genesee Country Village, Flint Hill Rd.,
Mumford, 538-6822
High Falls (museum & historic district),
Platt St. & Brown's Race, 325-2030
High Falls of the Genesee, Pont deRennes
bridge over the Genesee River
International Museum of Photography &

Mission Inn, 3649 7th St., 784-0300;
FAX 683-1342
✈ Red Lion, 222 N. Vineyard Ave.,
Ontario, 983-0909; FAX 983-8851
Super 8, 1199 University Ave., 682-9011;
FAX 369-6645

SELECTED RESTAURANTS: RIVERSIDE

Signature Food: Mexican
Carlos O'Brien's, 3667 Riverside Plaza,
686-5860
Cask 'N Cleaver, 1333 University Ave.,
682-4580
C.J.'s Beef and Produce, in the Sheraton
Riverside, 784-8000
El Gato Gordo, 1360 University Ave.,
787-8212
Mario's Place, 1725 Spruce St., 684-7755
Riverside Brewing Company, 3397 7th St.,
784-2739
The Spanish Patio, at the Mission Inn,
784-0300

SELECTED ATTRACTIONS: RIVERSIDE

California Citrus State Historic Park, 1879
Jackson St., 780-6222
California Museum of Photography, 3824
Main St., 787-4787
Castle Park (family recreational park),
3500 Polk St., 785-4140
Glen Ivy Hot Springs, 25000 Glen Ivy Rd.,
Corona, 277-3529
March Field Museum (aircraft), at March
Air Force Base, off I-215 and Van Buren
in Moreno Valley, 655-3725
Riverside Art Museum, 3425 Seventh St.,
684-7111
Riverside Municipal Museum, 3720
Orange St., 782-5273

INFORMATION SOURCES: RIVERSIDE

Riverside Visitors & Convention Bureau
3443 Orange St.
Riverside, California 92501
(909) 787-7950
Greater Riverside Chambers of Commerce
3685 Main St., Suite 350

Riverside, California 92501
(909) 683-7100

SELECTED HOTELS: SAN BERNARDINO

Best Western Sands, 606 North H St.,
889-8391; FAX 889-8394
Inland Empire Hilton, 285 E. Hospitality
Lane, 889-0133; FAX 381-4299
La Quinta Inn, 205 E. Hospitality Ln., 888-
7571; FAX 884-3864
Oak Creek Inn—Villa Viejo Super 8, 777
W. 6th St., 889-3561; FAX 384-7127
Radisson Hotel & Convention Center, 295
North E St., 381-6181; FAX 381-5288

SELECTED RESTAURANTS: SAN BERNARDINO

Signature Food: Mexican
La Potiniere, in the Inland Empire Hilton,
889-0133
Marie Callender's, 800 E. Highland,
882-1754
Red Lobster, 195 E. Hospitality Ln.,
888-2288
Spencer's, in the Radisson Hotel &
Convention Center, 381-6181
Stuart Anderson's Black Angus, 290 E.
Hospitality Ln., 885-7551
TGI Friday's, 390 E. Hospitality Ln.,
888-9934

SELECTED ATTRACTIONS: SAN BERNARDINO

Assistencia Mission, 26930 Barton Rd.,
Redlands, 793-5402
Heritage House, corner of 8th & D sts.,
384-5114
Historical Glass Museum (open
weekends; weekdays for groups by
appointment, call 2 weeks in advance),
1157 Orange St., 797-1528
San Bernardino Civic Light Opera, 562 W.
4th, 386-7353
San Bernardino County Museum, 2024
Orange Tree Lane, Redlands, 798-8570
San Manuel Indian Bingo, 5797 N.
Victoria Ave., Highland, 864-5050

INFORMATION SOURCES: SAN BERNARDINO

San Bernardino Convention & Visitors
Bureau

George Eastman House, 900 East Ave.,
271-3361
Memorial Art Gallery of the University of
Rochester, 500 University Ave., 473-7720
Rochester Museum & Science Center, 657
East Ave., 271-4320
Rochester Philharmonic Orchestra, Gibbs
& Main, 454-2620
Seabreeze Amusement Park, 4600 Culver
Rd., 323-1900
The Strong Museum, 1 Manhattan Square,
263-2700

INFORMATION SOURCES:
Greater Rochester Visitors Association
126 Andrews St.
Rochester, New York 14604-1102
(716) 546-3070
Greater Rochester Metro Chamber of
Commerce
55 Saint Paul St.
Rochester, New York 14604
(716) 454-2220

Sacramento, California

Population: (*1,481,102) 369,365 (1990C)
Altitude: 25 feet
Average Temp.: Jan., 45°F.; July, 75°F.
Telephone Area Code: 916
Time: 767-8900 **Weather:** 646-2000
Time Zone: Pacific

AIRPORT TRANSPORTATION:
Twelve miles from Sacramento
Metropolitan to downtown Sacramento.
Taxicab, airporter limousine and hotel
limousine service.

SELECTED HOTELS:
Best Western Sandman, 236 Jibboom St.,
443-6515; FAX 443-8346
Beverly Garland Hotel, 1780 Tribute Rd.,
929-7900; FAX 921-9147
Canterbury Inn, 1900 Canterbury Rd., 927-
3492; FAX 641-8594
Capitol Plaza Holiday Inn, 300 J St.,
446-0100; FAX 446-0100
Clarion Hotel, 700 16th St., 444-8000;
FAX 442-8129
Fountain Suites, 321 Bercut Dr., 441-1444;
FAX 441-6530
✈ Host Hotel by Marriott, 6945 Airport
Blvd., 922-8071; FAX 929-8636
Hyatt Regency Hotel, 1209 L St.,
443-1234; FAX 321-6699
Radisson Hotel, Hwy. 160 and Canterbury
Rd., 922-2020; FAX 649-9463
Red Lion Hotel, 2001 Point West Way,
929-8855; FAX 924-4913
Red Lion Sacramento Inn, 1401 Arden
Way, 922-8041; FAX 922-0386
Sacramento Hilton, 2200 Harvard,
922-4700; FAX 922-8418
Sheraton Rancho Cordova, 11211 Point
East Dr., Rancho Cordova, 638-1100;
FAX 638-5803

SELECTED RESTAURANTS:
Signature Food: almonds
A Shot of Class, 1020 11th St., 447-5340
Aldo's Restaurant, in the Town & Country
Village Shopping Center, Fulton &
Marconi, 483-5031
Biba, 2801 Capitol Ave., 455-2422
California Fat's, 1015 Front St., 441-7966

The Firehouse, 1112 2nd St., 442-4772
Frank Fat's, 806 L St., 442-7092
John Q's Penthouse, in the Capitol Plaza
Holiday Inn, 446-0100
Pheasant Club, 2525 Jefferson Blvd.,
371-9530
Terrace Grill, 544 Pavillion Ln., 920-3800

SELECTED ATTRACTIONS:
California State Railroad Museum, 2nd & I
St., 448-4466
Crocker Art Museum, 3rd & O St., 264-5423
Discovery Museum, 101 I St., Old
Sacramento, 264-7057
Discovery Museum Learning Center, 3615
Auburn Blvd., 277-6180
Historic Governor's Mansion, 16th & H
St., 323-3047
Sacramento Zoo, in William Land Park,
corner of Sutterville Rd. & Land Park
Dr., 264-5885
State Capitol, 10th St. & Capitol Mall,
324-0333
Sutter's Fort, 27th & L sts., 445-4422
Waterworld USA, 1600 Exposition Blvd.,
924-0556

INFORMATION SOURCES:
Sacramento Convention & Visitors Bureau
1421 K St.
Sacramento, California 95814
(916) 264-7777
Visitor Information Center
1104 Front St.
Old Sacramento, California 95814
(916) 442-7644 (7 days a week)
Sacramento Metropolitan Chamber of
Commerce
917 7th St.
Sacramento, California 95814
(916) 552-6800

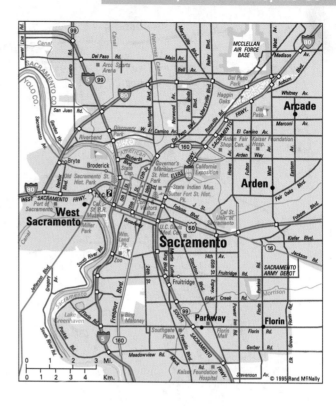

St. Louis, Missouri

Population: (*2,444,099) 396,685 (1990C)
Altitude: 470 feet
Average Temp.: Jan., 32°F.; July, 79°F.
Telephone Area Code: 314
Time: 321-2522 **Weather:** 321-2222
Time Zone: Central

AIRPORT TRANSPORTATION:
Fifteen miles to downtown St. Louis.
Taxicab and limousine bus service.

SELECTED HOTELS:
Adam's Mark, 4th & Chestnut, 241-7400;
FAX 241-6618
Cheshire Inn & Lodge, 6300 Clayton Rd.,
647-7300; FAX 647-0442
The Frontenac Hilton, 1335 S. Lindbergh
Blvd., 993-1100; FAX 993-8546
Hotel Majestic, 1019 Pine St., 436-2355;
FAX 436-2355
Hyatt Regency, 1 St. Louis Union Station,
231-1234; FAX 923-3970
Marriott's Pavilion Hotel, 1 S. Broadway,
421-1776; FAX 331-9029
Regal River Front Hotel, 200 S. 4th St.,
241-9500; FAX 241-9977
Ritz Carlton St. Louis, 100 Carondelet
Plaza, 863-6300; FAX 863-3525
✈ St. Louis Airport Hilton, 10330 Natural
Bridge Rd., 426-5500; FAX 426-3429
✈ St. Louis Airport Marriott, I-70 at
Lambert Int'l Airport, 423-9700;
FAX 423-0213
Seven Gables Inn, 26 N. Meramec,
Clayton, 863-8400; FAX 863-8846
✈ Stouffer Concourse Hotel, 9801 Natural
Bridge Rd., 429-1100; FAX 429-3625

SELECTED RESTAURANTS:
Signature Food: toasted ravioli;
concreates (thick ice cream)

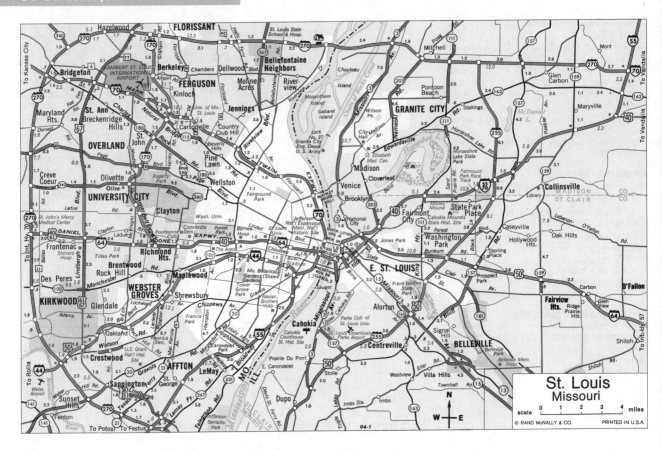

St. Louis
Missouri

scale 0 1 2 3 4 miles

© RAND McNALLY & CO.   PRINTED IN U.S.A.

94-1

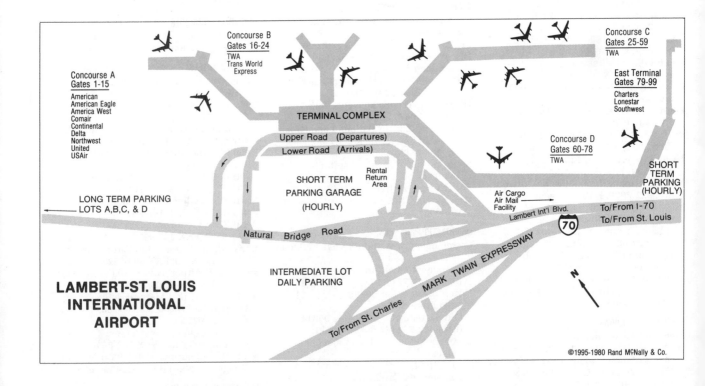

Concourse A
Gates 1-15

American
American Eagle
America West
Comair
Continental
Delta
Northwest
United
USAir

Concourse B
Gates 16-24
TWA
Trans World
Express

Concourse C
Gates 25-59
TWA

East Terminal
Gates 79-99

Charters
Lonestar
Southwest

TERMINAL COMPLEX

Upper Road   (Departures)
Lower Road   (Arrivals)

Rental
Return
Area

Concourse D
Gates 60-78
TWA

**SHORT
TERM
PARKING
(HOURLY)**

**SHORT TERM
PARKING GARAGE
(HOURLY)**

Air Cargo
Air Mail
Facility

Lambert Int'l Blvd.

To/From I-70
To/From St. Louis

LONG TERM PARKING
LOTS A,B,C, & D

Natural   Bridge   Road

INTERMEDIATE LOT
DAILY PARKING

To/From St. Charles

MARK TWAIN EXPRESSWAY

N

**LAMBERT-ST. LOUIS
INTERNATIONAL
AIRPORT**

©1995-1980 Rand McNally & Co.

Dierdorf & Hart's, at St. Louis Union
Station, 421-1772
Dominic's, 5101 Wilson Ave., 771-1632
Giovanni's on the Hill, 5201 Shaw,
772-5958
Henry VIII, in the Ramada Henry VIII
Hotel, 4690 N. Lindbergh Blvd.,
Bridgeton, 731-3040
Kemoll's Restaurant, 211 N. Broadway,
421-0555
Tony's, 410 Market St., 231-7007

SELECTED ATTRACTIONS:

Anheuser-Busch Brewery, 12th & Lynch,
577-2626
Gateway Arch, 425-4465
Gateway Riverboat Cruises, 621-4040
Grant's Farm, 10501 Gravois, 843-1700
Missouri Botanical Garden, 4344 Shaw,
577-5100
St. Louis Art Museum, One Fine Arts Dr.,
Forest Park, 721-0067
St. Louis Science Center, 5050 Oakland
Ave., 289-4400
St. Louis Union Station, Market St.
between 18th & 20th, 421-6655
St. Louis Zoo, US 40 & Hampton Ave.,
Forest Park, 781-0900
Six Flags Over Mid-America, Eureka,
938-5300

INFORMATION SOURCES:

St. Louis Convention & Visitors
Commission
10 S. Broadway, Suite 1000
St. Louis, Missouri 63102
(314) 421-1023
Business Info. (800) 325-7962;
Visitor Info. (800) 888-3861
St. Louis Regional Commerce & Growth
Association
100 S. 4th St., Suite 500
St. Louis, Missouri 63102
(314) 231-5555

Salt Lake City, Utah

Population: (*1,072,227) 159,936 (1990C)
Altitude: 4,260 feet
Average Temp.: Jan., 27°F.; July, 77°F.
Telephone Area Code: 801
Time: 975-213   **Weather:** 524-5133
Time Zone: Mountain

AIRPORT TRANSPORTATION:

Six miles to downtown Salt Lake City.
Taxicab, bus, and limousine bus service.

SELECTED HOTELS:

✈ Airport Hilton, 5151 Wiley Post Way,
539-1515; FAX 539-1113
Doubletree Hotel, 215 W. South Temple,
531-7500; FAX 531-7442
Holiday Inn, Ninth S. & Main St.,
359-8600; FAX 359-7186
Little America Hotel, 500 S. Main St., 363-
6781; FAX 596-5911
Red Lion Salt Lake, 255 S. West Temple,
328-2000; FAX 532-1953
Salt Lake City Marriott Hotel, 75 S. West
Temple, 531-0800; FAX 532-4127
Salt Lake Hilton, 150 W. 500 South,
532-3344; FAX 532-3344

SELECTED RESTAURANTS:

Signature Food: Utah scones

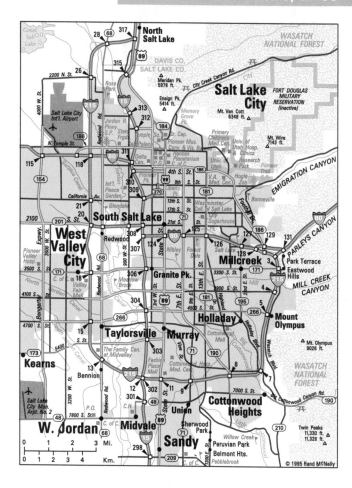

Benihana of Tokyo, 165 S. West Temple
St., 322-2421
Cafe Pierpont, 122 W. Pierpont Ave.,
364-1222
Cowboy Grub, 2350½ Foothill Blvd.,
466-8334
La Caille at Quail Run, 9565 Wasatch
Blvd., 942-1751
La Fleur de Lys, 39 Market St., 359-5753
Market Street Grill, 54 Market St.,
322-4668
Mikado Japanese Restaurant, 67 W. 100
South St., 328-0929
The New Yorker, 60 Market St., 363-0166
Ristorante Della Fontana, 336 S. 4th East,
328-4243

SELECTED ATTRACTIONS:

Beehive House, State St. & South Temple,
240-2671
Family History Library, 35 N. West
Temple, 240-2331
Hansen Planetarium, 15 S. State St.,
538-2098
Hogle Zoo, 2600 E. Sunnyside Ave.,
582-1631
Lagoon Amusement Park & Pioneer
Village, 17 mi. north on I-15, 451-8000
Museum of Church History & Art, 45 N.
West Temple, 240-3310
Pioneer Trail State Park, 2601 E.
Sunnyside Ave., 584-8392
Raging Waters, 1200 West 1700 South,
973-4020
Temple Square, 50 W. North Temple,
240-2534

Utah Museum of Fine Arts, south of the
Marriott Library, Univ. of Utah campus,
581-7049
Utah State Capitol, 350 N. Main, 538-3000
Violin Making School of America (tours
by appointment), 308 East 200 South,
364-3651

INFORMATION SOURCES:

Salt Lake Convention and Visitors Bureau
180 S. West Temple St.
Salt Lake City, Utah 84101
(801) 521-2822
Salt Lake Area Chamber of Commerce
175 E. 400 S., Suite 600
Salt Lake City, Utah 84111
(801) 364-3631

San Antonio, Texas

Population: (*1,302,099) 935,933 (1990C)
Altitude: 505 to 1,000 feet
Average Temp.: Jan., 51°F.; July, 84°F.
Telephone Area Code: 210
Time: 226-3232   **Weather:** 609-2033
Time Zone: Central

AIRPORT TRANSPORTATION:

Eight miles to downtown San Antonio.
Taxicab and limousine bus service.

SELECTED HOTELS:

Best Western Continental Inn, 9735 I-35
N., 655-3510; FAX 655-0778
Embassy Suites Northwest, 7750
Briaridge, 340-5421; FAX 340-1843

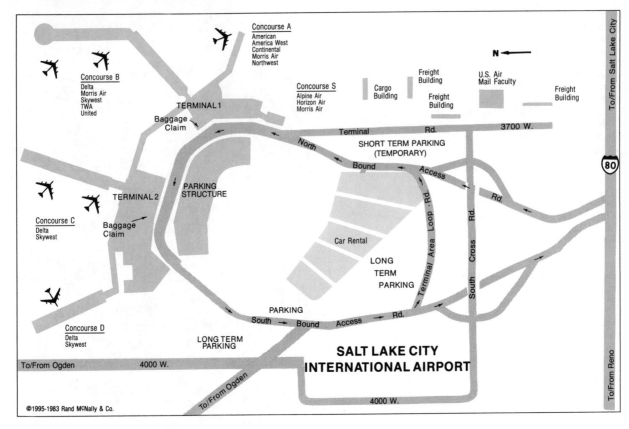

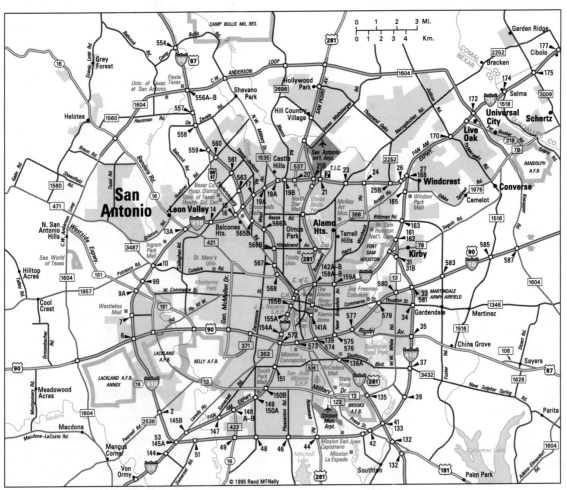

Hilton Palacio del Rio, 200 S. Alamo St.,
222-1400; FAX 226-4018
Hyatt Regency San Antonio, 123 Losoya,
222-1234; FAX 227-4925
La Mansion del Rio, 112 College St.,
225-2581; FAX 226-1365
Plaza San Antonio Hotel, 555 S. Alamo
St., 229-1000; FAX 229-1418
St. Anthony Hotel, 300 E. Travis St.,
227-4392; FAX 227-0915
San Antonio Marriott on the Riverwalk,
711 E. Riverwalk, 224-4555;
FAX 224-2754
✈ The Sheraton Fiesta, 37 NE. Loop 410,
366-2424; FAX 341-0410

SELECTED RESTAURANTS:

Signature Food: Tex-Mex
Chez Ardid Restaurant Gastronomique,
1919 San Pedro, 732-3203
Crystal Baking Co., 1039 NE. Loop 410,
826-2371
Fig Tree, 515 Paseo la Villita, 224-1976
Grey Moss Inn, 19010 Scenic Loop Rd.,
695-8301
Las Canarias, in the La Mansion del Rio,
225-2581
Paesano's, 1715 McCullough Ave.,
226-9541
Restaurant Biga, 206 E. Locust, 225-0722

SELECTED ATTRACTIONS:

The Alamo, Alamo Plaza, 225-1391
Brackenridge Park, 3500 block of St.
Mary's St., 299-8480
La Villita Historical District, on the River
Walk between S. Alamo & E. Nueva sts.,
299-8610
Market Square (El Mercado), 514 W.
Commerce St., 299-8600
The River Walk, downtown San Antonio
San Antonio Missions National Historical
Park, 2202 Roosevelt Ave., 229-5701
San Antonio Zoo & Aquarium, 3903 N. St.
Mary's St., 734-7183
Texas Adventure, 307 Alamo Plaza,
227-8224
Tower of the Americas, in HemisFair Park,
299-8615
University of Texas Institute of Texan
Cultures at San Antonio, 801 S. Bowie
St., 226-7651
Witte Museum, 3801 Broadway St.,
820-2169

INFORMATION SOURCES:

San Antonio Convention and Visitors
Bureau
121 Alamo Plaza
P.O. Box 2277
San Antonio, Texas 78298
(210) 270-8700; (800) 447-3372
Visitor Information Center
317 Alamo Plaza
San Antonio, Texas 78205
(210) 299-8155; (800) 447-3372
Chamber of Commerce of Greater San
Antonio
602 E. Commerce St.
San Antonio, Texas 78205
(210) 229-2100

San Diego, California

Population: (*2,498,016) 1,110,549
(1990C)

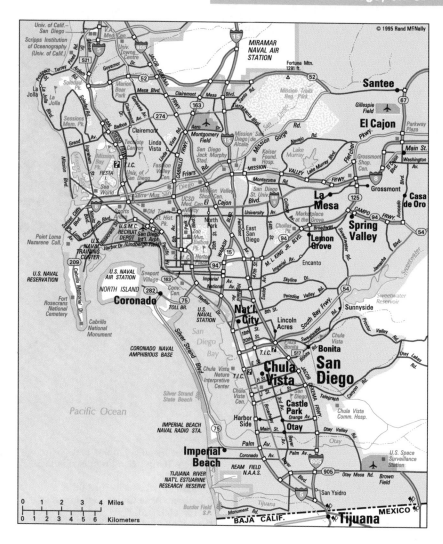

Altitude: Sea level to 823 feet
Average Temp.: Jan., 57°F.; July, 71°F.
Telephone Area Code: 619
Time: 853-1212 **Weather:** 289-1212
Time Zone: Pacific

AIRPORT TRANSPORTATION:

Three miles to downtown San Diego.
Taxicab and bus service.

SELECTED HOTELS:

Bristol Court, 1055 First Ave., 232-6141;
FAX 232-0118
Embassy Suites, 601 Pacific Hwy.,
239-2400; FAX 239-1520
Glorietta Bay Inn, 1630 Glorietta Blvd.,
435-3101; FAX 435-6182
Hanalei Hotel, 2270 Hotel Circle N.,
297-1101; FAX 297-6049
✈ Holiday Inn on the Bay, 1355 N. Harbor
Dr., 232-3861; FAX 232-4924
Hotel del Coronado, 1500 Orange Ave.,
Coronado, 435-6611; FAX 522-8262
Hyatt Islandia, 1441 Quivira Rd.,
224-1234; FAX 224-0348
Hyatt Regency San Diego, 1 Market Pl.,
232-1234; FAX 233-6464
Mission Valley Inn, 875 Hotel Circle S.,
298-8281; FAX 295-5610
Rancho Bernardo Inn, 17550 Bernardo
Oaks Dr., 487-1611; FAX 673-0311

San Diego Hilton, 1775 E. Mission Bay
Dr., 276-4010; FAX 275-7991
San Diego Marriott Hotel & Marina, 333
W. Harbor Dr., 234-1500; FAX 234-8678
✈ Sheraton Grand, 1590 Harbor Island
Dr., 291-6400; FAX 291-4847
✈ Sheraton Harbor Island Hotel—East,
1380 Harbor Island Dr., 291-2900;
FAX 294-3279
Town & Country Hotel, 500 Hotel Circle
N., 291-7131; FAX 291-3584
Westgate Hotel, 1055 2nd Ave., 238-1818;
FAX 557-3737

SELECTED RESTAURANTS:

Signature Food: fish tacos; Mexican
Anthony's Star of the Sea Room, Harbor
Dr. & Ash, 232-7408
El Bizcocho, in the Rancho Bernardo Inn,
487-1611
Gourmet Room, in the Town & Country
Hotel, 291-7131
The Marine Room, 2000 Spindrift Dr., La
Jolla, 459-7222
Mister A's, 2550 5th Ave., Financial
Center, 239-1377
Old Trieste, 2335 Morena Blvd., 276-1841
Thee Bungalow, 4996 W. Point Loma
Blvd., 224-2884
Tom Ham's Lighthouse, 2150 Harbor
Island Dr., 291-9110

SAN DIEGO INTERNATIONAL AIRPORT / LINDBERGH FIELD

WEST TERMINAL

EAST TERMINAL

Aeromexico
Air LA
Alaska
America West
TWA
USAir
USAir Express

American
American Eagle
Delta

Midwest
Express
Northwest
Reno Air
Skywest

Continental
Markair
Morris Air
Southwest
United
United Express

N

PARKING

PARKING

PARKING

PARKING

PARKING

CAR RENTAL

CAR RENTAL

CAR RENTAL

RENTAL CAR AREA

Harbor Dr.

Harbor Dr.

Harbor Island Dr

©1995-1983 Rand McNally & Co.

Top O' the Cove, 1216 Prospect, La Jolla, 454-7779

SELECTED ATTRACTIONS:
Balboa Park (museums, theaters, gardens, galleries), Center of city Cabrillo National Monument, on the tip of Point Loma, 557-5450
Museum of Contemporary Art, 1001 Kettner Blvd. at Broadway, 234-1001
Old Town State Historic Park, (Visitor Center) 4002 Wallace St., 220-5422
San Diego Zoo, off Park Blvd. in Balboa Park, 234-3153
Sea World of California, on Sea World Dr., 226-3901
Wild Animal Park, 15500 San Pasqual Valley Rd., Escondido, 234-6541

INFORMATION SOURCES:
San Diego Convention and Visitors Bureau
401 B St., Suite 1400
San Diego, California 92101-4237
(619) 232-3101
International Visitor Information Center
11 Horton Plaza
1st Ave. & F St.
San Diego, California 92101
(619) 236-1212
Greater San Diego Chamber of Commerce
402 W. Broadway, Suite 1000
San Diego, California 92101
(619) 232-0124

San Francisco-Oakland, California

Population: San Francisco (*1,603,678) 723,959; Oakland (*2,082,914) 372,242 (1990C)

Altitude: Sea level to 934 feet
Average Temp.: San Francisco Jan., 50°F.; July, 59°F.
Oakland Jan., 48°F.; July, 63°F.
Telephone Area Code: (San Francisco) 415; (Oakland) 510
Time: 767-8900 **Weather:** San Francisco 364-7974; Oakland 936-1212
Time Zone: Pacific

AIRPORT TRANSPORTATION:
Fifteen miles to downtown San Francisco from San Francisco International; 25 miles to downtown Oakland from San Francisco International Airport. Eight miles to downtown Oakland from Oakland International.
Taxicab, bus, and limousine bus service to downtown San Francisco. Taxicab, limousine, bus, and BART service to downtown Oakland.

SELECTED HOTELS: SAN FRANCISCO
The Donatello, 501 Post St., 441-7100; FAX 885-8842
The Fairmont Hotel, 950 Mason St., 772-5000; FAX 781-4027
Four Seasons-Clift, 495 Geary St., 775-4700; FAX 441-4621
Grand Hyatt on Union Square, 345 Stockton St., 398-1234; FAX 391-1780
Hilton Square, O'Farrell St. between Taylor & Mason, 771-1400; FAX 771-6807
The Holiday Inn Union Square, 480 Sutter St., 398-8900; FAX 989-8823
Huntington Hotel, 1075 California St., 474-5400; FAX 474-6227
Hyatt Regency San Francisco, 5 Embarcadero Center, 788-1234; FAX 398-2567

Mark Hopkins Inter-Continental, 999 California St., 392-3434; FAX 421-3302
Miyako Hotel, 1625 Post St., 922-3200; FAX 921-0417
The Phoenix Hotel, 601 Eddy St., 776-1380; FAX 885-3109
Queen Anne Hotel, 1590 Sutter St., 441-2828; FAX 775-5212
The Stouffer Stanford Court Hotel, 905 California St., 989-3500; FAX 391-0513
The Westin St. Francis, 335 Powell St., 397-7000; FAX 774-0124

SELECTED RESTAURANTS: SAN FRANCISCO
Signature Food: sourdough bread
Amelio's, 1630 Powell St., North Beach, 397-4339
Empress of China, 838 Grant Ave., 434-1345
Ernie's Restaurant, 847 Montgomery St., 397-5969
Fleur de Lys, 777 Sutter St., 673-7779
Fournou's Ovens, in The Stouffer Stanford Court Hotel, 989-1910
Palio D'Asti, 640 Sacramento St., 395-9800
Tommy Toy's Haute Cuisine Chinoise, 655 Montgomery St., 397-4888
The Waterfront Restaurant, Pier 7, The Embarcadero, 391-2696

SELECTED ATTRACTIONS: SAN FRANCISCO
Alcatraz Island, San Francisco Bay, 546-2700
Chinatown, Grant Ave. & Stockton St. (begins at Grant & Bush)
Coit Tower/Telegraph Hill, northeast San Francisco, reached via Lombard St.
Exploratorium, 3601 Lyon St., 561-0360
Fisherman's Wharf, on Jefferson St. between Hyde & Powell sts.

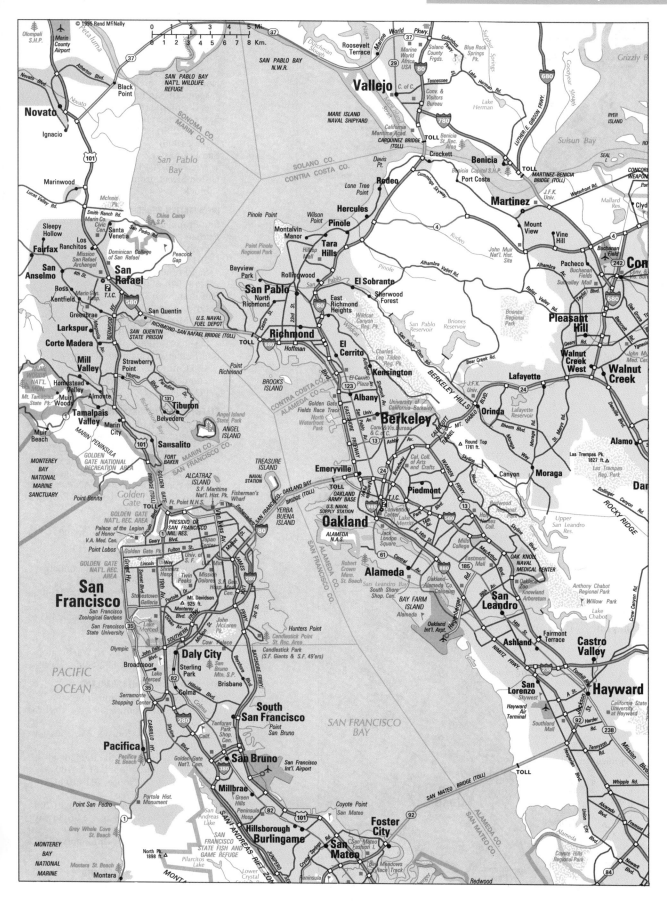

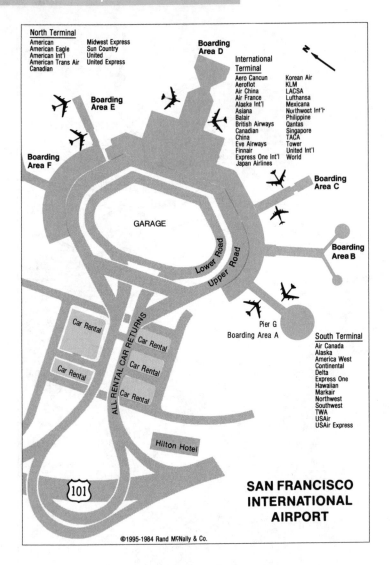

SAN FRANCISCO INTERNATIONAL AIRPORT

North Terminal
American
American Eagle
American Int'l
American Trans Air
Canadian
Midwest Express
Sun Country
United
United Express

Boarding Area D

International Terminal
Aero Cancun
Aeroflot
Air China
Air France
Alaska Int'l
Asiana
Balair
British Airways
Canadian
China
Eve Airways
Finnair
Express One Int'l
Japan Airlines
Korean Air
KLM
LACSA
Lufthansa
Mexicana
Northwest Int'l
Philippine
Qantas
Singapore
TACA
Tower
United Int'l
World

Boarding Area E

Boarding Area F

Boarding Area C

Boarding Area B

GARAGE

Lower Road
Upper Road

ALL RENTAL CAR RETURNS

Car Rental

Pier G
Boarding Area A

South Terminal
Air Canada
Alaska
America West
Continental
Delta
Express One
Hawaiian
Markair
Northwest
Southwest
TWA
USAir
USAir Express

101

Hilton Hotel

©1995-1984 Rand McNally & Co.

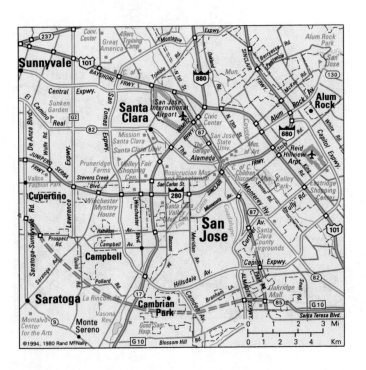

Golden Gate Bridge, spans the entrance to San Francisco Bay

Golden Gate Park, from Stanyan St. to the Pacific Ocean, between Fulton St. & Lincoln Blvd.

North Beach (night-life area), northeast sector of San Francisco, Columbus Ave. is the main artery

San Francisco Museum of Modern Art, 151 3rd St., 357-4000 (due to open in Jan., 1995)

San Francisco Zoo, Sloat Blvd. & 45th Ave., 753-7083

Union Street, from the 1600 to the 2200 block

INFORMATION SOURCES: SAN FRANCISCO

San Francisco Convention & Visitors Bureau
  Convention Plaza
  201 Third St., Suite 900
  San Francisco, California 94103
  (415) 974-6900

San Francisco Visitor Information Center
  Lower Level, Hallidie Plaza
  900 Market St. (at Powell)
  San Francisco, California 94102
  (415) 391-2000

San Francisco Chamber of Commerce
  465 California St., 9th Floor
  San Francisco, California 94104
  (415) 392-4511

SELECTED HOTELS: OAKLAND

Best Western Park Plaza, 150 Hegenberger Rd., 635-5300; FAX 635-9661

Best Western Thunderbird Inn, 233 Broadway, 452-4565; FAX 452-4634

Claremont Resort & Spa, Domingo & Ashby aves., 843-3000; FAX 848-6208

Holiday Inn—Bay Bridge, 1800 Powell St., Emeryville, 658-9300; FAX 547-8166

Holiday Inn Oakland Airport, 500 Hegenberger Rd., 562-5311; FAX 636-1539

Oakland Airport Hilton, 1 Hegenberger Rd., 635-5000; FAX 635-0244

Parc Oakland, 1001 Broadway, 451-4000; FAX 835-3466

Ramada at Oakland International, 455 Hegenberger Rd., 562-6100; FAX 569-5681

Waterfront Plaza Hotel, 10 Washington St., 836-3800; FAX 832-5695

SELECTED RESTAURANTS: OAKLAND

Signature Food: sourdough bread

Bay Wolf Restaurant, 3853 Piedmont Ave., 655-6004

Crogan's Seafood House & Bar, 1601 La Salle Ave., 339-2098

Ducks & Company, in the Ramada at Oakland International, 562-6100

The Gingerbread House, 741 5th St., 444-7373

La Brasserie, 542 Grand Ave., 893-6206

Scott's Seafood Grill & Bar, 2 Broadway, 444-3456

Trader Vic's, 9 Anchor Dr., Emeryville 653-3400

SELECTED ATTRACTIONS: OAKLAND

Camron-Stanford House (1876; tours), 1418 Lakeside Dr., 836-1976

Children's Fairyland, in Lakeside Park at Grand & Bellevue aves., 452-2259

Lakeside Park/Lake Merritt, 238-FUNN
Marine World Africa U.S.A., Marine
World Pkwy., Vallejo, (707) 643-ORCA
Oakland Museum, 10th & Oak St.,
834-2413
Rotary Nature Center, in Lakeside Park at
Bellevue & Perkins aves., 238-3739
Takara Sake Company (weekday tours
with advance reservations), 708
Addison St., Berkeley, 540-8250
Trial & Show Gardens, in Lakeside Park,
238-3208

INFORMATION SOURCES: OAKLAND

Oakland Convention and Visitors Bureau
1000 Broadway, Suite 200
Oakland, California 94607-4020
(510) 839-9000; (800) 262-5526
Oakland Chamber of Commerce
475 14th St.
Oakland, California 94612-1928
(510) 874-4800

San Jose, California

Population: (*1,497,577) 782,248 (1990C)
Altitude: 87 feet
Average Temp.: Jan., 51°F.; July, 65°F.
Telephone Area Code: 408
Time: 767-8900   **Weather:** 976-9847
Time Zone: Pacific

AIRPORT TRANSPORTATION:

Five miles to downtown San Jose.
Taxicab and bus service.

SELECTED HOTELS:

Best Western Le Baron, 1350 North 1st St.,
453-6200; FAX 437-9558
Fairmont Hotel, 170 S. Market St.,
998-1900; FAX 287-1648
Holiday Inn Park Center Plaza, 282
Almaden Blvd., 998-0400;
FAX 289-9081
✈ Howard Johnson's, 1755 North 1st St.,
453-3133; FAX 452-1849
✈ Hyatt San Jose, 1740 North 1st St.,
993-1234; FAX 453-0259
✈ The Red Lion Hotel, 2050 Gateway Pl.,
453-4000; FAX 437-2898
Rodeway Inn, 2112 Monterey Hwy.,
294-1480; FAX 947-0343
San Jose Hilton & Towers, 300 Almaden
Blvd., 287-2100; FAX 947-4489
✈ Super 8—Airport, 1355 North 4th St.,
453-5340; FAX 453-5340
✈ Vagabond Inn, 1488 North 1st St.,
453-8822; FAX 453-0559

SELECTED RESTAURANTS:

Signature Food: American
Amalfi's, in the Hyatt San Jose, 993-1234
Emile's, 545 S. 2nd St., 289-1960
Eulipia, 374 S. 1st St., 280-6161
L'Horizon Restaurant, 1250 Aviation Ave.,
at the San Jose Jet Center, 295-1771
Lou's Village, 1465 W. San Carlos St.,
293-4570
Ninth Floor, in the Best Western Le Baron
Hotel, 453-6200
Paolo's Continental Restaurant, 333 W.
San Carlos, first floor of the Riverfront
Tower, 294-2558

SELECTED ATTRACTIONS:

Children's Discovery Museum of San Jose,
180 Woz Way, 298-5437

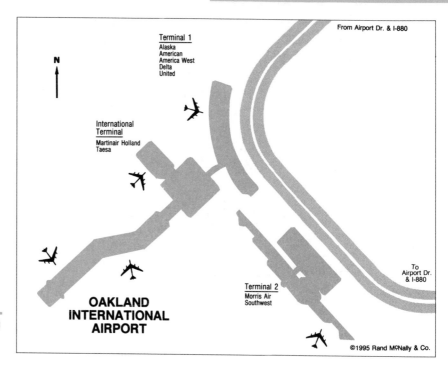

OAKLAND
INTERNATIONAL
AIRPORT

©1995 Rand McNally & Co.

Fallon House, 175 W. St. John, 993-8182
Japanese Friendship Garden, 1500 Senter
Rd. in Kelley Park, 287-2290
Monterey Bay Aquarium, 886 Cannery
Row, Monterey, 648-4800
Roaring Camp & Big Trees Narrow Gauge
Railroad, Felton, 335-4484
Rosicrucian Egyptian Museum, Park,
Planetarium & Art Gallery, Park &
Naglee aves., 947-3636
San Jose Historical Museum (25 acres of
original and replica 1890s buildings),
1600 Senter Rd. in Kelley Park,
287-2290
TECH Museum of Innovation, 145 W. San
Carlos St., 279-7150
Winchester Mystery House, 525 S.
Winchester Blvd., 247-2000

INFORMATION SOURCES:

San Jose Convention & Visitors Bureau
333 W. San Carlos, Suite 1000
San Jose, California 95110
(408) 295-9600; FYI Events
Line (408) 295-2265; (800) SAN-JOSE
Visitor & Business Center
San Jose McEnery Convention Center
150 W. San Carlos
San Jose, California 95113
(408) 283-8833; (800) SAN-JOSE
San Jose Chamber of Commerce
180 S. Market St.
San Jose, California 95113
(408) 291-5250

Seattle, Washington

Population: (*1,972,961) 516,259 (1990C)
Altitude: Sea level to 520 feet
Average Temp.: Jan., 41°F.; July, 66°F.
Telephone Area Code: 206
Time: 361-8463   **Weather:** 526-6087
Time Zone: Pacific

AIRPORT TRANSPORTATION:

Fourteen miles to downtown Seattle.
Taxicab, limousine bus, Metro Transit,
and Airporter service.

SELECTED HOTELS:

Doubletree Inn, 205 Strander Blvd.,
246-8220; FAX 575-4749
Four Seasons Olympic, 411 University
St., 621-1700; FAX 682-9633
Holiday Inn Crowne Plaza, 6th Ave. and
Seneca St., 464-1980; FAX 340-1617
Hyatt Regency—Bellevue, Bellevue Way &
NE 8th St., Bellevue, 462-1234;
FAX 646-7567
✈ Radisson Seattle Airport, 17001
International Blvd., 244-6000;
FAX 246-6835
✈ Red Lion Inn/Sea-Tac, 18740 Pacific
Hwy. S., 246-8600; FAX 248-8072
✈ Sea-Tac Holiday Inn Airport, 17338
Pacific Hwy. S., 248-1000;
FAX 242-7089
Seattle Hilton, 6th & University, 624-0500;
FAX 682-9029
✈ Seattle Marriott Sea-Tac Airport, 3201
S. 176th St., 241-2000; FAX 248-0789
Seattle Sheraton Hotel & Towers, 1400 6th
Ave., 621-9000; FAX 621-8441
Stouffer Madison, 515 Madison,
583-0300; FAX 622-8635
Westin Hotel Seattle, 1900 5th Ave.,
728-1000; FAX 728-2259

SELECTED RESTAURANTS:

Signature Food: seafood—clams, oysters,
salmon, scallops
Carvery, Seattle-Tacoma International
Airport, 433-5622
Fuller's, in the Seattle Sheraton Hotel &
Towers, 621-9000
The Hunt Club, in the Sorrento Hotel, 900
Madison, 622-6400
Jonah's, in the Best Western Bellevue Inn,
11211 Main St., Bellevue, 455-5240

Maximilien in the Market, 93 Pike St., Suite 316, 682-7270

Nikko, in the Westin Hotel Seattle, 728-1000

The Painted Table, in the Alexis Hotel, 1007 1st Ave. at Madison, 624-4844

Ray's Boathouse, 6049 Seaview NW, 789-3770

Reinor's, 1106 8th Ave., 624-2222

SELECTED ATTRACTIONS:

Hiram M. Chittenden Locks, located where NW. Market turns left, Ballard, 783-7059

Museum of Flight, 9404 East Marginal Way S., 764-5700

Pacific Science Center, 200 Second Ave. N. in Seattle Center, 443-2001

Pike Place Market, First & Pike sts., 682-7453

Seattle Aquarium, Pier 59 in Waterfront Park, 386-4320

Seattle Art Museum, 100 University St., 654-3100

Seattle Underground Tour, 682-1511

Space Needle, 219 Fourth Ave. N. in Seattle Center, 443-2100

Tillicum (Indian) Village Tour, Blake Island State Park (tour starts from Pier 56 on the Seattle Waterfront), 443-1244

Woodland Park Zoo, 50th & Fremont Ave. N., 684-4800

INFORMATION SOURCES:

Seattle-King County Convention & Visitors Bureau/Convention Sales
520 Pike St., Suite 1300
Seattle, Washington 98101
(206) 461-5800

Seattle-King County Convention & Visitors Bureau/Visitor Information
Level 1, Galleria
Washington State Convention & Trade Center
800 Convention Pl.
Seattle, Washington 98101
(206) 461-5840

Greater Seattle Chamber of Commerce
600 University St., Suite 1200
Seattle, Washington 98101
(206) 389-7200

Spokane, Washington

Population: (*361,364) 177,196 (1990C)
Altitude: 1,898 feet
Average Temp.: Jan., 26°F.; July, 70°F.
Telephone Area Code: 360
Time: 976-1616   **Weather:** 624-8905
Time Zone: Pacific

AIRPORT TRANSPORTATION:

Five miles to downtown Spokane.
Taxicab, airporter limousine and limousine service.

SELECTED HOTELS:

Best Western Trade Winds "Downtown," W. 907 3rd Ave., 838-2091; FAX 838-2094

Best Western Trade Winds—North, N. 3033 Division St., 326-5500; FAX 328-1357

Cavanaugh's Inn at the Park, W. 303 N. River Dr., 326-8000; FAX 325-7329

Cavanaugh's River Inn, N. 700 Division St., 326-5577; FAX 326-1120

Courtyard by Marriott, N. 401 Riverpoint
  Blvd., 456-7600; FAX 456-0969
Friendship Inn, 4301 W. Sunset Hwy.,
  838-1471; FAX 838-1705
Holiday Inn Spokane West, 4212 W.
  Sunset Blvd., 747-2021; FAX 747-5950
✈ Ramada Inn, at Spokane International
  Airport, 838-5211; FAX 838-1074
Red Lion Motor Inn, N. 1100 Sullivan Rd.,
  924-9000; FAX 922-4965
Sheraton—Spokane, N. 322 Spokane Falls
  Ct., 455-9600; FAX 455-6285
Shilo Inn, E. 923 3rd Ave., 535-9000;
  FAX 535-5740
West Coast Ridpath, 515 W. Sprague Ave.,
  838-2711; FAX 747-6970

SELECTED RESTAURANTS:

Signature Food: huckleberries
Ankeny's, in the West Coast Ridpath,
  838-6311
Chapter Eleven, E. 105 Mission Ave.,
  326-0466
1881, in the Sheraton—Spokane,
  455-9600
The Mustard Seed, W. 245 Spokane Falls
  Blvd., 747-2689
The Onion Bar and Grill, W. 302
  Riverside, 747-3852
Patsy Clark's Mansion, W. 2208 2nd Ave.,
  838-8300
Sea Galley, N. 1221 Howard Ave.,
  327-3361
Spokane House, in the Friendship Inn,
  838-1475
Stockyard Inn, 3827 E. Boone Ave.,
  534-1212
Town & Country, 5615 Trent Ave.,
  534-3868

SELECTED ATTRACTIONS:

Antique 1909 Loof Carousel, on Spokane
  Falls Blvd. between Washington &
  Howard, in Riverfront Park, 625-6600
Centennial Trail, follows the Spokane
  River from downtown to the
  Washington/Idaho state line
Cheney Cowles Memorial Museum
  (Eastern Washington Historical Society),
  W. 2316 First, 456-3931
Downtown Skywalk Shopping System
Manito Park/Japanese Gardens, 21st &
  Bernard, 625-6622
Riverfront Park, N. 507 Howard, 625-6600

INFORMATION SOURCES:

Spokane Regional Convention and
Visitors Bureau
  W. 926 Sprague, Suite 180
  Spokane, Washington 99204
  Business Info. (509) 624-1341;
  Visitor Info. (509) 747-3230;
  (800) 248-3230
Spokane Area Chamber of Commerce
  W. 1020 Riverside
  Spokane, Washington 99201
  (509) 624-1393

Tampa-St. Petersburg, Florida

Population: (*2,067,959—Tampa-St.
  Petersburg-Clearwater Metro Area)
  Tampa 280,150; St. Petersburg 238,629
  (1990C)
Altitude: (Tampa) 57 feet; (St. Petersburg)
  44 feet

SEA-TAC INTERNATIONAL AIRPORT, SEATTLE

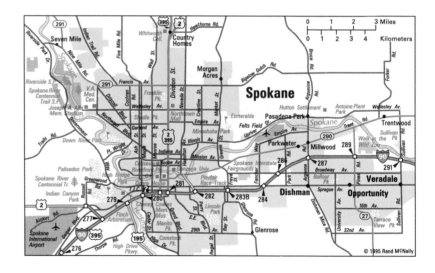

Average Temp.: Jan., 61°F.; July, 82°F.
Telephone Area Code: 813
Time and Weather: 888-9700
Time Zone: Eastern

AIRPORT TRANSPORTATION:

Five miles from Tampa International to
  downtown Tampa; 25 miles to
  downtown St. Petersburg. Ten miles
  from St. Petersburg/Clearwater Airport
  to St. Petersburg.
Taxicab and limousine service to Tampa
  and St. Petersburg; also bus service to
  Tampa.

SELECTED HOTELS: TAMPA

Best Western, 9331 Adamo Dr., 621-5511;
  FAX 626-6032
Guest Quarters, 3050 N. Rocky Point Dr.
  W., 888-8800; FAX 888-8743
Holiday Inn, 4500 W. Cypress St.,
  879-4800; FAX 873-1832
Holiday Inn Downtown, 111 W. Fortune
  St., 223-1351; FAX 221-2000
Howard Johnson, 4139 E. Busch Blvd.,
  988-9191; FAX 988-9195
Hyatt Regency—Tampa, 211 N. Tampa St.,
  225-1234; FAX 273-0234

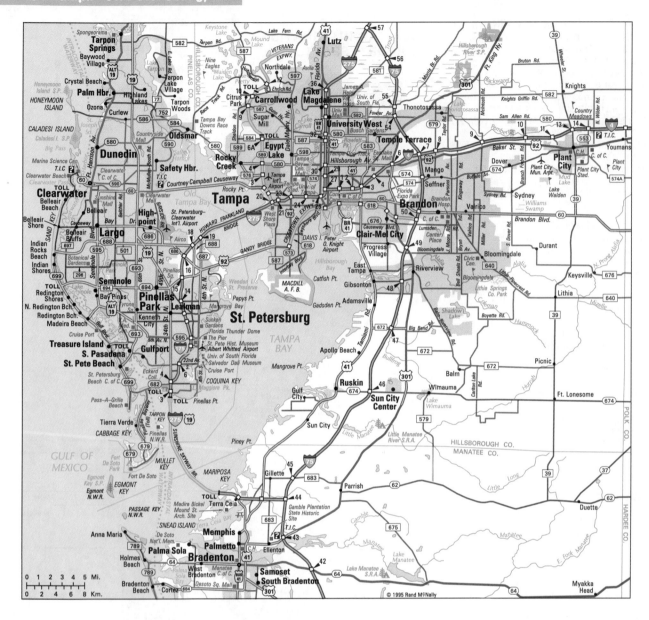

✈ Hyatt Regency—Westshore, 6200 Courtney Campbell Causeway, 874-1234; FAX 281-9168

Omni Hotel—Tampa, 700 N. Westshore Blvd., 289-8200; FAX 289-9166

✈ Radisson Bay Harbor Inn, 7700 Courtney Campbell Causeway, 281-8900; FAX 281-0189

Ramada Inn, 400 E. Bearss Ave., 961-1000; FAX 961-5704

Sheraton Grand, 4860 W. Kennedy Blvd., 286-4400; FAX 286-4053

✈ Tampa Airport Marriott, Tampa Int'l Airport, 879-5151; FAX 873-0945

Tampa Marriott Westshore, 1001 N. Westshore Blvd., 287-2555; FAX 289-5464

SELECTED RESTAURANTS: TAMPA

Signature Food: Cuban sand; paella; black bean soup

Armani's, in the Hyatt Regency—Westshore, 874-1234

Bern's Steak House, 1208 S. Howard Ave., 251-2421

C.K.'s Revolving Rooftop Restaurant, in the Tampa Airport Marriott, 879-5151

The Colonnade, 3401 Bayshore Blvd., 839-7558

The Columbia, 2117 E. 7th Ave., 248-4961

Cypress Room, in the Saddlebrook Resort, 25 miles north of Tampa on I-75, 973-1111

J. Fitzgerald's, in the Sheraton Grand, 286-4444

Old Spaghetti Warehouse, 1911 N. 13th St., in Ybor Square, 248-1720

Oyster Catchers, in the Hyatt Regency—Westshore, 874-1234

SELECTED ATTRACTIONS: TAMPA

Adventure Island, 4545 Bougainvillea Ave., 987-5660

Busch Gardens, Busch Blvd. & 40th St., 987-5082

Children's Museum of Tampa, 7550 North Blvd., 935-8441

The Florida Aquarium, 300 S. 13th St., (800) 44-TAMPA (due to open in March, 1995)

Henry B. Plant Museum, 401 W. Kennedy Blvd., 254-1891

Lowry Park Zoo, 7530 North Blvd., 935-8552

Museum of Science & Industry, 4801 E. Fowler Ave., 987-6300

Tampa Museum of Art, 601 Doyle Carlton Dr., 223-8130

Ybor City State Museum, 1818 Ninth Ave., 247-6323

Ybor Square, 8th Ave. & 13th St., 247-4497

INFORMATION SOURCES: TAMPA

Tampa/Hillsborough Convention & Visitors Association
111 Madison St., Suite 1010
Tampa, Florida 33602-4706
(813) 223-1111; (800) 44-TAMPA

Greater Tampa Chamber of Commerce
801 E. Kennedy
P.O. Box 420
Tampa, Florida 33601
(813) 228-7777

SELECTED HOTELS: ST. PETERSBURG

Days Inn, 2595 54th Ave. N., 522-3191;
FAX 527-5120
Days Inn Marina Beach Resort, 6800 34th
St. S., 867-1151; FAX 864-4494
The Don CeSar, 3400 Gulf Blvd., St. Pete
Beach, 360-1881; FAX 367-6952
Holiday Inn—Stadium, 4601 34th St. S.,
867-3131; FAX 867-2025
Lamp Post Inn, 1200 34th St. N.,
323-1221; FAX 323-1259
La Quinta, 4999 34th St. N., (US 19),
527-8421; FAX 527-8851
The St. Petersburg Hilton and Towers, 333
First St. S., 894-5000; FAX 894-7655
The Tradewinds, 5500 Gulf Blvd., St. Pete
Beach, 367-6461; FAX 367-7496

SELECTED RESTAURANTS: ST. PETERSBURG

Signature Food: grouper
Alessi's, at The Pier, 800 2nd Ave. NE,
894-1133
Billy's Tierra Verde Seafood, 1 Colony
Rd., Tierra Verde, 866-2115
The Columbia, at The Pier, 800 2nd Ave.
NE, 822-8000
The Lobster Pot, 17814 Gulf Blvd.,
Reddington Shores, 391-8592
Parker's Landing, in the Days Inn Marina
Beach Resort, 867-1151

SELECTED ATTRACTIONS: ST. PETERSBURG

Great Explorations (hands-on museum),
1120 4th St. S., 821-8992
Museum of Fine Arts, 255 Beach Dr. NE,
896-2667
P. Buckley Moss Gallery, 190 Fourth Ave.
NE, 894-2899
The Pier, on the waterfront at the end of
2nd Ave. NE, 821-6164
St. Petersburg Museum of History, 335
Second Ave. NE, 894-1052
Salvador Dali Museum, 1000 Third St. S.,
823-3767
Sunken Gardens, 1825 Fourth St. N.,
896-3186

INFORMATION SOURCES: ST. PETERSBURG

St. Petersburg/Clearwater Area
Convention & Visitors Bureau
ThunderDome
1 Stadium Dr., Suite A
St. Petersburg, Florida 33705
(813) 582-7892; visitor information:
(800) 345-6710 (in the U.S.) or
(800) 688-2235 (in Canada)
Suncoast Welcome Center
2001 Ulmerton Rd.
Clearwater, Florida 34622
(813) 573-1449
St. Petersburg Area Chamber of
Commerce
100 2nd Ave. N.
P.O. Box 1371
St. Petersburg, Florida 33731
Business Info. (813) 821-4069
Visitor Info. (813) 821-4715

Toronto, Ontario, Canada

Population: (*3,427,168) 599,217 (1986C)
Altitude: 275 ft.
Average Temp.: Jan., 23°F.; July, 69°F.
Telephone Area Code: 416
Time: (none) **Weather:** (905) 676-3066
Time Zone: Eastern

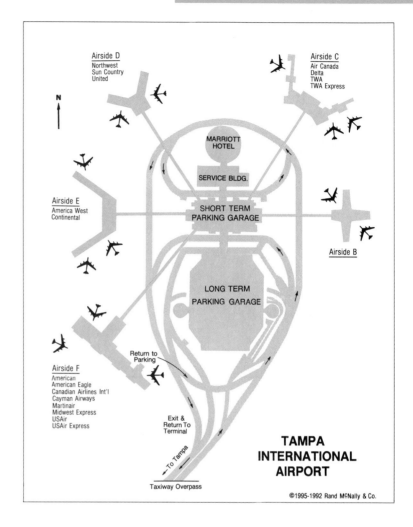

TAMPA INTERNATIONAL AIRPORT

Airside D — Northwest, Sun Country, United
Airside C — Air Canada, Delta, TWA, TWA Express
Airside E — America West, Continental
Airside B
Airside F — American, American Eagle, Canadian Airlines Int'l, Cayman Airways, Martinair, Midwest Express, USAir, USAir Express
MARRIOTT HOTEL
SERVICE BLDG.
SHORT TERM PARKING GARAGE
LONG TERM PARKING GARAGE
Return to Parking
Exit & Return To Terminal
To Tampa
Taxiway Overpass
©1995-1992 Rand McNally & Co.

LESTER B. PEARSON INTERNATIONAL AIRPORT, TORONTO

Highway 409
Highway 427
Airport Road
Dixon Rd.
PARKING GARAGE
SWISSÔTEL
Cargo Facilities
SHUTTLE PARKING
ADMINISTRATION
PARKING GARAGE

Terminal 3
Air New Zealand
American
British Airways
Canadian Airlines
Canadian Partners
Japan
KLM Royal Dutch
Lufthansa
Pemair
Qantas
United
United Express
Varig Brazilian

Terminal 1
Aerolines
Argentines
Aeromexico
Air Columbus
Air Europe Italy
Air Niagara
Air Transat
Air Ukraine
Alitalia
American Trans Air
Balkan Bulgarian
Bradley First
Air Services
Business Express
BWIA Int'l
Caledonian Airways
Canada 3000

PARKING GARAGE

Comair
Delta
Express One
Jetstream Int'l Martinair

Northwest
Olympic Airways
Piedmont
Royal
TAP Air Portugal
USAir

Terminal 2
Air Canada
Air France
Air India
Air Jamaica
Air Ontario
Czechoslovak
Cubana
El Al Israel
Finnair
Guyana Airways
Iberia
Korean
LOT Polish
Pakistan
Royal
Jordanian
Swissair
VIASA
Venezuelan

©1995-1992 Rand McNally & Co.

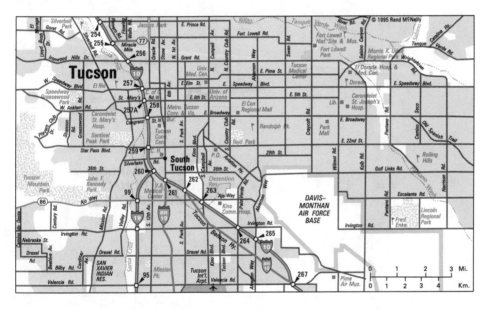

AIRPORT TRANSPORTATION:
Eighteen miles to downtown Toronto.
Taxicab and bus service.

SELECTED HOTELS:
Four Seasons Hotel, 21 Avenue Rd.,
  964-0411; FAX 964-2301
Four Seasons Inn-on-the-Park, 1100
  Eglinton Ave. E., 444-2561;
  FAX 446-3312
Hilton International, 145 Richmond St.
  W., 869-3456; FAX 869-3187
King Edward Hotel, 37 King St. E.,
  863-9700; FAX 367-5515
Park Plaza, 4 Avenue Rd., 924-5471;
  FAX 924-4933
Ramada—Downtown, 89 Chestnut St.,
  977-0707; FAX 977-1136

Royal York, 100 Front St. W., 368-2511;
  FAX 368-2884
Sheraton Centre, 123 Queen St. W.,
  361-1000; FAX 947-4854
✈ Sheraton Gateway, at the Lester B.
  Pearson International Airport,
  (905) 672-7000; fax (905) 672-7100
Sutton Place, 955 Bay St., 924-9221;
  FAX 924-1778
Westbury Howard Johnson, 475 Yonge St.,
  924-0611; FAX 924-5338

SELECTED RESTAURANTS:
Signature Food: seafood
The Acadian Room, in the Royal York
  Hotel, 368-2511
Chopstix and Rice, 1 Adelaide St. E.,
  363-7423

Fisherman's Wharf, 145 Adelaide St. W.,
  364-1346
Hy's Steakhouse, 73 Richmond St. W.,
  364-1792
Lighthouse, in the Westin Harbour Castle,
  1 Harbour Sq., 869-1600
Old Spaghetti Factory, 54 Esplanade,
  864-9761
The Old Mill, 21 Old Mill Rd., 236-2641
Splendido, 88 Harbord St., 929-7788

SELECTED ATTRACTIONS:
Art Gallery of Ontario, 317 Dundas St. W.,
  977-0414
Black Creek Pioneer Village, 1000 Murray
  Ross Pkwy., Downsview, 736-1733
Casa Loma, 1 Austin Terrace, 923-1171

CN Tower, 301 Front St. W., 360-8500
Harbourfront Centre, 235 Queen's Quay, 973-3000
Metro Zoo, Meadowvale Rd. N., Scarborough, 392-5900
Ontario Place, 955 Lakeshore Blvd. W., 314-9900
Ontario Science Centre, 770 Don Mills Rd., 429-4100
Royal Ontario Museum, 100 Queen's Park, 586-5549
SkyDome, 1 Blue Jays Way, 341-3663

INFORMATION SOURCES:
Metropolitan Toronto Convention & Visitors Association
207 Queens Quay W., Suite 590
Box 126
Toronto, Ontario M5J 1A7
(416) 203-2600; (800) 363-1990
Canadian Consulate General
1251 Avenue of the Americas
New York, New York 10020-1175
(212) 596-1600

Tucson, Arizona

Population: (*666,880) 405,390 (1990C)
Altitude: 2,386 feet
Average Temp.: Jan., 50°F.; July, 86°F.
Telephone Area Code: 520 † as of 3/19/95; 602 until 3/19/95
Time: 676-1676 **Weather:** 294-2522
Time Zone: Mountain Standard

AIRPORT TRANSPORTATION:
Ten miles to downtown Tucson.
Taxicab, limousine, van and bus service.

SELECTED HOTELS:
Arizona Inn, 2200 E. Elm St., 325-1541; FAX 881-5830
Aztec Inn, 102 N. Alvernon Way, 795-0330; FAX 326-2111
Best Western Executive Inn, 333 W. Drachman St., 791-7551; FAX 628-7803
✈ Best Western Inn at the Airport, 7060 S. Tucson Blvd., 746-0271; FAX 889-7391
Best Western Tucson Innsuites, 6201 N. Oracle Rd., 297-8111; FAX 297-2935
Country Suites, 7411 N. Oracle Rd., 575-9255; FAX 575-8671
Discovery Inn, 1010 S. Freeway, 622-5871; FAX 620-0097
Doubletree Hotel, 445 S. Alvernon Way, 881-4200; FAX 323-5225
Embassy Suites, 5335 E. Broadway, 745-2700; FAX 790-9232
✈ The Embassy Suites—Airport, 7051 S. Tucson Blvd., 573-0700; FAX 741-9645
Holiday Inn, 4550 S. Palo Verde Blvd., 746-1161; FAX 741-1170
Pueblo Inn, 350 S. Freeway, 622-6611; FAX 622-8143
Quality Inn University, 1601 N. Oracle Rd., 623-6666; FAX 884-7422
Radisson Suite Hotel, 6555 E. Speedway Blvd., 721-7100; FAX 721-1991
Ramada Inn Downtown, 475 N. Granada, 624-8341; FAX 623-8922
The Viscount Suite Hotel, 4855 E. Broadway Blvd., 745-6500; FAX 790-5114
Wayward Winds Lodge, 707 W. Miracle Mile, 791-7526; FAX 791-9502

SELECTED RESTAURANTS:
Signature Food: southwestern; Mexican
Charles, 6400 E. El Dorado Circle, 296-7173
Daniel's, in the St. Philip's Plaza, 4340 N. Campbell Ave., 742-3200
Gold Room, at the Westward Look Resort, 245 E. Ina Rd., 297-1151
Janos, 150 N. Main, 884-9426
Le Rendez-Vous, 3844 E. Fort Lowell, 323-7373
Palomino, 2959 N. Swan, 325-0413
Saguaro Corners, 3750 S. Old Spanish Trail, 886-5424
Scordato's Restaurant, 4405 W. Speedway, 792-3055
The Tack Room, 2800 N. Sabino Canyon Rd., 722-2800

SELECTED ATTRACTIONS:
Arizona-Sonora Desert Museum, 2021 N. Kinney Rd., 883-1380
Colossal Cave, Old Spanish Trail, 17 miles southeast of Tucson, 647-7275
Grace H. Flandrau Planetarium and Science Center, at the University of Arizona, Cherry Ave. & University Blvd., 621-4515
International Wildlife Museum, 4800 W. Gates Pass Rd., 624-4024 or 629-0100
Old Tucson (famous movie location and "Old West" town), 201 S. Kinney Rd., 883-0100
Pima Air & Space Museum, 6000 E. Valencia Rd., 574-9658
Sabino Canyon Tours, 749-2861
Tucson Botanical Gardens, 2150 N. Alvernon Way, 326-9255

INFORMATION SOURCES:
Metropolitan Tucson Convention and Visitors Bureau
130 S. Scott
Tucson, Arizona 85701
(602)† 624-1817; (800) 638-8350
Tucson Metropolitan Chamber of Commerce
465 W. St. Mary's Rd.
Tucson, Arizona 85702
(602)† 792-1212

Tulsa, Oklahoma

Population: (*708,954) 367,302 (1990C)
Altitude: 711 feet
Average Temp. Jan., 39°F.; July, 83°F.
Telephone Area Code: 918
Time: 477-1000 **Weather:** 743-3311
Time Zone: Central

AIRPORT TRANSPORTATION:
Eight miles to downtown Tulsa.
Taxicab and limousine service.

SELECTED HOTELS:
Adams Mark Hotel, 10 E. 2nd, 582-9000; FAX 560-2261
Best Western Trade Winds Central Inn, 3141 E. Skelly Dr., 749-5561; FAX 749-6312
Best Western Trade Winds East Inn, 3337 E. Skelly Dr., 743-7931; FAX 743-4308
Doubletree Downtown, 616 W. 7th St., 587-8000; FAX 587-1642
Econo Lodge, 11620 E. Skelly Dr., 437-9200; FAX 437-2935

Holiday Inn—Tulsa Holidome, 8181 E. Skelly Dr., 663-4541; FAX 665-7109
✈ Radisson—Tulsa Airport, 2201 N. 77 East Ave., 835-9911; FAX 838-2452
Ramada Hotel, 5000 E. Skelly Dr., 622-7000; FAX 664-9353
Southern Hills Marriott, 1902 E. 71st St., 493-7000; FAX 481-7147
Tulsa Marriott, 10918 E. 41st, 627-5000; FAX 627-4003

SELECTED RESTAURANTS:
Signature Food: chicken-fried steak; steak; barbeque
Bravos, in the Adams Mark Hotel, 582-9000
Charlie Mitchells, 81st St. & Lewis, 481-3250
Fountains Restaurant, 6540 S. Lewis Ave., 749-9915
Interurban Restaurant, 717 S. Houston, 585-3134
The Olive Garden, 7019 S. Memorial Dr., 254-0082
Rosie's Rib Joint, 8125 E. 49th St., 663-2610

SELECTED ATTRACTIONS:
Allen Ranch, 196th & S. Memorial, 366-3010
Bell's Amusement Park, 3901 E. 21st, 744-1991
Creek Nation Bingo, 81st St. & Riverside, 299-0100
Discoveryland, W. 41st St., 5 mi. west of OK Hwy. 97, 245-6552
Gilcrease Museum, 1400 Gilcrease Rd., 596-2700
Oral Roberts University campus, 7777 S. Lewis, 495-6807
Perryman Wrangler Ranch, 11524 S. Elwood, 299-2997
Philbrook Museum of Art, 27th Pl. & Peoria, 749-7941
Tulsa Rose Garden/Woodward Park, 23rd & Peoria, 746-5125
Tulsa Zoo, in Mohawk Park, 5701 E. 36th St. N., 669-6200

INFORMATION SOURCE:
Convention & Visitors Bureau, Metropolitan Tulsa Chamber of Commerce
616 S. Boston Ave., Suite 100
Tulsa, Oklahoma 74119
(918) 585-1201

Vancouver, British Columbia, Canada

Population: (*1,380,729) 414,281 (1986C)
Altitude: Sea level to 40 ft.
Average Temp.: Jan., 36.5°F.; July, 72°F.
Telephone Area Code: 604
Time: (none) **Weather:** 664-9010
Time Zone: Pacific

AIRPORT TRANSPORTATION:
Eleven miles to downtown Vancouver.
Taxicab and bus service.

SELECTED HOTELS:
Coast Plaza Hotel, 1733 Comox St., 688-7711; FAX 688-5934

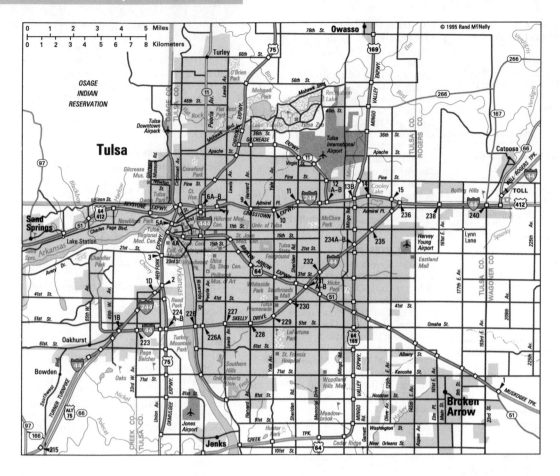

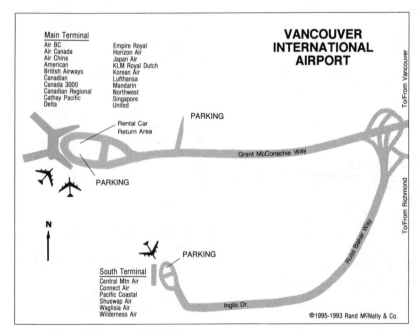

SELECTED RESTAURANTS:

Signature Food: seafood; salmon

The Cannery, 2205 Commissioner St., 254-9606

Captain's Palace, 309 Belleville St., 388-9191

Cavalier Room, in the Hotel Georgia, 682-5566

Hy's Encore, 637 Hornby St., 683-7671

Monk McQueens Fresh Seafood & Oyster Bar, 601 Stamps Landing, 877-1351

Mulvaney's Restaurant, 1535 Johnston St., on Granville Island, 685-6571

1066 Restaurant, 1066 W. Hastings St., 689-1066

Umberto al Porto, 321 Water St., 683-8376

SELECTED ATTRACTIONS:

Chinatown (includes the Dr. Sun Yat-Sen Classical Chinese Garden), bounded by Hastings, Georgia, Carrall & Princess sts.

Gastown (1880s district; includes the Gastown Steam Clock), bounded by Water, Seymour, Cordova, & Columbia sts.

Granville Island, underneath the south end of Granville St. bridge, 666-5784

Grouse Mountain (skiing, all-year recreation), 6400 Nancy Greene Way, North Vancouver, 984-0661

Museum of Anthropology, 6393 NW Marine Dr., University of British Columbia, 822-3825

Planetarium, 1100 Chestnut St., 736-3656

Robsonstrasse (European district), Robson St. between Howe & Broughton sts.

✈ Delta Vancouver Airport Hotel, 3500 Cessna Dr., 278-1241; FAX 276-1975

Hotel Georgia, 801 W. Georgia St., 682-5566; FAX 682-8192

Hotel Vancouver, 900 W. Georgia St., 684-3131; FAX 662-1929

Hyatt Regency, 655 Burrard St., 687-6543; FAX 643-5812

Pan Pacific Hotel, 999 Canada Pl., 662-8111; FAX 685-8690

Waterfront Centre Hotel, 900 Canada Place Way, 691-1991; FAX 691-1838

Westin Bayshore, 1601 W. Georgia St., 682-3377; FAX 687-3102

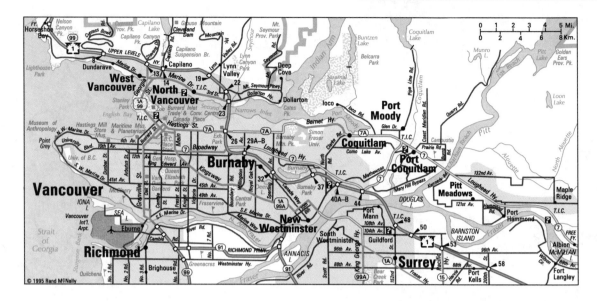

Science World, Quebec St. & Terminal
Ave., 268-6363

Vancouver Maritime Museum/St. Roch
National Historic Site, 1905 Ogden
Ave., 257-8300

Vancouver Public Aquarium, located in
Stanley Park (which includes numerous
other major attractions), 682-1118

INFORMATION SOURCES:

Greater Vancouver Convention & Visitors
Bureau
Suite 210, Waterfront Centre
200 Burrard St.
Vancouver, British Columbia V6C 3L6
(604) 682-2222

Vancouver Tourist Info Centre
Plaza Level, Waterfront Centre
200 Burrard St.
Vancouver, British Columbia V6C 3L6
(604) 683-2000

Canadian Consulate General
1251 Avenue of the Americas
New York, New York 10020-1175
(212) 596-1600

Washington, D.C.

Population: (*3,923,574) 606,900 (1990C)
Altitude: 1 to 410 feet
Average Temp.: Jan., 36°F.; July, 79°F.
Telephone Area Code: 202
Time: 844-2525   **Weather:** 936-1212
Time Zone: Eastern

AIRPORT TRANSPORTATION:

Three miles from Washington National
Airport to downtown Washington; 26
miles from Dulles International Airport
to downtown Washington; 37 miles
from Baltimore-Washington
International Airport to downtown
Washington.

Taxicab, bus, and rapid service between
National Airport and downtown
Washington; taxicab and bus service to
Dulles International Airport; taxicab,
bus, and train service to Baltimore-
Washington International Airport.

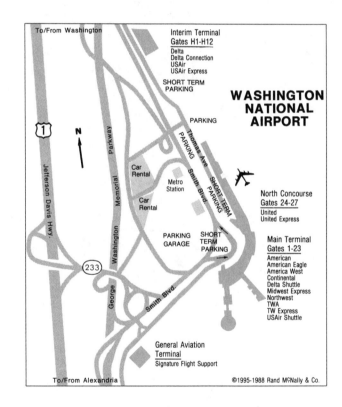

SELECTED HOTELS:

Capital Hilton, 16th & K sts. NW,
393-1000; FAX 639-5784

Four Seasons Hotel, 2800 Pennsylvania
Ave. NW, 342-0444; FAX 944-2076

Georgetown Inn, 1310 Wisconsin Ave.
NW, 333-8900; FAX 625-1744

Grand Hyatt Washington, 1000 H St. NW,
582-1234; FAX 637-4781

The Hay-Adams Hotel, 800 16th St. NW,
638-6600; FAX 638-2716

Loew's L'Enfant Plaza, 480 L'Enfant Plaza
SW, 484-1000; FAX 646-4456

Madison Hotel, 1177 15th St. NW,
862-1600; FAX 785-1255

One Washington Circle Hotel, 23rd St. &
New Hampshire, 872-1680;
FAX 887-4989

The Ritz-Carlton, 2100 Massachusetts
Ave. NW, 293-2100; FAX 293-0641

Sheraton-Carlton, 923 16th St. NW,
638-2626; FAX 638-4231

Washington Hilton and Towers, 1919
Connecticut Ave. NW, 483-3000;
FAX 265-8221

The Watergate Hotel, 2650 Virginia Ave.
NW, 965-2300; FAX 337-7915

SELECTED RESTAURANTS:

Signature Food: Senate bean soup
Cantina Romana Ristorante, 3251 Prospect
St., 337-5130

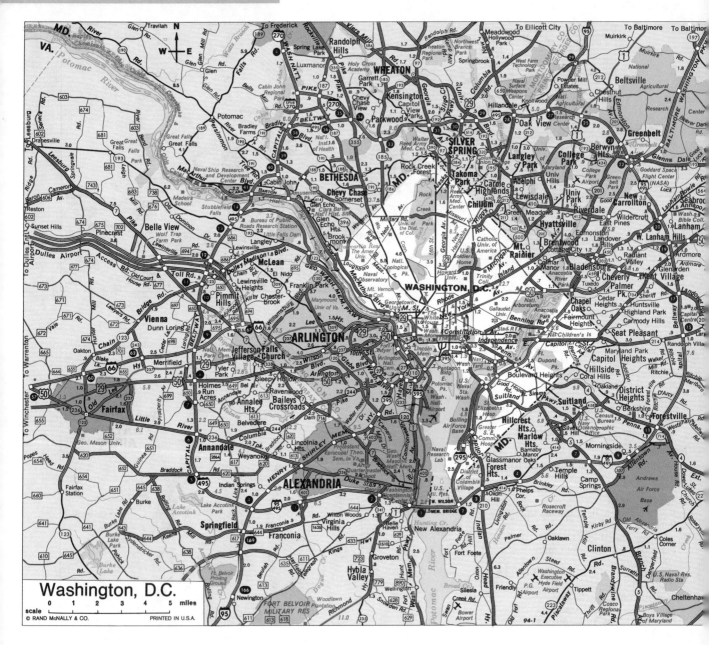

Washington, D.C.

scale
0 1 2 3 4 5 miles
© RAND McNALLY & CO.   PRINTED IN U.S.A.

Jockey Club Restaurant, in The Ritz-
  Carlton, 293-2100
Le Lion D'or, 1150 Connecticut Ave. NW,
  18th St. entrance, 296-7972
Maison Blanche Restaurant, 1725 F St.
  NW, 842-0070
Montpelier, in the Madison Hotel,
  862-1600
1789 Restaurant, 1226 36th St. NW,
  965-1789
Seasons, in the Four Seasons Hotel,
  342-0444
Trader Vic's, in the Capital Hilton,
  347-7100

SELECTED ATTRACTIONS:

The Jefferson Memorial, Tidal Basin
  (South Bank), East Potomac Park,
  426-6841
The Lincoln Memorial, 23rd St. NW &
  Constitution Ave., 426-6841

National Air & Space Museum, 6th St. &
  Indepedence Ave. SW, 357-2700
National Gallery of Art, 4th St. &
  Constitution Ave. NW, 737-4215
National Museum of American History,
  14th St. & Constitution Ave. NW,
  357-2700
National Museum of Natural History, 10th
  St. & Constitution Ave. NW, 357-2700
United States Capitol, National Mall (East
  End), 225-6827
Vietnam Veterans Memorial, Constitution
  Ave. & 22nd St. NW, 426-6841
The Washington Monument, National
  Mall at 15th St. NW & Constitution
  Ave., 426-6841
The White House, 1600 Pennsylvania Ave.
  NW, 456-7041

INFORMATION SOURCE:

Washington, D.C. Convention and Visitors
Association

1212 New York Ave. NW Washington,
D.C. 20005
(202) 789-7000

Wichita, Kansas

Population: (*485,270) 304,011 (1990C)
Altitude: 1,305 feet
Average Temp.: Jan., 31°F.; July, 81°F.
Telephone Area Code: 316
Time: 686-6621 **Weather:** 942-3102
Time Zone: Central

AIRPORT TRANSPORTATION:

Six miles to downtown Wichita.
Taxicab, bus and limousine service.

SELECTED HOTELS:

Best Western Tudor Inn, 9100 E. Kellogg,
  685-0371; FAX 685-4668
Century Plaza Hotel, 250 W. Douglas,
  264-1181; FAX 263-0135

✈ Hilton Airport and Executive
Conference Center, 2098 Airport Rd.,
945-5272; FAX 945-7620
✈ Holiday Inn—Airport, 5500 W. Kellogg,
943-2181; FAX 943-6587
Holiday Inn—East, 7335 E. Kellogg,
685-1281; FAX 685-1281
Ramada Hotel, 400 W. Douglas, 262-5000;
FAX 262-5000
Ramada Inn, 5805 W. Kellogg, 942-7911;
FAX 942-0854
Wichita East Hotel, 549 S. Rock Rd.,
686-7131; FAX 686-0018
Wichita Marriott, 9100 Corporate Hills
Dr., 651-0333; FAX 651-0990
Wichita Royale, 125 N. Market St.,
263-2101; FAX 263-0135

SELECTED RESTAURANTS:
Signature Food: steak
Amarillo Grill, 5730 E. Central, 685-5050
The American Cafe, in the Wichita East
Hotel, 686-7131
Chateaubriand, 9100 E. Kellogg, 682-4212
The Olive Tree, 2949 N. Rock Rd.,
636-1100
The Portobello Rd., 504 S. Bluff, 684-5591
Scotch & Sirloin, 3941 E. Kellogg,
685-8701

SELECTED ATTRACTIONS:
Botanica—The Wichita Gardens, 701 N.
Amidon, 264-0448
Joyland Amusement Park, 2801 S.
Hillside, 684-0179
Mid-America All-Indian Center &
Museum, 650 N. Seneca, 262-5221
Old Cowtown Museum, 1871 Sim Park
Dr., 264-0671
The Sedgwick County Zoo & Botanical
Gardens, Exit 10 on I-235, 942-2212
or -2213
Wichita Art Museum, 619 N. Stackman
Dr., 268-4921
Wichita Greyhound Park, 1500 E. 77th St.
N., 755-4000

Wichita Omnisphere & Science Center,
220 S. Main, 264-6178 or -3174
Wichita-Sedgwick County Historical
Museum, 204 S. Main, 265-9314

INFORMATION SOURCE:
Wichita Convention & Visitors Bureau
100 S. Main, Suite 100
Wichita, Kansas 67202
(316) 265-2800; (800) 288-9424

Winnipeg, Manitoba, Canada

Population: (*625,304) 594,551 (1986C)
Altitude: 915 feet
Average Temp.: Jan., 3°F.; July, 67°F.
Telephone Area Code: 204
Time: 783-2119 **Weather:** 983-2050
Time Zone: Central

AIRPORT TRANSPORTATION:
Eight miles to downtown Winnipeg.
Taxicab, bus, airport limousine, and hotel
shuttle service.

SELECTED HOTELS:
The Charter House Hotel, 330 York Ave.,
942-0101; FAX 956-0665
The Days Inn Marlborough Inn, 331 Smith
St., 942-6411; FAX 942-2017
The Delta Winnipeg, 288 Portage Ave.,
956-0410; FAX 947-1129
Holiday Inn—Crowne Plaza, 350 St. Mary
Ave., 942-0551; FAX 943-8702
Holiday Inn South—Conference Centre,
1330 Pembina Hwy., 452-4747;
FAX 284-2751
Hotel Fort Garry, 222 Broadway,
942-8251; FAX 956-2351
✈ International Inn—Best Western, 1808
Wellington Ave., 786-4801;
FAX 786-1329
Place Louis Riel All-Suite Hotel, 190
Smith St., 947-6961; FAX 947-3029
✈ Radisson Suite Hotel, 1800 Wellington
Ave., 783-1700; FAX 786-6588

Sheraton Winnipeg, 161 Donald St.,
942-5300; FAX 943-7975
Travelodge Hotel, 360 Colony St.,
786-7011; FAX 772-1443
The Westin Hotel, Portage & Main,
957-1350; FAX 956-1791

SELECTED RESTAURANTS:
Signature Food: gold eye (smoked fish)
Café Jardin, 340 Provencher Blvd., in the
Cultural Center, 233-9515
Fork & Cork Bistro, 218 Sherbrook,
783-5754
Hy's Steak Loft, 216 Kennedy St.,
942-1000
The Olive Garden, 1544 Portage Ave.,
774-9725
The Round Table, 800 Pembina Hwy.,
453-3631
La Taverna, 771 St. Mary's Rd., 257-2116
La Vieille Gare, 630 Des Meurons,
237-5015

SELECTED ATTRACTIONS:
Assiniboine Park & Zoo, Corydon Ave. &
Shaftsbury Rd., 986-6921
The Forks Market, downtown Winnipeg,
Main St. & The Forks Rd., 942-6302;
The Forks Info. Line 957-7618
IMAX Theatre, 3rd Floor, Portage Place,
393 Portage Ave., 956-4629 or 956-2400
Lower Fort Garry National Historic Site,
20 miles north of Winnipeg on MB
Hwy. 9, 983-3600
Manitoba Children's Museum, in the
Kinsmen Building, 45 Forks Market Rd.,
956-KIDS or 956-1888
Manitoba Museum of Man & Nature, 190
Rupert Ave., 956-2830
Paddlewheel/River Rouge Tours, Water
Ave. at Gilroy St., 947-6843
Royal Canadian Mint, 520 Lagimodiere
Blvd., 257-3359
Touch the Universe/Manitoba
Planetarium, 190 Rupert Ave., 943-3142
or 956-2830

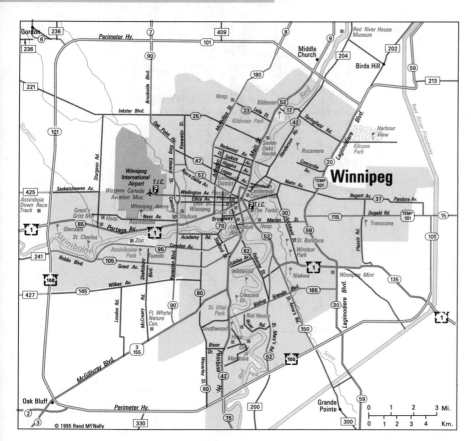

© 1995 Rand McNally

Winston-Salem

© 1995 Rand McNally

Winnipeg Art Gallery, 300 Memorial
Blvd., 786-6641

INFORMATION SOURCES:
Tourism Winnipeg
320-25 Forks Market Rd.
Winnipeg, Manitoba, Canada R3C 4S8
(204) 943-1970; (800) 665-0204
Canadian Consulate General
1251 Avenue of the Americas
New York, New York 10020-1175
(212) 596-1600

Winston-Salem, North Carolina

Population: (*942,091) (Greensboro-
Winston Salem-High Point Metro Area)
143,485 (1990C)
Altitude: 912 feet
Average Temp.: Jan., 41°F.; July, 78°F.
Telephone Area Code: 910
Time: 773-0000   **Weather:** 761-8411
Time Zone: Eastern

AIRPORT TRANSPORTATION:
Nineteen miles to downtown Winston-
Salem.
Taxicab, bus and limousine service.

SELECTED HOTELS:
The Adams Mark Winston Plaza, 425 N.
Cherry St., 725-3500; FAX 722-6475
Hampton Inn, 1990 Hampton Inn Ct.,
760-1660; FAX 768-9168
Hawthorne Inn & Conference Center, 420
High St., 777-3000; FAX 777-3282
Marque Winston-Salem, 460 N. Cherry St.,
725-1234; FAX 722-9182
Ramada Inn North, 531 Akron Dr.,
767-8240; FAX 661-9513
Salem Inn, 127 S. Cherry St., 725-8561;
FAX 725-2318
Tanglewood Park Lodge, U.S. 158, West
Clemmons, 766-0591; FAX 766-8723

SELECTED RESTAURANTS:
Signature Food: barbeque
The Dirtwater Fox, in the Marque
Winston-Salem, 725-1234
The Maze, 120 Reynolda Village,
748-0269

Old Salem Tavern, 736 S. Main St.,
748-8585
Ryan's, 719 Coliseum Dr., 724-6132
Winston Bar & Grill, in the Stouffer
Winston Plaza, 725-3500

SELECTED ATTRACTIONS:
Historic Bethabara Park, 2147 Bethabara
Rd., 924-8191
Museum of Early Southern Decorative
Arts, 924 S. Main St., 721-7360
Old Salem, S. Main St. & Academy,
721-7300
Piedmont Craftsmen, 1204 Reynolda Rd.,
725-1516
R.J. Reynolds Tobacco Company, Whitaker
Park/Reynolds Blvd.,
741-5718
Reynolda House Museum of American
Art, 1900 Reynolda Rd., 725-5325
SciWorks, I-52 at Hanes Mill Rd.,
767-6730
Southeastern Center for Contemporary
Art, 750 Marguerite Dr., 725-1904
Stroh Brewery, 4791 Schlitz Ave.,
788-6710
Tanglewood Park, U.S. 158, West
Clemmons, 766-0591

INFORMATION SOURCES:
Winston-Salem Convention and Visitors
Bureau
601 W. 4th St.
P.O. Box 1408
Winston-Salem, North Carolina 27102
(910) 777-3787; (800) 331-7018
Winston-Salem Visitor Center
601 N. Cherry St., Suite 100
Winston-Salem, North Carolina 27101
(910) 777-3796; (800) 331-7018
Greater Winston-Salem Chamber of
Commerce
601 W. 4th
P.O. Box 1408
Winston-Salem, North Carolina 27102
(910) 777-3787

Maps: United States, Canada, and Mexico

Introduction

One of the most important parts of any driving trip is knowing the best way to get to your destination. Using the detailed, accurate maps found in this atlas, you can easily plan your trip to avoid the all-too-common frustrations of wrong turns and mis-judged mileage.

The maps in this section include a United States map; maps of the 50 states; a Canada map; and maps of Alberta, Atlantic provinces, British Columbia, Manitoba, Ontario, Qué-bec, and Saskatchewan, and Mexico.

Among the many features to be found on the maps, the following are of special use in trip planning:

Detailed Transportation Routes. The atlas differentiates between toll roads, freeways, routes under construction, principal highways and various sec-ondary roads.

Place Indexes. An index to cities and towns is conveniently located with each state and province map, as well

as the map of Mexico. Capitals are noted at the head of each index; land area and total population are also provided.

The United States map includes indexes to National Parks and National Monuments.

Distance Guides. Mileage can be computed by using either the map scale included with each map, or by referring to the accumulated mileage numbers that appear between the red pointers located along major highways.

Trip Planning

A key element to a successful trip is determining the exact location of your destination and the most direct way to get there.

Locating the Destination. To locate your destination, look in the alphabet-ical listing of cities and towns accom-panying the appropriate map. Each town or city has a map reference key, composed of a letter and a number. To

find a town on the map with a refer-ence key of B-5, for example, look down the side of the map for the letter B, and draw an imaginary line across the map until it intersects with an imaginary line drawn up or down from the number 5. Your destination will be within an inch-and-a-quarter square surrounding this point.

Marking the Route. Once you have located your destination, determine the most direct route by considering both distance and type of road. A secondary road may seem to provide a shortcut, while in actuality it takes more time because of lower speed limits, unimproved surface condi-tions, etc.

If time is a consideration, toll roads and freeways most often provide the quickest ride and should be used wherever possible.

To ease your map reading while driving, mark your chosen route with a see-through, felt-tip marker in a color that won't conflict with the lines, names, or symbols shown on the map.

Legend

Roads and Related Symbols

	Free Limited Access Highways:
	Under Construction
	Toll Limited-Access Highways:
	Under Construction
	Other Four-Lane Divided Highways
	Principal Highways
	Other Through Highways
	Other Roads (conditions vary — local inquiry suggested)
	Unpaved Roads (conditions vary — local inquiry suggested)
90 190 80/90	Interstate Highways
ALT 17 183 18	U.S. Highways
8 18 14/83	State and Provincial Highways
4 43 147	Secondary State, Provincial, and County Highways
N NM	County Trunk Highways
1 20	Trans-Canada Highway, Canadian Autoroutes
5	Mexican and Central American Highways

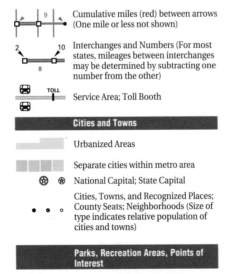

Cumulative miles (red) between arrows (One mile or less not shown)

Interchanges and Numbers (For most states, mileages between interchanges may be determined by subtracting one number from the other)

Service Area; Toll Booth

Cities and Towns

Urbanized Areas

Separate cities within metro area

National Capital; State Capital

Cities, Towns, and Recognized Places; County Seats; Neighborhoods (Size of type indicates relative population of cities and towns)

Parks, Recreation Areas, Points of Interest

U.S. and Canadian National, State and Provincial Parks; Recreation Areas:

with camping facilities

without camping facilities

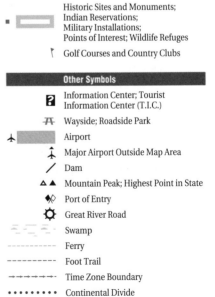

National Forests and Grasslands; City Parks

Historic Sites and Monuments; Indian Reservations; Military Installations; Points of Interest; Wildlife Refuges

Golf Courses and Country Clubs

Other Symbols

Information Center; Tourist Information Center (T.I.C.)

Wayside; Roadside Park

Airport

Major Airport Outside Map Area

Dam

Mountain Peak; Highest Point in State

Port of Entry

Great River Road

Swamp

Ferry

Foot Trail

Time Zone Boundary

Continental Divide

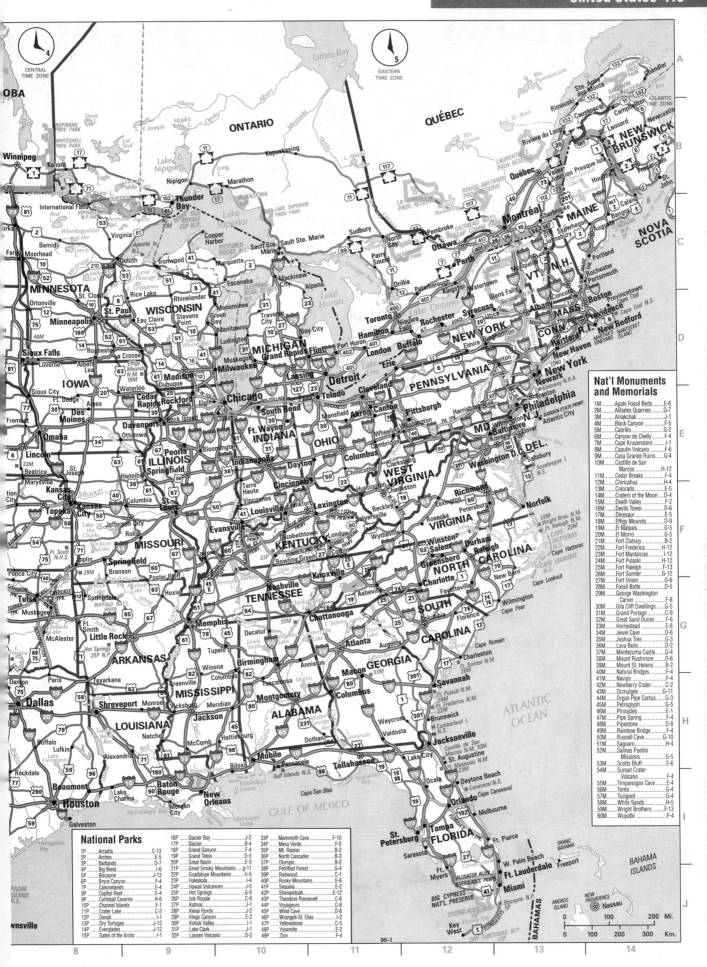

4 CENTRAL TIME ZONE

5 EASTERN TIME ZONE

ONTARIO

QUÉBEC

NEW BRUNSWICK

NOVA SCOTIA

MAINE

MINNESOTA

WISCONSIN

MICHIGAN

NEW YORK

VT. N.H.

MASS.

CONN. R.I.

IOWA

ILLINOIS

INDIANA

OHIO

PENNSYLVANIA

N.J.

MD. DEL.

MISSOURI

KENTUCKY

WEST VIRGINIA

VIRGINIA

ARKANSAS

TENNESSEE

NORTH CAROLINA

SOUTH CAROLINA

MISSISSIPPI

ALABAMA

GEORGIA

LOUISIANA

ATLANTIC OCEAN

FLORIDA

GULF OF MEXICO

BAHAMA ISLANDS

Nat'l Monuments and Memorials

1M	Agate Fossil Beds	E-6
2M	Alibates Quarries	G-7
3M	Aniakchak	
4M	Black Canyon	G-2
5M	Cabrillo	G-2
6M	Canyon de Chelly	F-4
7M	Cape Krusenstern	I-1
8M	Capulin Volcano	F-6
9M	Casa Grande Ruins	G-4
10M	Castillo de San Marcos	H-12
11M	Cedar Breaks	F-4
12M	Chiricahua	H-4
13M	Colorado	E-5
14M	Craters of the Moon	D-4
15M	Death Valley	F-2
16M	Devils Tower	D-6
17M	Dinosaur	D-5
18M	Effigy Mounds	D-9
19M	El Malpais	G-5
20M	El Morro	G-5
21M	Fort Clatsop	B-2
22M	Fort Frederica	H-12
23M	Fort Mantanzas	I-12
24M	Fort Pulaski	H-12
25M	Fort Raleigh	F-13
26M	Fort Sumter	G-12
27M	Fort Union	D-5
28M	Fossil Butte	D-5
29M	George Washington Carver	F-8
30M	Gila Cliff Dwellings	G-5
31M	Grand Portage	C-9
32M	Great Sand Dunes	F-6
33M	Homestead	E-6
34M	Jewel Cave	D-6
35M	Joshua Tree	G-3
36M	Lava Beds	D-2
37M	Montezuma Castle	G-4
38M	Mount Rushmore	D-6
39M	Mount St. Helens	B-2
40M	Natural Bridges	F-4
41M	Navajo	F-4
42M	Newberry Crater	C-2
43M	Ocmulgee	G-11
44M	Organ Pipe Cactus	G-3
45M	Petroglyph	G-5
46M	Pinnacles	F-1
47M	Pipe Spring	F-4
48M	Pipestone	D-8
49M	Rainbow Bridge	F-4
50M	Russell Cave	G-10
51M	Saguaro	G-3
52M	Salinas Pueblo Missions	G-5
53M	Scotts Bluff	E-6
54M	Sunset Crater Volcano	F-4
55M	Timpanogos Cave	E-4
56M	Tonto	G-4
57M	Tuzigoot	G-4
58M	White Sands	H-5
59M	Wright Brothers	F-13
60M	Wupatki	F-4

National Parks

1P	Arcadia	C-13
2P	Arches	E-5
3P	Badlands	D-7
4P	Big Bend	I-6
5P	Biscayne	J-13
6P	Bryce Canyon	F-4
7P	Canyonlands	E-4
8P	Capitol Reef	H-6
9P	Carlsbad Caverns	H-6
10P	Channel Islands	F-1
11P	Crater Lake	C-2
12P	Denali	I-1
13P	Dry Tortugas	J-12
14P	Everglades	J-12
15P	Gates of the Arctic	I-1
16P	Glacier Bay	J-2
17P	Glacier	B-4
18P	Grand Canyon	F-4
19P	Grand Teton	D-5
20P	Great Basin	E-3
21P	Great Smoky Mountains	g-11
22P	Guadalupe Mountains	I-5
23P	Haleakala	I-4
24P	Hawaii Volcanoes	J-5
25P	Hot Springs	G-8
26P	Isle Royale	C-9
27P	Katmai	J-2
28P	Kenai Fjords	I-2
29P	Kings Canyon	E-2
30P	Kobuk Valley	I-1
31P	Lake Clark	J-2
32P	Lassen Volcanic	D-2
33P	Mammoth Cave	F-10
34P	Mesa Verde	F-5
35P	Mt. Rainier	B-3
36P	North Cascades	B-3
37P	Olympic	B-2
38P	Petrified Forest	G-4
39P	Redwood	C-1
40P	Rocky Mountains	E-6
41P	Sequoia	E-2
42P	Shenandoah	E-12
43P	Theodore Roosevelt	C-6
44P	Voyageurs	C-8
45P	Wind Cave	D-6
46P	Wrangell-St. Elias	I-2
47P	Yellowstone	C-5
48P	Yosemite	E-2
49P	Zion	F-4

0 100 200 Mi.
0 100 200 300 Km.

95-1

Alabama

Population: 4,062,608
(1990 Census)
Land Area: 50,766 sq. mi.
Capital: Montgomery

Cities and Towns

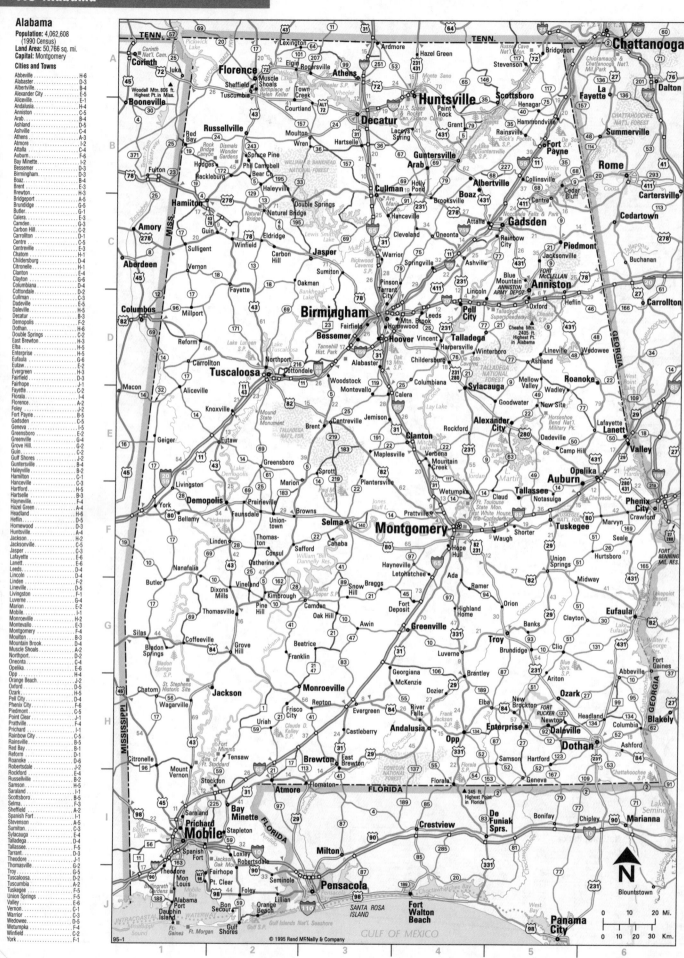

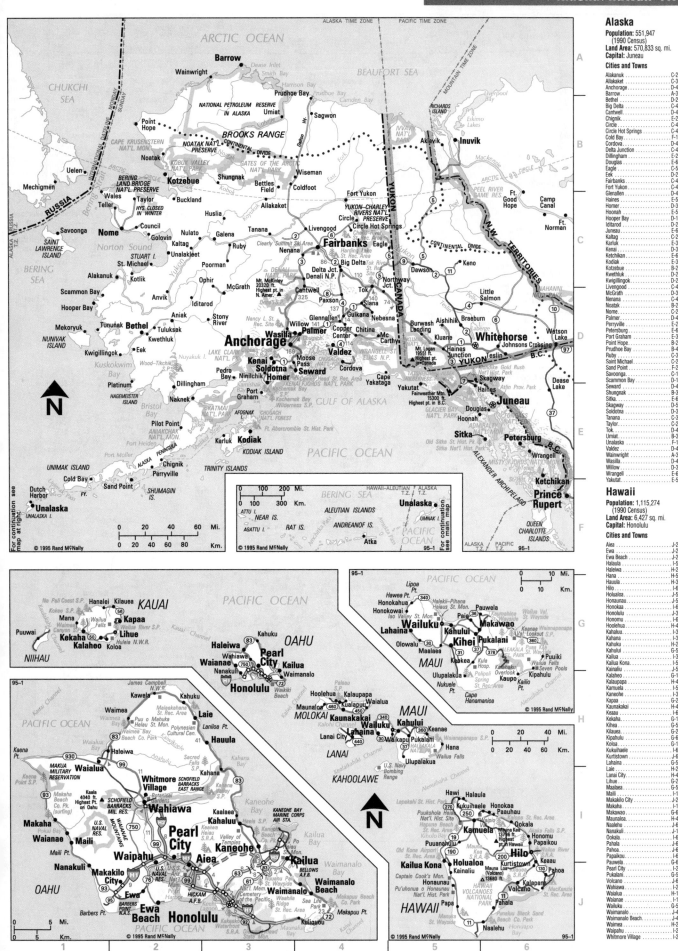

Alaska

Population: 551,947
(1990 Census)
Land Area: 570,833 sq. mi.
Capital: Juneau

Cities and Towns

Hawaii

Population: 1,115,274
(1990 Census)
Land Area: 6,427 sq. mi.
Capital: Honolulu

Cities and Towns

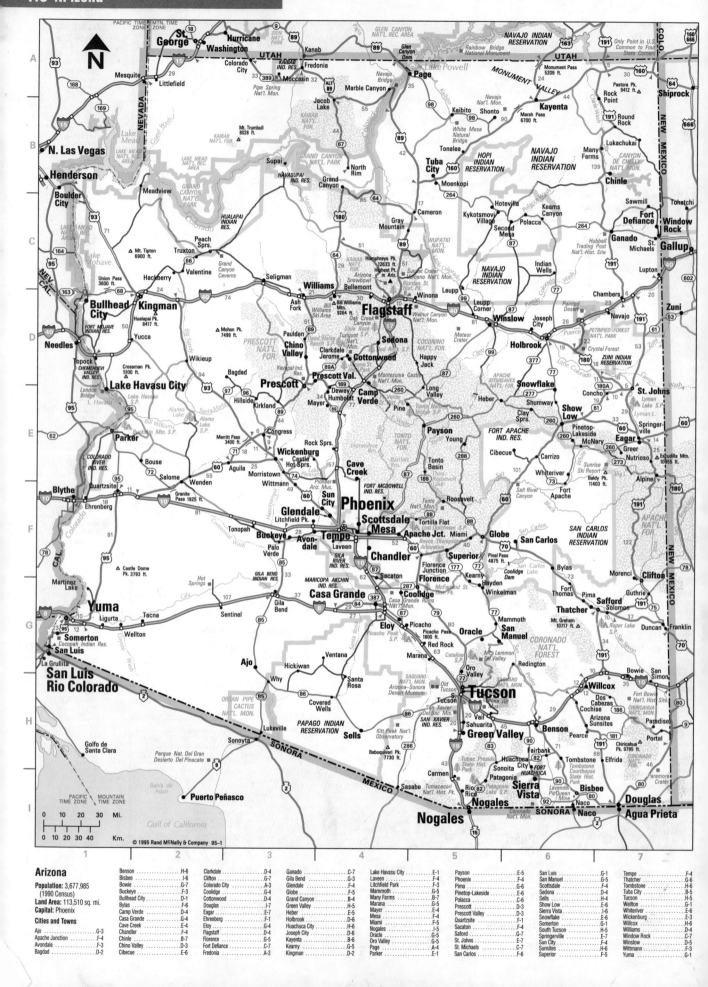

Arizona

Population: 3,677,985
(1990 Census)
Land Area: 113,510 sq. mi.
Capital: Phoenix

Cities and Towns

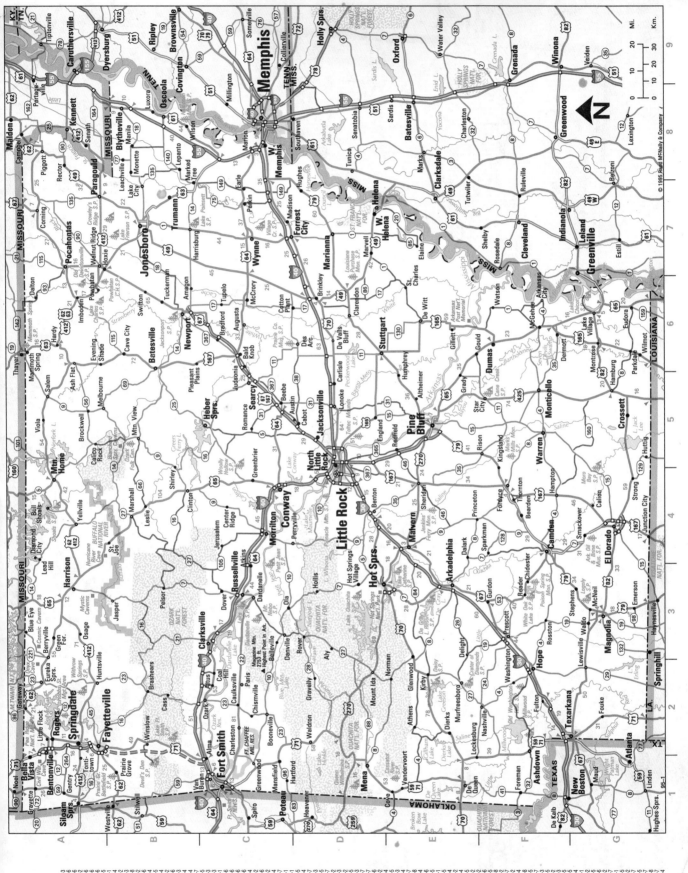

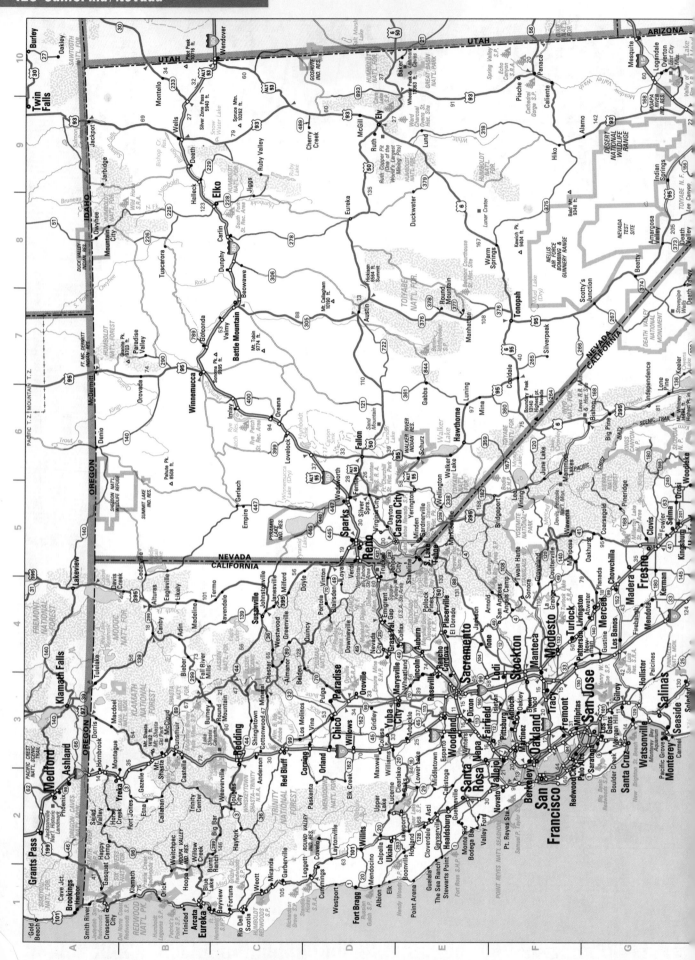

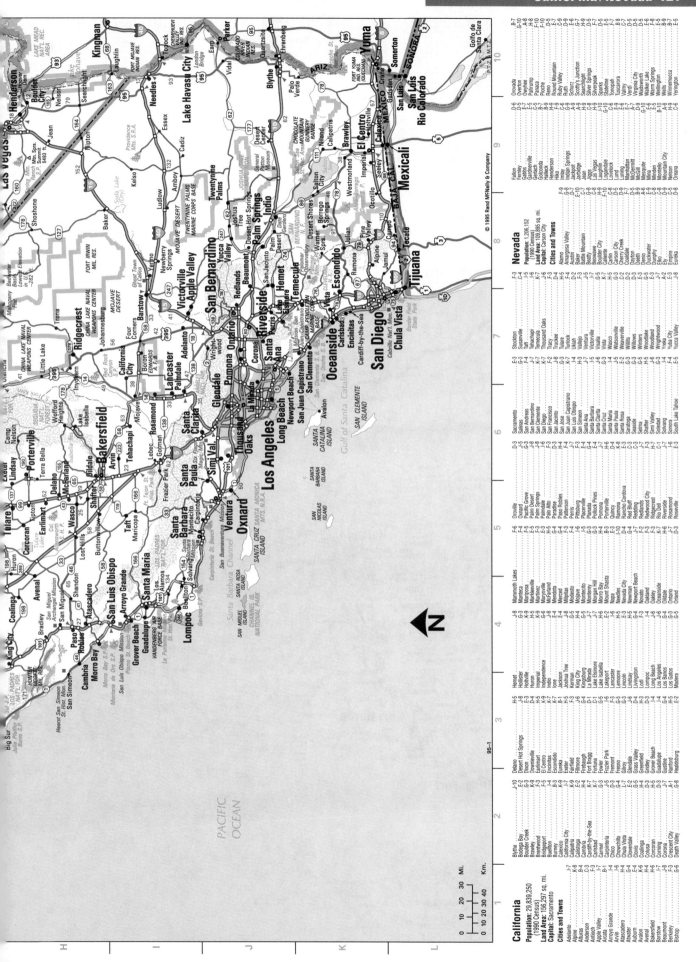

© 1995 Rand McNally & Company

California

Population: 29,839,250
(1990 Census)
Land Area: 156,297 sq. mi.
Capital: Sacramento

Cities and Towns

Adelanto	J-7
Alpine	K-8
Alturas	C-3
Anaheim	J-6
Anderson	D-2
Antioch	F-3
Apple Valley	J-7
Arcata	C-1
Arroyo Grande	H-4
Arvin	H-5
Atascadero	H-4
Atwater	F-4
Auburn	E-3
Avalon	K-6
Avenal	G-4
Bakersfield	H-5
Barstow	J-7
Beaumont	J-7
Berkeley	F-3
Bishop	F-5

Blythe	J-9
Bodega Bay	E-2
Boulder Creek	F-3
Brawley	K-9
Brentwood	F-3
Bridgeport	E-5
Buellton	H-4
Burney	C-2
Calexico	K-9
California City	H-6
Calipatria	K-9
Calistoga	E-2
Cambria	H-3
Cardiff-by-the-Sea	K-7
Carlsbad	K-7
Carmel	G-3
Carpinteria	H-5
Chico	D-3
Chowchilla	F-4
Chula Vista	K-7
Coalinga	G-4
Colusa	D-3
Corcoran	G-5
Corning	D-2
Corona	J-7
Crescent City	B-1
Death Valley	G-6

Delano	G-5
Desert Hot Springs	J-8
Dixon	E-3
Downieville	D-4
Earlimart	G-5
El Centro	K-9
Encinitas	K-7
Escondido	K-7
Eureka	C-1
Exeter	G-5
Fairfield	E-3
Fillmore	H-5
Fort Bragg	D-1
Fortuna	C-1
Fowler	G-4
Fremont	F-3
Fresno	G-4
Gilroy	F-3
Glendale	J-6
Grass Valley	D-3
Greenfield	G-3
Gridley	D-3
Grover Beach	H-4
Guadalupe	H-4
Gustine	F-4
Hanford	G-4
Healdsburg	E-2

Hemet	J-7
Hollister	F-3
Holtville	K-9
Huron	G-4
Imperial	K-9
Independence	F-5
Indio	J-8
Ione	E-4
Jackson	E-4
Joshua Tree	J-8
Kerman	G-4
King City	G-3
Kingsburg	G-4
La Mirada	J-6
Lake Elsinore	J-7
Lake Isabella	H-5
Lakeport	D-2
Lancaster	H-6
Lemoore	G-4
Lincoln	E-3
Lindsay	G-5
Livingston	F-4
Lodi	E-3
Lompoc	H-4
Long Beach	J-6
Los Angeles	J-6
Los Banos	F-4
Los Gatos	F-3
Madera	F-4

Mammoth Lakes	F-5
Manteca	F-3
Mariposa	F-4
Marysville	D-3
Martinez	F-3
McFarland	G-5
Mendota	G-4
Merced	F-4
Milpitas	F-3
Modesto	F-4
Mojave	H-5
Montecito	H-5
Monterey	G-3
Morgan Hill	F-3
Morro Bay	H-4
Mount Shasta	C-2
Napa	E-2
Needles	I-9
Nevada City	D-3
Newport Beach	J-6
Novato	E-2
Oakdale	F-4
Oakley	F-3
Oceanside	K-7
Ojai	H-5
Ontario	J-6

Oroville	D-3
Oxnard	H-5
Pacific Grove	G-3
Palm Desert	J-8
Palm Springs	J-8
Palmdale	H-6
Palo Alto	F-3
Paradise	D-3
Paso Robles	H-4
Patterson	F-4
Perris	J-7
Pittsburg	F-3
Placerville	E-4
Planada	F-4
Pleasanton	F-3
Pomona	J-6
Porterville	G-5
Quincy	D-4
Ramona	K-7
Rancho Cordova	E-3
Red Bluff	D-2
Redding	C-2
Redwood City	F-3
Rio Dell	C-1
Rosamond	H-5
Roseville	E-3

Sacramento	E-3
Salinas	G-3
San Andreas	E-4
San Bernardino	J-7
San Clemente	J-6
San Diego	K-7
San Francisco	F-2
San Jacinto	J-8
San Jose	F-3
San Juan Capistrano	J-6
San Luis Obispo	H-4
San Simeon	H-3
Santa Ana	J-6
Santa Barbara	H-5
Santa Cruz	F-3
Santa Maria	H-4
Santa Rosa	E-2
Saratoga	F-3
Seaside	G-3
Selma	G-4
Shafter	G-5
Simi Valley	H-5
Soledad	G-3
Sonora	F-4
South Lake Tahoe	E-5

Stockton	F-3
Susanville	C-3
Taft	H-5
Temecula	J-7
Thousand Oaks	H-5
Tracy	F-3
Truckee	D-4
Tulare	G-5
Turlock	F-4
Ukiah	D-2
Vacaville	E-3
Vallejo	E-3
Ventura	H-5
Victorville	J-7
Visalia	G-4
Vista	K-7
Wasco	G-5
Weaverville	C-2
Westmorland	K-9
Willits	D-2
Willows	D-2
Woodland	E-3
Wrightwood	J-7
Yreka	B-2
Yuba City	D-3
Yucca Valley	J-8

Nevada

Population: 1,206,152
Land Area: 109,889 sq. mi.
Capital: Carson City

Cities and Towns

Alamo	H-5
Austin	G-3
Baker	F-3
Battle Mountain	F-3
Beatty	G-5
Boulder City	I-7
Caliente	H-5
Carlin	F-3
Carson City	E-4
Cherry Creek	F-3
Dayton	E-4
Death	D-3
Duckwater	G-3
Elko	E-3
Ely	G-4
Empire	E-3
Eureka	G-3

Fallon	F-3
Fernley	E-3
Gabbs	F-4
Gardnerville	E-4
Gerlach	E-3
Golconda	F-3
Hawthorne	F-4
Henderson	I-6
Hiko	H-5
Imlay	F-3
Indian Springs	H-5
Jackpot	D-4
Jarbidge	D-4
Jean	I-6
Jiggs	E-4
Las Vegas	I-6
Laughlin	I-7
Logandale	H-6
Lovelock	F-3
Lund	G-4
Manhattan	G-4
McDermitt	D-3
McGill	G-3
Mina	F-4
Minden	E-4
Montello	D-4
Mountain City	D-3
Nelson	I-6
Oreana	F-3

Orovada	D-3
Overton	H-6
Owyhee	D-3
Pahrump	H-5
Panaca	H-5
Pioche	H-5
Reno	E-4
Round Mountain	G-4
Ruby Valley	E-4
Ruth	G-3
Schurz	F-4
Scotty's Junction	G-5
Searchlight	I-7
Silver Springs	E-4
Silverpeak	G-5
Stateline	E-4
Tonopah	G-4
Tuscarora	E-3
Valmy	F-3
Virginia City	E-4
Wadsworth	E-3
Walker Lake	F-4
Warm Springs	G-4
Washington	E-4
Wells	E-4
Winnemucca	E-3
Yerington	E-4

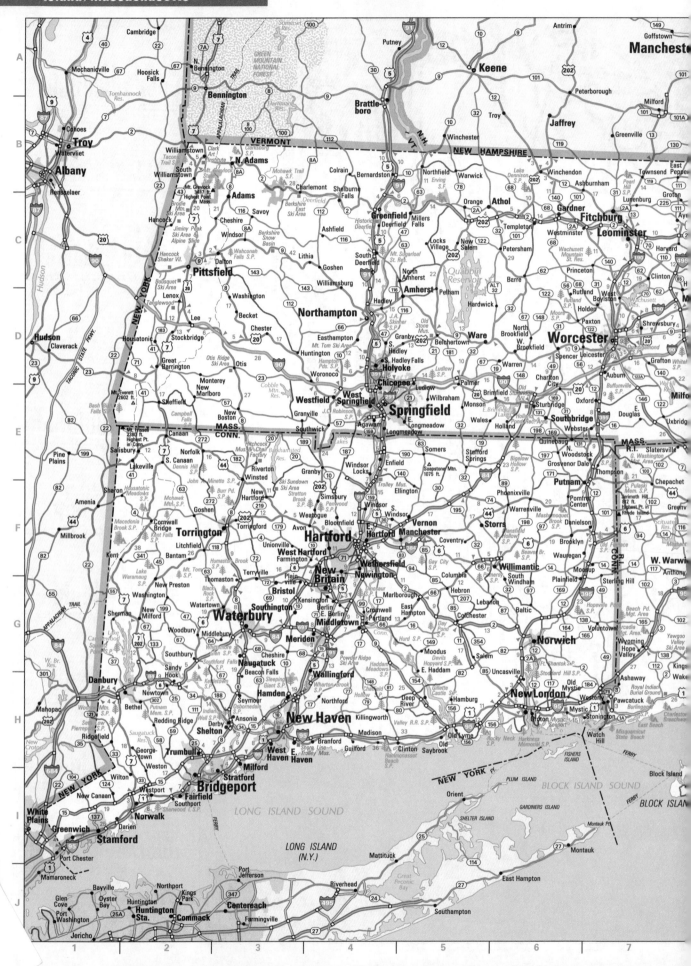

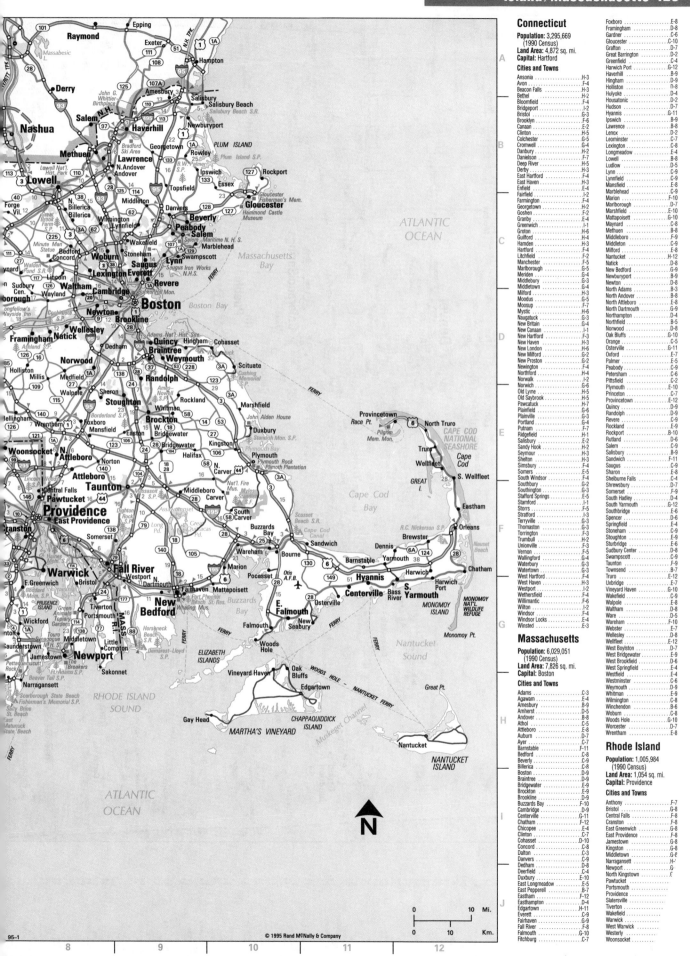

Connecticut

Population: 3,295,669 (1990 Census)
Land Area: 4,872 sq. mi.
Capital: Hartford

Cities and Towns

Ansonia	H-3
Avon	F-4
Beacon Falls	H-3
Bethel	H-2
Bloomfield	F-4
Bridgeport	I-2
Bristol	G-3
Brooklyn	F-6
Canaan	E-2
Clinton	H-5
Colchester	G-5
Cromwell	G-4
Danbury	H-2
Danielson	F-7
Deep River	H-5
Derby	H-3
East Hartford	F-4
East Haven	H-3
Enfield	E-4
Fairfield	I-2
Farmington	F-4
Georgetown	H-2
Goshen	F-2
Granby	E-4
Greenwich	I-1
Groton	H-6
Guilford	H-4
Hamden	H-3
Hartford	F-4
Litchfield	F-2
Manchester	F-5
Marlborough	G-5
Meriden	G-4
Middlebury	G-3
Middletown	G-4
Milford	I-3
Moodus	G-5
Moosup	F-7
Mystic	H-6
Naugatuck	G-3
New Britain	G-4
New Canaan	I-1
New Hartford	F-3
New Haven	H-3
New London	H-6
New Milford	G-2
New Preston	G-2
Newington	F-4
Northford	H-4
Norwalk	I-2
Norwich	G-6
Old Lyme	H-5
Old Saybrook	H-5
Pawcatuck	H-7
Plainfield	G-6
Plainville	G-4
Portland	G-4
Putnam	F-7
Ridgefield	H-1
Salisbury	E-2
Sandy Hook	H-2
Seymour	H-3
Shelton	H-3
Simsbury	F-4
Somers	E-5
South Windsor	F-4
Southbury	G-2
Southington	G-3
Stafford Springs	E-5
Stamford	I-1
Storrs	F-5
Stratford	I-3
Terryville	G-3
Thomaston	G-3
Torrington	F-3
Trumbull	H-2
Unionville	F-5
Vernon	F-5
Wallingford	G-4
Waterbury	G-3
Watertown	G-3
West Hartford	F-4
West Haven	H-3
Westport	I-2
Wethersfield	F-4
Willimantic	F-5
Wilton	I-2
Windsor	E-4
Windsor Locks	E-4
Winsted	F-3

Massachusetts

Population: 6,029,051 (1990 Census)
Land Area: 7,826 sq. mi.
Capital: Boston

Cities and Towns

Adams	C-3
Agawam	E-4
Amesbury	B-9
Amherst	D-4
Andover	B-8
Athol	C-5
Attleboro	E-8
Auburn	D-7
Ayer	C-7
Barnstable	F-11
Bedford	C-8
Beverly	C-9
Billerica	C-8
Boston	D-9
Braintree	D-9
Bridgewater	E-9
Brockton	E-9
Brookline	D-9
Buzzards Bay	F-10
Cambridge	D-9
Centerville	G-11
Chatham	F-12
Chicopee	E-4
Clinton	C-7
Cohasset	D-10
Concord	C-8
Dalton	C-3
Danvers	C-9
Dedham	D-8
Deerfield	D-4
Duxbury	E-10
East Longmeadow	E-5
East Pepperell	B-7
Eastham	F-12
Easthampton	D-4
Edgartown	H-11
Everett	C-9
Fairhaven	F-9
Fall River	F-9
Falmouth	G-10
Fitchburg	C-7

Foxboro	E-8
Framingham	D-8
Gardner	C-6
Gloucester	C-10
Grafton	D-7
Great Barrington	D-2
Greenfield	C-4
Harwich Port	G-12
Haverhill	B-9
Hingham	D-9
Holliston	D-8
Holyoke	D-4
Housatonic	D-2
Hudson	D-7
Hyannis	G-11
Ipswich	B-9
Lawrence	B-8
Lenox	D-2
Leominster	C-7
Lexington	C-8
Longmeadow	E-4
Lowell	B-8
Ludlow	D-5
Lynn	C-9
Lynnfield	C-9
Mansfield	E-8
Marblehead	C-9
Marion	F-10
Marlborough	D-7
Marshfield	E-10
Mattapoisett	G-10
Maynard	C-8
Methuen	B-8
Middleboro	F-9
Middleton	C-9
Milford	D-7
Nantucket	H-12
Natick	D-8
New Bedford	G-9
Newburyport	B-9
Newton	D-8
North Adams	C-3
North Andover	B-8
North Attleboro	E-8
North Dartmouth	G-9
Northampton	D-4
Northfield	B-5
Norwood	D-8
Oak Bluffs	G-10
Orange	C-5
Osterville	G-11
Oxford	E-7
Palmer	E-5
Peabody	C-9
Petersham	C-6
Pittsfield	D-2
Plymouth	E-10
Princeton	C-7
Provincetown	E-12
Quincy	D-9
Randolph	D-9
Revere	C-9
Rockland	E-9
Rockport	B-10
Rutland	D-6
Salem	C-9
Salisbury	B-9
Sandwich	F-11
Saugus	C-9
Sharon	E-8
Shelburne Falls	C-4
Shrewsbury	D-7
Somerset	F-9
South Hadley	D-4
South Yarmouth	G-12
Southbridge	E-6
Spencer	D-6
Springfield	E-4
Stoneham	C-9
Stoughton	E-9
Sturbridge	E-6
Sudbury Center	D-8
Swampscott	C-9
Taunton	F-9
Townsend	B-7
Truro	E-12
Uxbridge	E-7
Vineyard Haven	G-10
Wakefield	C-9
Walpole	D-8
Waltham	C-8
Ware	D-5
Wareham	F-10
Webster	E-7
Wellesley	D-8
Wellfleet	E-12
West Boylston	D-7
West Bridgewater	E-9
West Brookfield	D-6
West Springfield	E-4
Westfield	E-4
Westminster	C-6
Weymouth	D-9
Whitman	E-9
Wilmington	C-8
Winchendon	B-6
Woburn	C-8
Woods Hole	G-10
Worcester	D-7
Wrentham	E-8

Rhode Island

Population: 1,005,984 (1990 Census)
Land Area: 1,054 sq. mi.
Capital: Providence

Cities and Towns

Anthony	F-7
Bristol	G-8
Central Falls	F-8
Cranston	F-8
East Greenwich	G-8
East Providence	F-8
Jamestown	G-8
Kingston	G-8
Middletown	G-8
Narragansett	H-7
Newport	G-8
North Kingstown	G-7
Pawtucket	
Portsmouth	
Providence	
Slatersville	
Tiverton	
Wakefield	
Warwick	
West Warwick	
Westerly	
Woonsocket	

© 1995 Rand McNally & Company

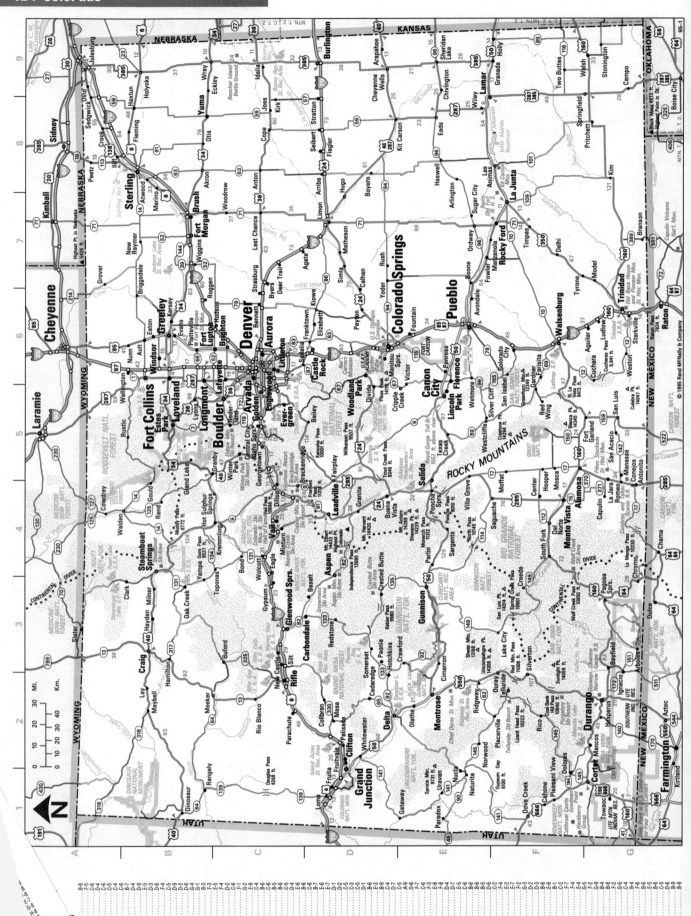

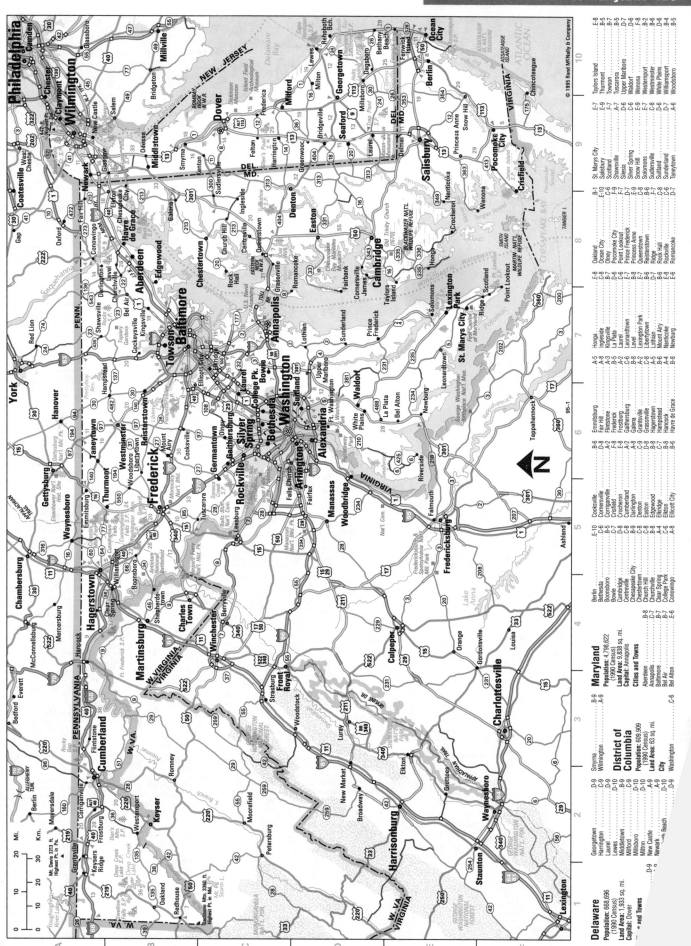

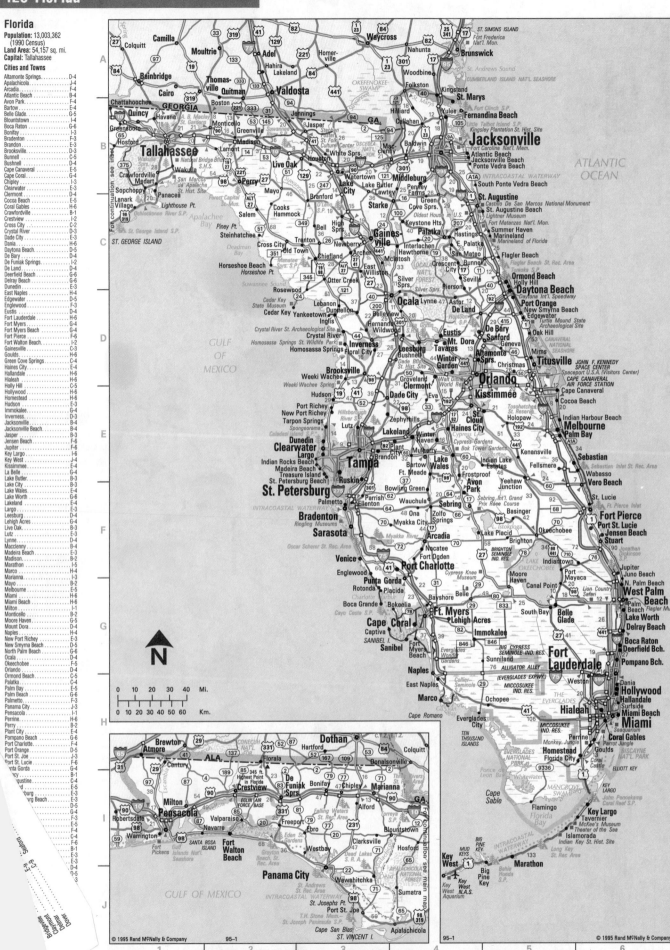

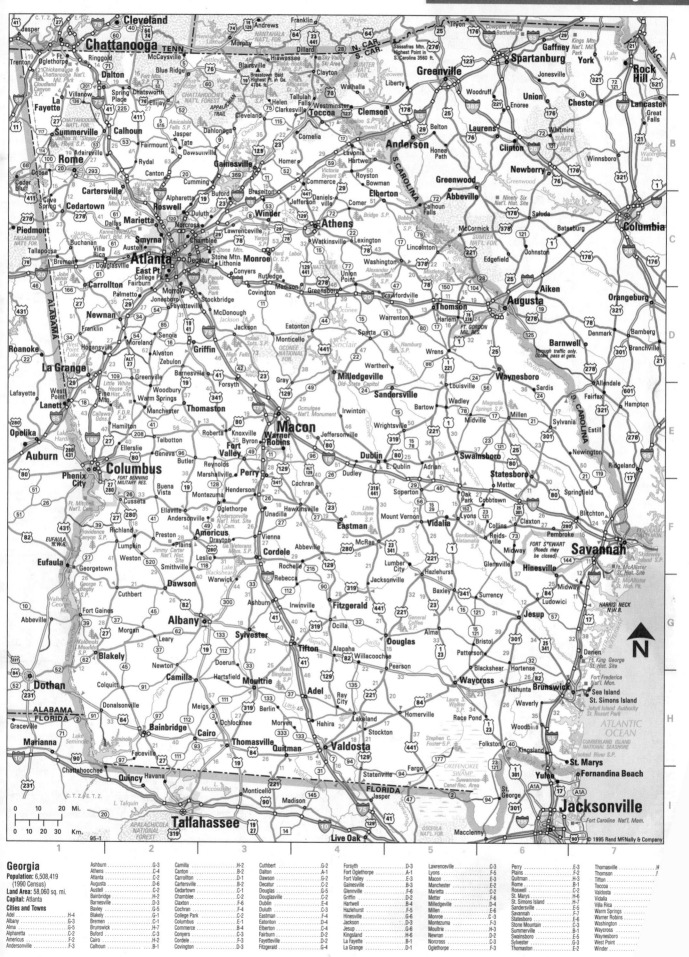

Idaho

Population: 1,011,986
(1990 Census)
Land Area: 82,413 sq. mi.
Capital: Boise

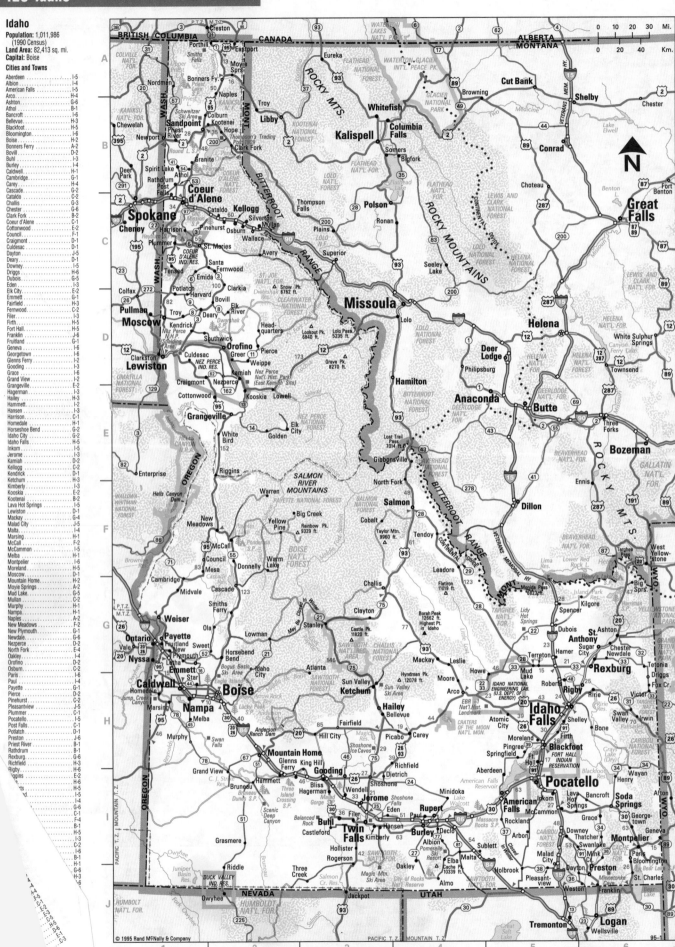

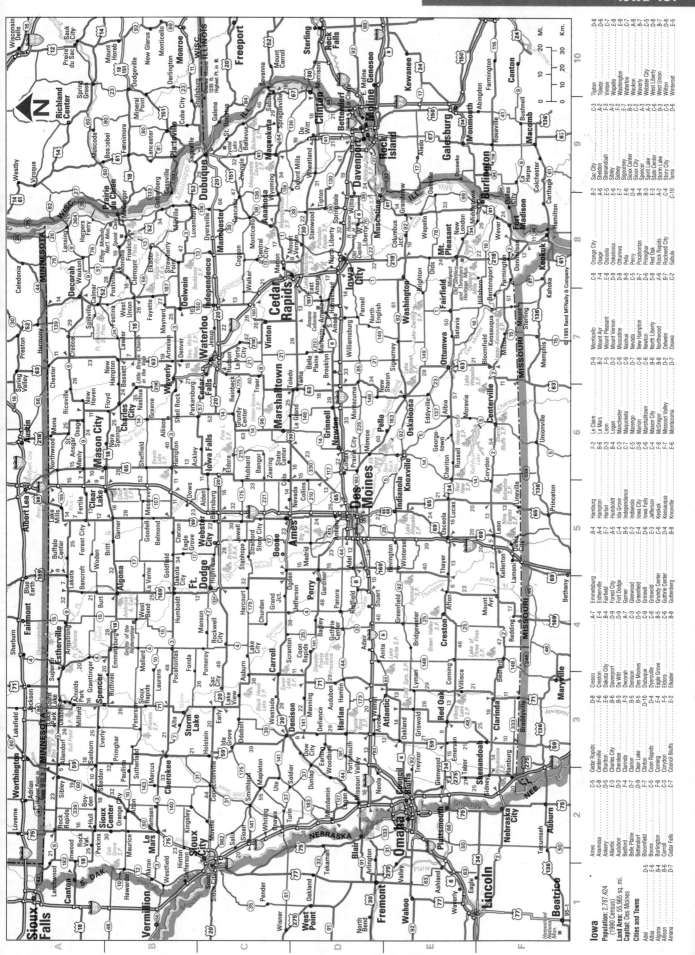

© 1995 Rand McNally & Company

95–1

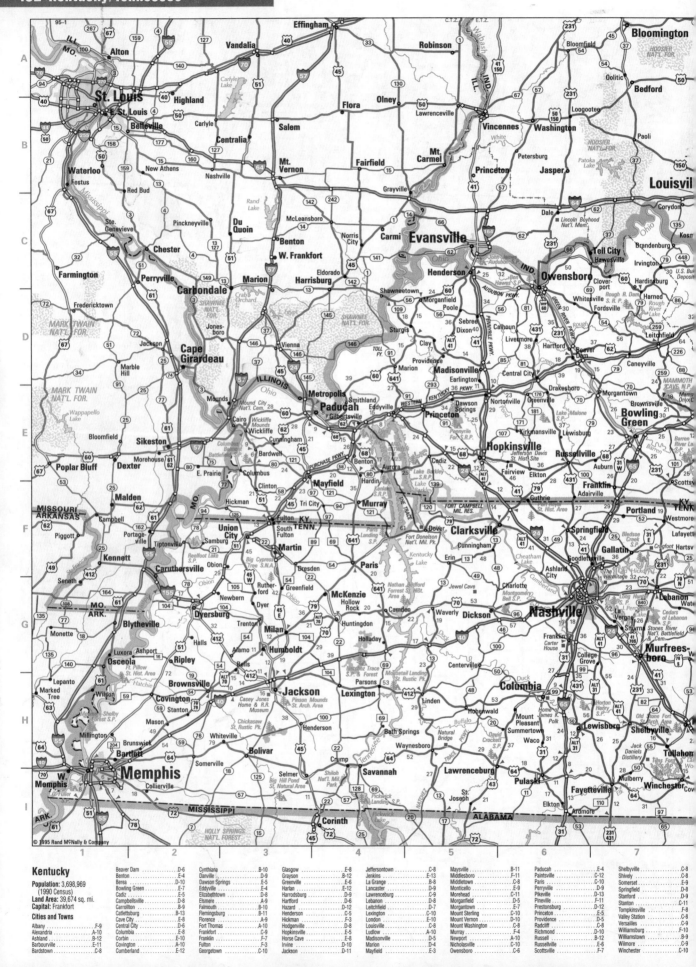

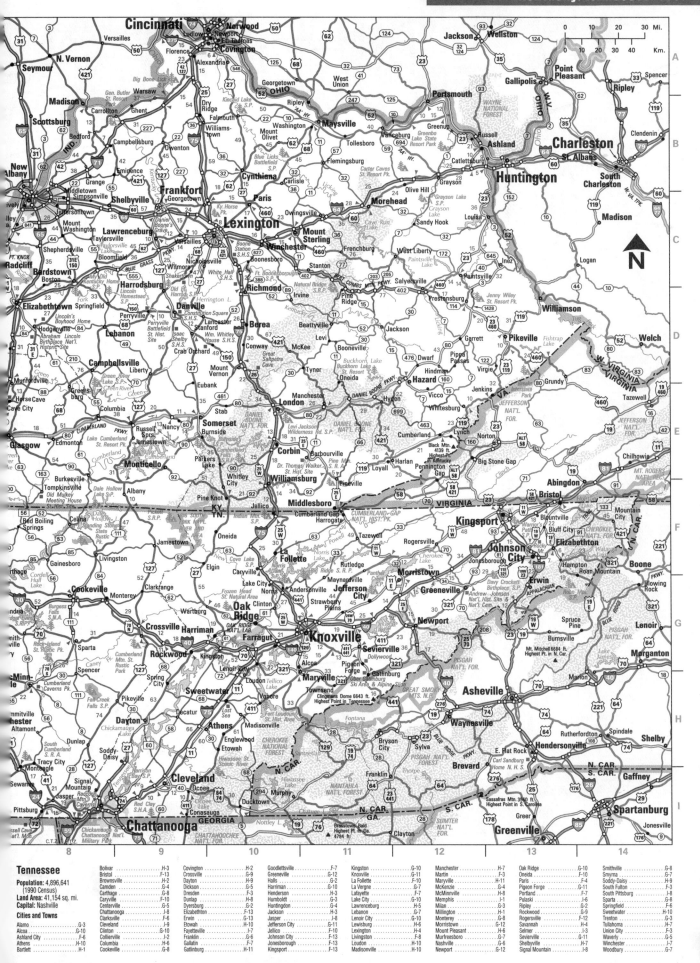

Kansas

Population: 2,485,600 (1990 Census)
Land Area: 81,783 sq. mi.
Capital: Topeka

Cities and Towns

Abilene	C-7
Anthony	E-6
Arkansas City	F-7
Ashland	F-4
Atchison	B-9
Atwood	B-2
Augusta	E-7
Baldwin City	C-9
Baxter Springs	B-6
Beloit	B-6
Burlington	D-9
Caney	F-9
Chanute	E-9
Chapman	C-7
Cherryvale	F-9
Cimarron	F-4
Clay Center	B-7
Coffeyville	F-9
Colby	B-2
Coldwater	F-4
Columbus	F-10
Concordia	B-6
Cottonwood Falls	D-8
Council Grove	D-8
Derby	E-7
Dighton	D-3
Dodge City	E-3
El Dorado	E-7
Elkhart	F-2
Ellinwood	D-5
Ellsworth	C-6
Emporia	D-8
Erie	E-9
Eureka	E-8
Ft. Scott	E-10
Fredonia	E-9
Frontenac	E-10
Garden City	E-2
Garnett	D-9
Girard	E-10
Goodland	C-1
Great Bend	D-5
Greensburg	E-4
Harper	E-6
Herington	D-7
Hays	C-4
Hiawatha	B-9
Hill City	C-4
Hillsboro	D-7
Hoisington	D-5
Holton	B-9
Horton	B-9
Howard	E-8
Hoxie	C-3
Hugoton	F-2
Humboldt	E-9
Hutchinson	D-6
Iola	E-9
Johnson	E-1
Junction City	C-7
Kinsley	E-4
Kiowa	F-5
La Crosse	D-4
Lakin	E-2
Larned	D-4
Lawrence	C-9
Leavenworth	B-9
Leoti	D-2
Liberal	F-2
Lincoln	C-6
Lyons	D-6
Manhattan	C-8
Mankato	B-6
Marysville	B-7
McPherson	D-6
Meade	F-3
Medicine Lodge	E-5
Minneapolis	C-6
Mound City	D-10
Neodesha	E-9
Ness City	D-4
Newton	D-7
Norton	B-4
Oakley	C-2
Oberlin	B-3
Olathe	C-10
Osage City	D-8
Osawatomie	D-10
Osborne	C-5
Oskaloosa	C-9
Oswego	F-10
Ottawa	C-9
Paola	C-10
Parsons	E-9
Peabody	D-7
Phillipsburg	B-4
Pittsburg	E-10
Plainville	C-4
Pratt	E-5
Russell	C-5
Sabetha	B-9
St. Francis	B-1
St. John	E-5
St. Marys	C-8
Salina	C-6
Scott City	D-2
Sedan	F-8
Seneca	B-8
Sharon Springs	C-1
Shawnee	B-5
Smith Center	B-5
South Hutchinson	D-6
Sterling	C-5
Stockton	C-4
Sublette	F-2
Syracuse	D-1
Tonganoxie	C-9
Topeka	C-9
Troy	B-9
Ulysses	E-2
WaKeeney	C-4
Wamego	C-8
Washington	B-7
Wellington	E-7
Wichita	E-7
Winfield	F-7
Yates Center	E-9

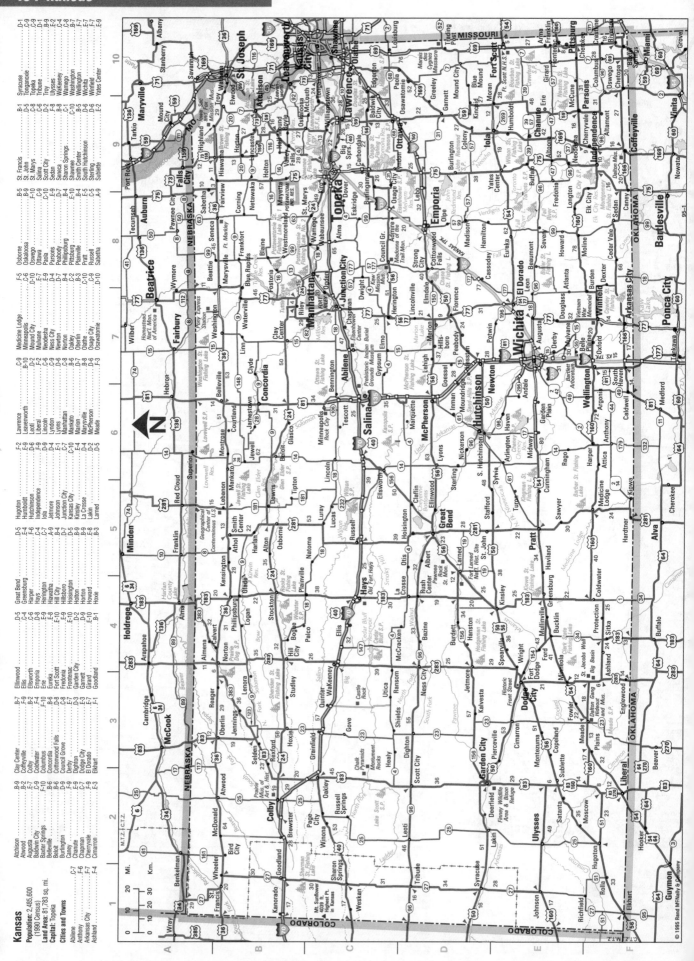

© 1995 Rand McNally & Company

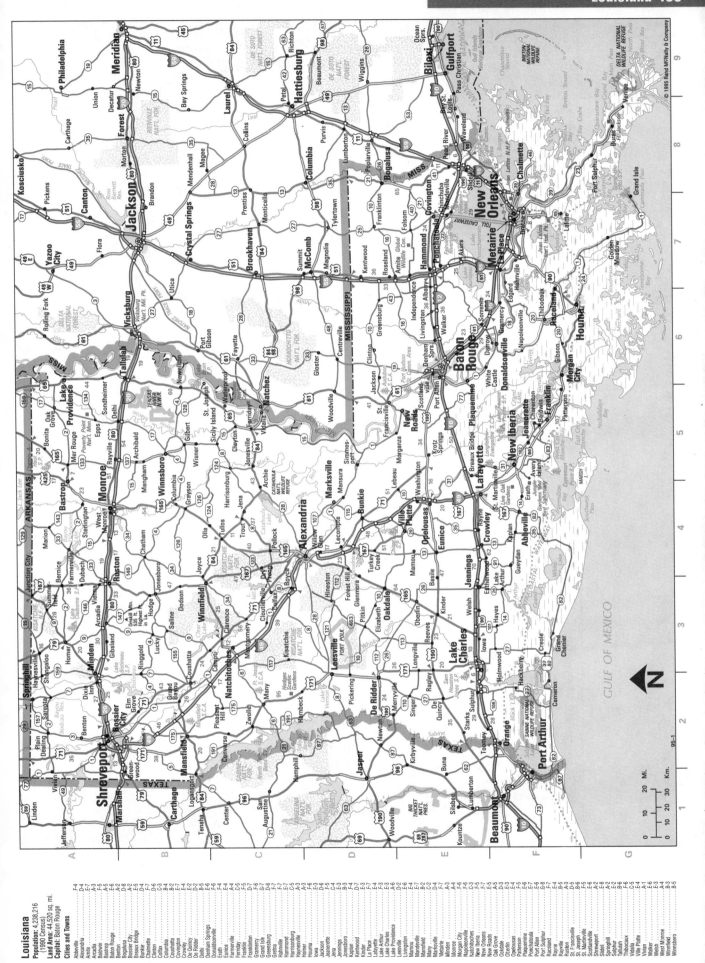

© 1995 Rand McNally & Company

Louisiana
Population: 4,238,216
(1990 Census)
Land Area: 44,520 sq. mi.
Capital: Baton Rouge

Cities and Towns

Maine

Population: 1,233,223
(1990 Census)
Land Area: 30,995 sq. mi.
Capital: Augusta

Cities and Towns

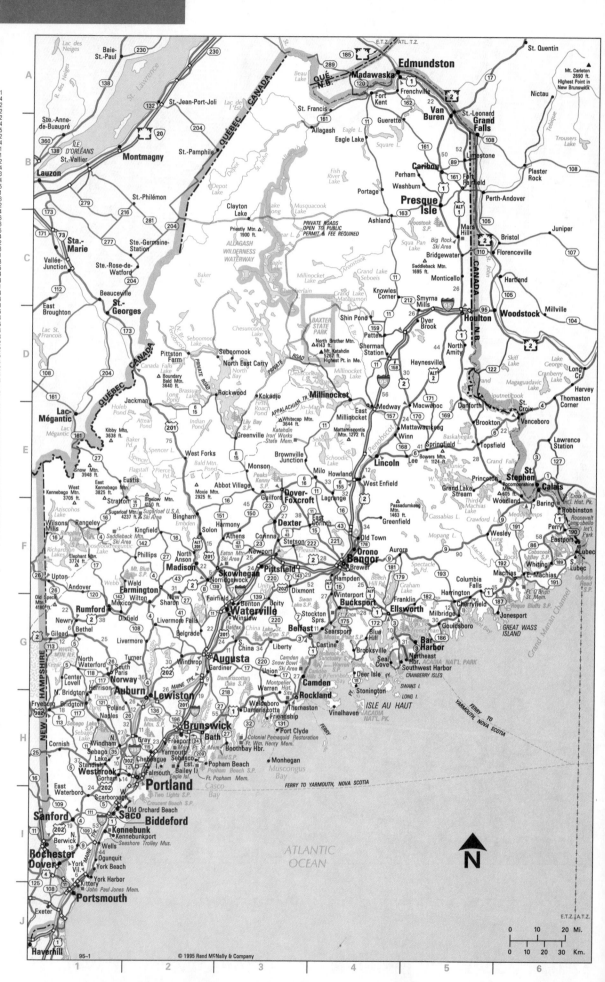

95–1

© 1995 Rand McNally & Company

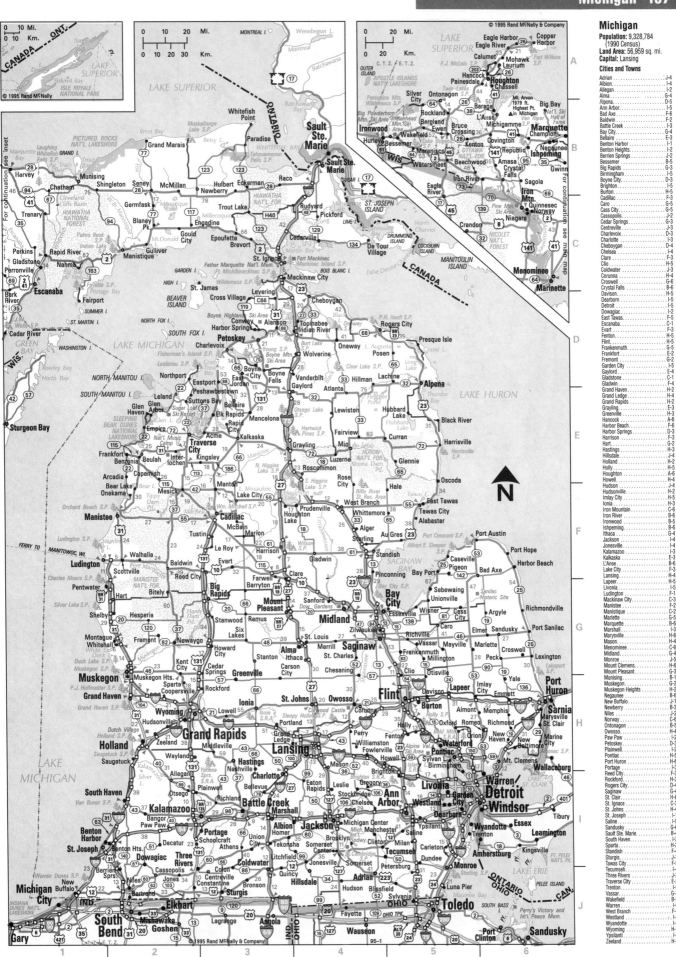

© 1995 Rand McNally & Company

Michigan
Population: 9,328,784
(1990 Census)
Land Area: 56,959 sq. mi.
Capital: Lansing
Cities and Towns

Minnesota

Population: 4,387,029 (1990 Census)
Land Area: 79,548 sq. mi.
Capital: St. Paul

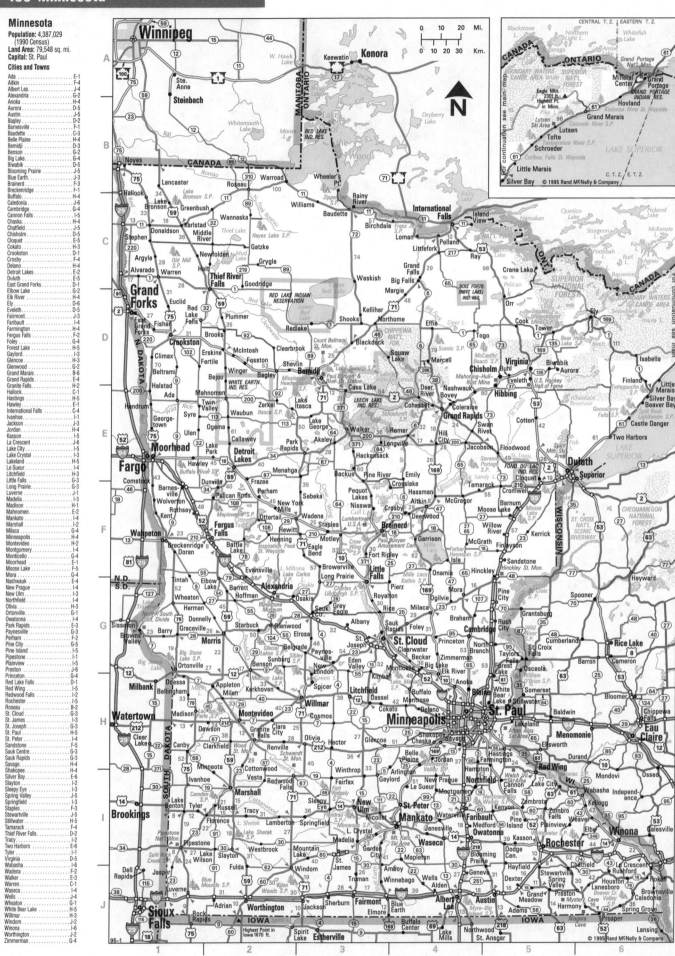

© 1995 Rand McNally & Company

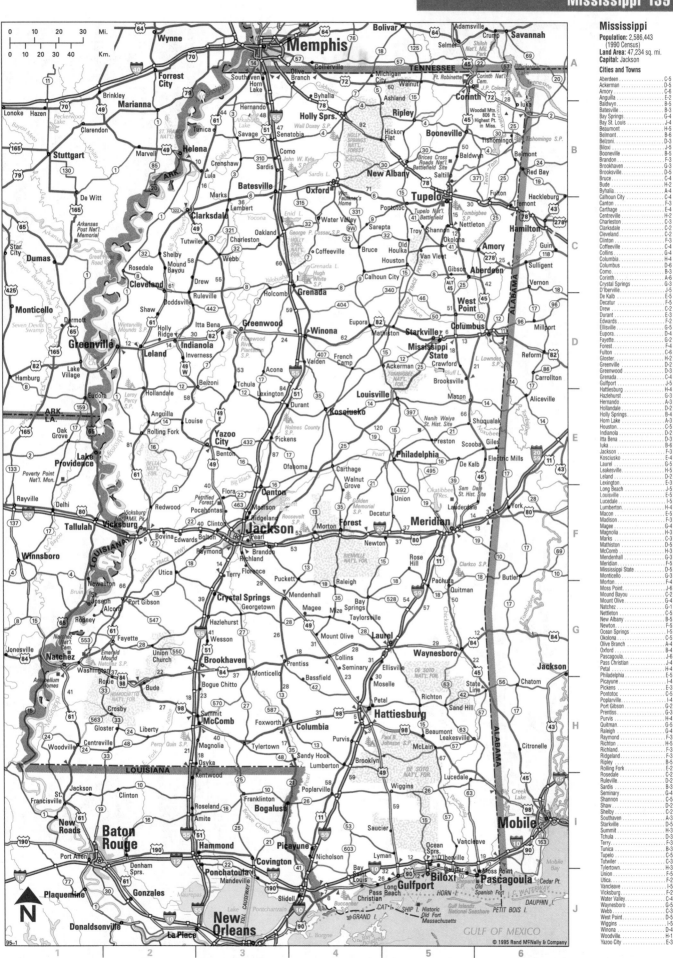

95-1

© 1995 Rand McNally & Company

Mississippi

Population: 2,586,443
(1990 Census)
Land Area: 47,234 sq. mi.
Capital: Jackson

Cities and Towns

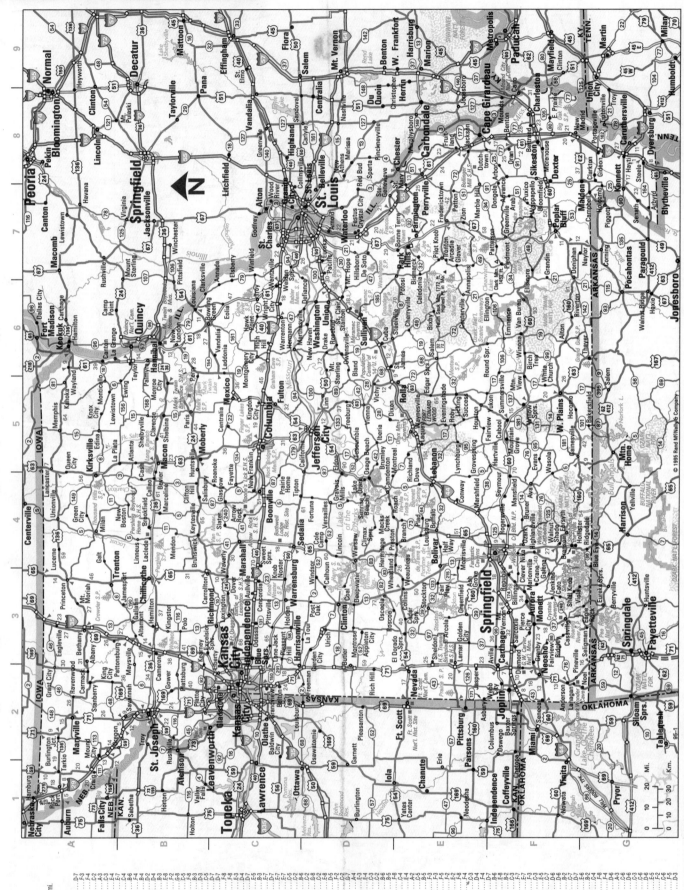

Missouri

Population: 5,137,804
(1990 Census)
Land Area: 68,945 sq. mi.
Capital: Jefferson City

Cities and Towns

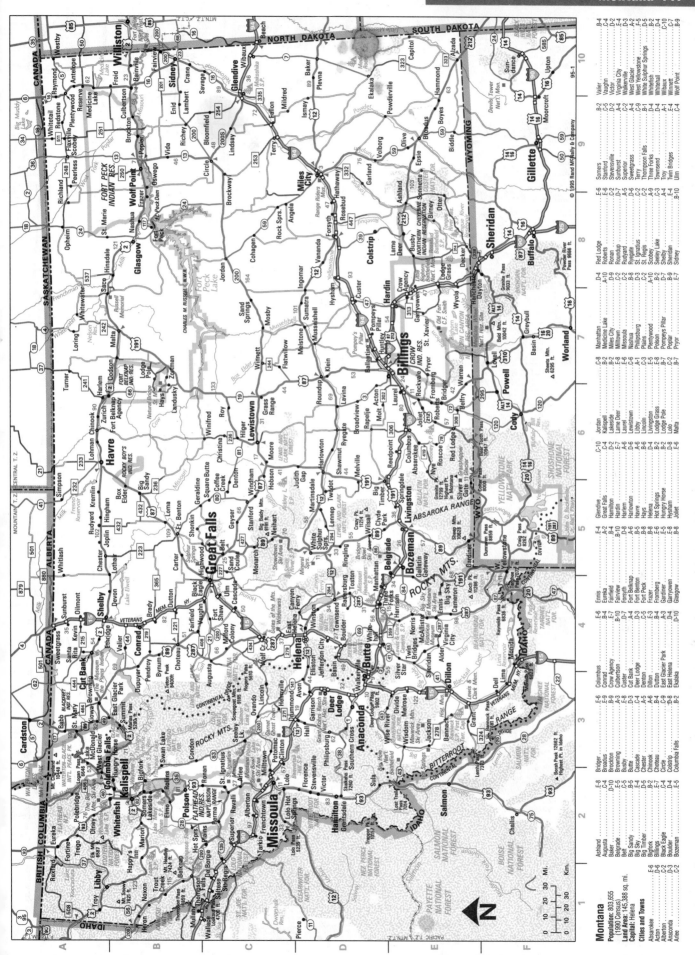

© 1995 Rand McNally & Company

95-1

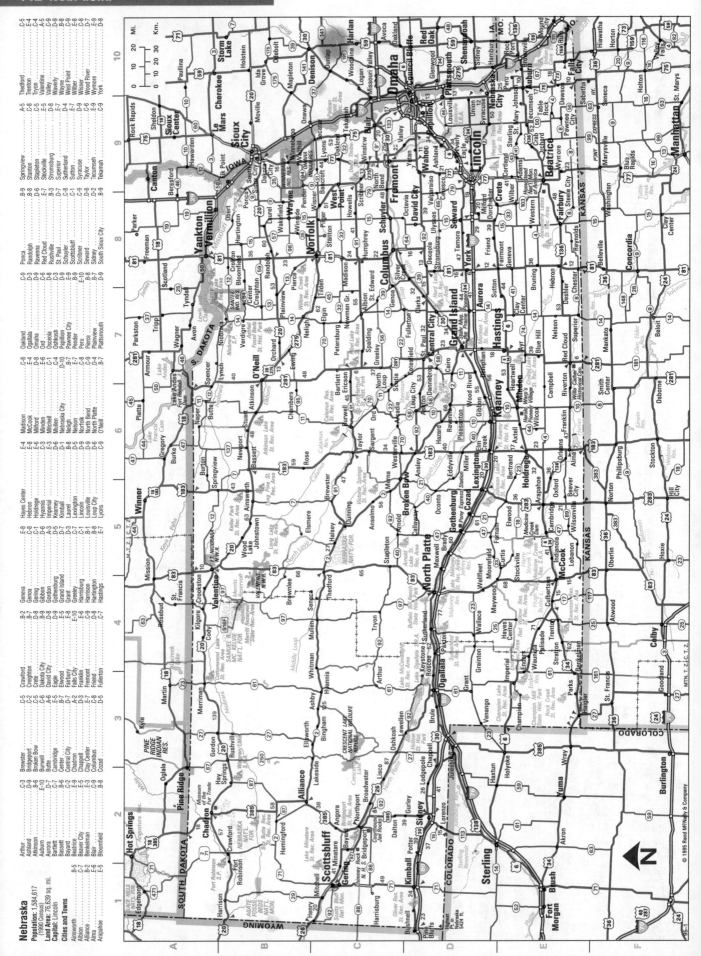

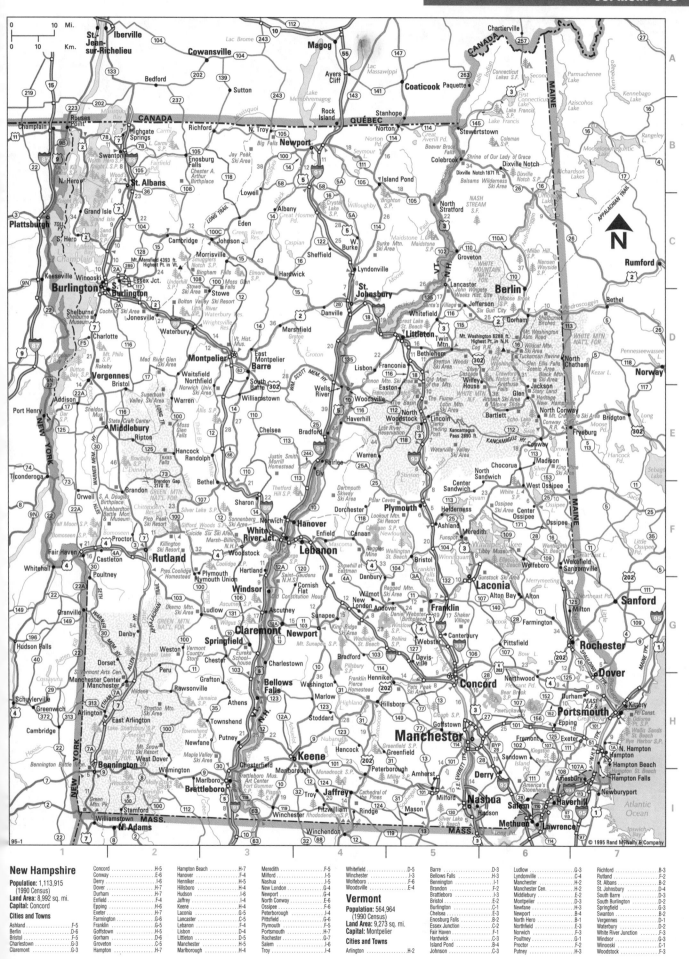

New Jersey

Population: 7,748,634
(1990 Census)
Land Area: 7,468 sq. mi.
Capital: Trenton

Cities and Towns

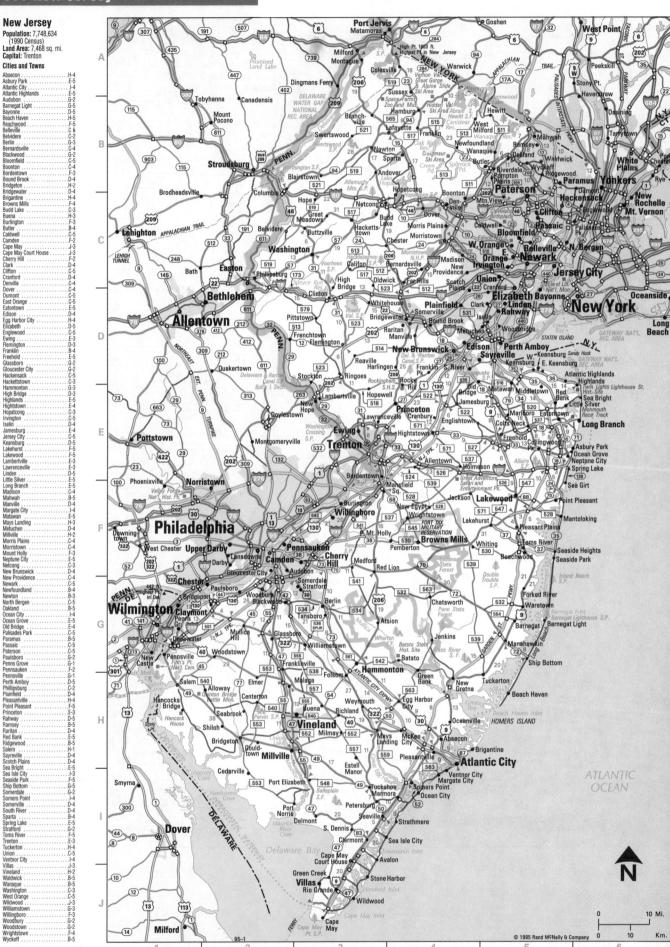

© 1995 Rand McNally & Company

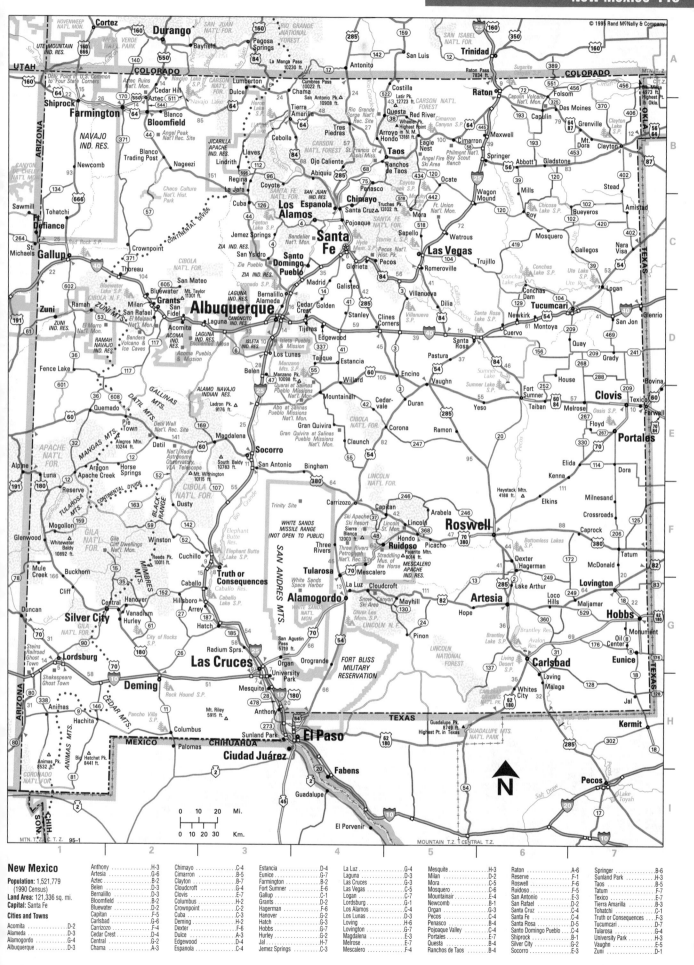

95-1

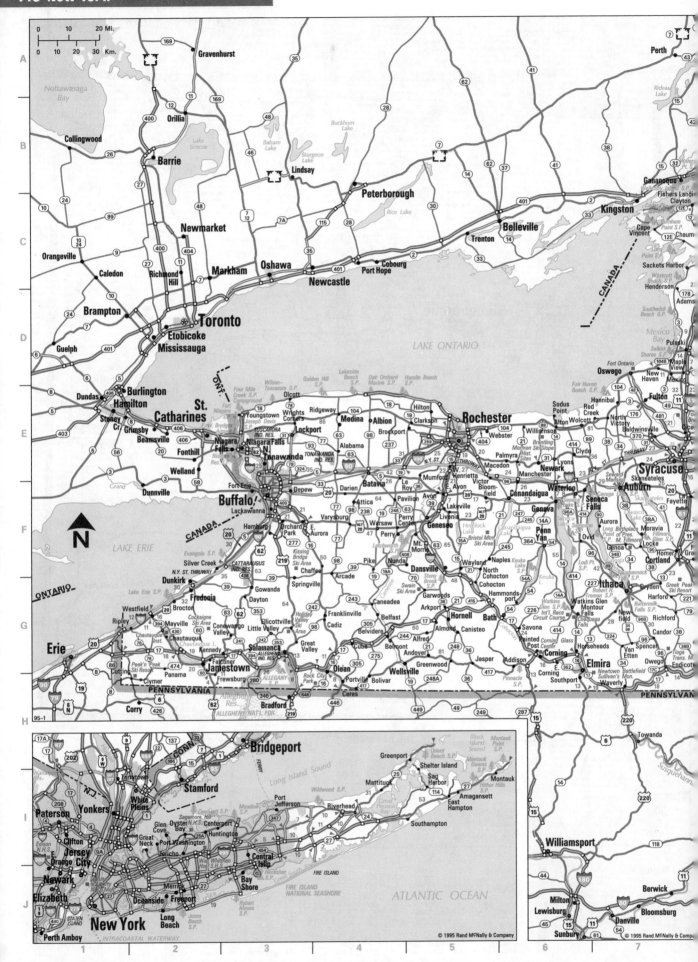

New York

Population: 18,044,505
(1990 Census)
Land Area: 47,379 sq. mi.
Capital: Albany

Cities and Towns

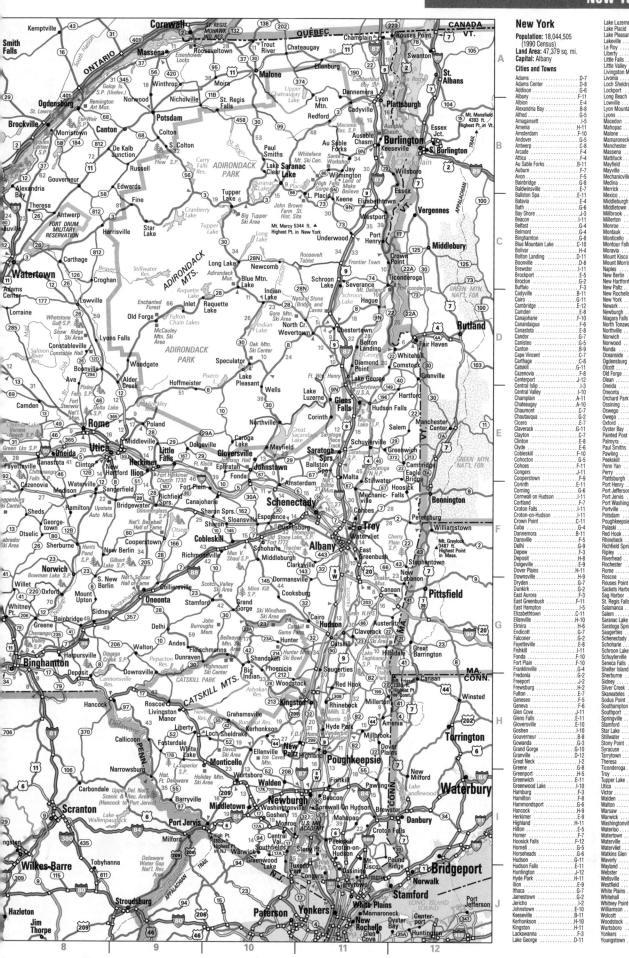

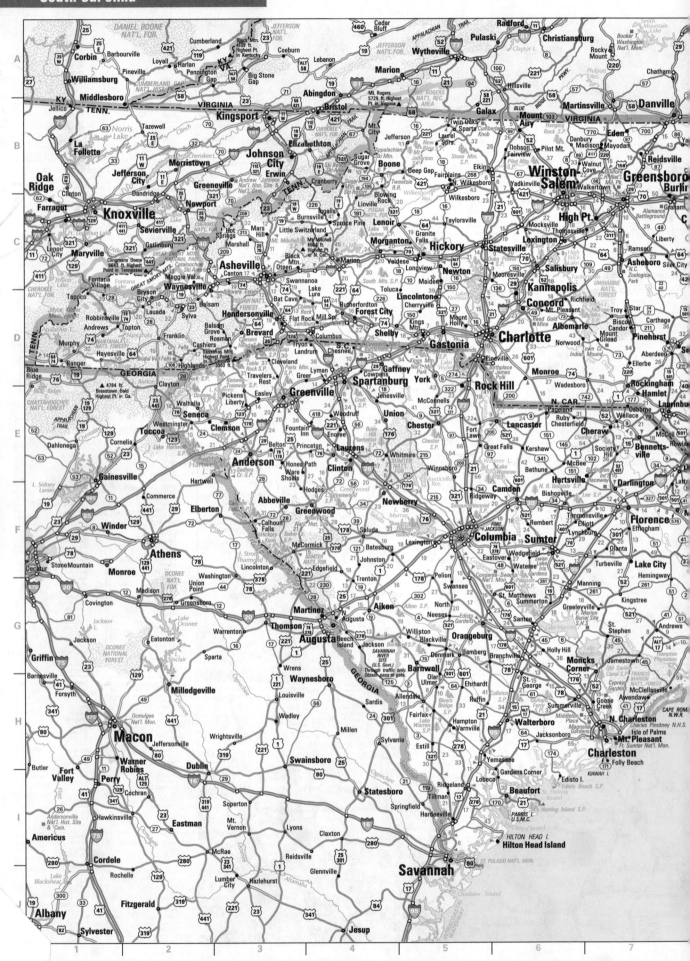

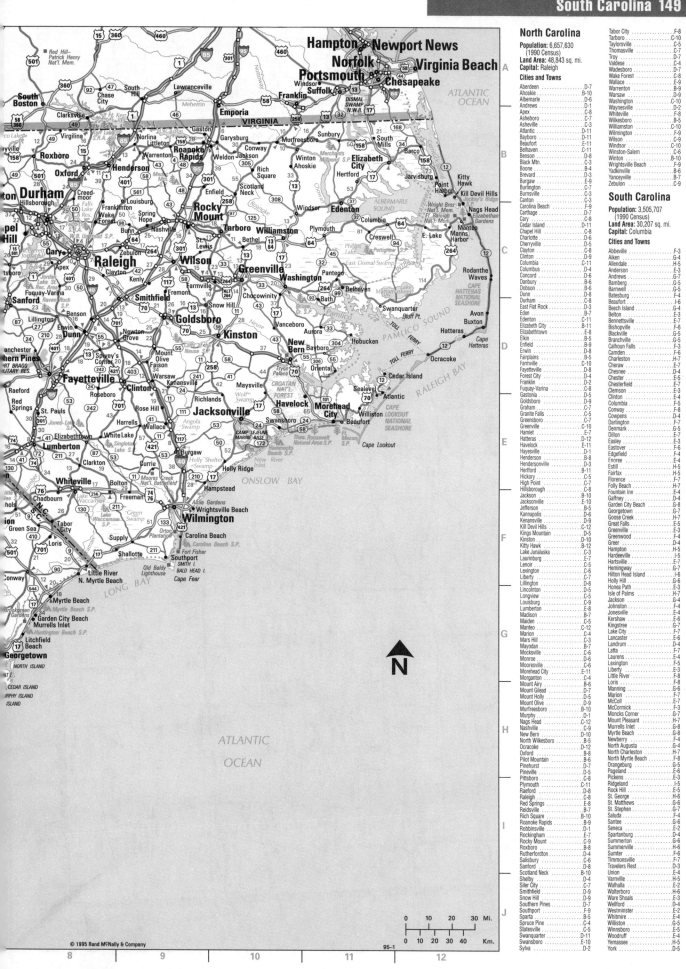

North Carolina

Population: 6,657,630
(1990 Census)
Land Area: 48,843 sq. mi.
Capital: Raleigh

Cities and Towns

Aberdeen	D-7
Ahoskie	B-10
Albemarle	D-6
Andrews	D-1
Apex	C-7
Asheboro	C-7
Asheville	C-3
Atlantic	D-11
Bayboro	D-11
Beaufort	E-11
Belhaven	C-11
Benson	D-7
Black Mtn.	C-3
Boone	B-4
Brevard	D-3
Burgaw	E-9
Burlington	C-7
Burnsville	C-3
Canton	C-2
Carolina Beach	F-9
Carthage	D-7
Cary	C-7
Cedar Island	D-11
Chapel Hill	C-7
Charlotte	D-5
Cherryville	D-5
Clayton	C-7
Clinton	D-8
Columbia	C-11
Columbus	D-4
Concord	D-5
Danbury	B-6
Dobson	B-6
Durham	C-8
East Flat Rock	D-3
Eden	B-6
Edenton	C-11
Elizabeth City	B-11
Elizabethtown	E-8
Elkin	B-5
Enfield	B-9
Fairplains	B-5
Farmville	C-10
Fayetteville	D-8
Forest City	D-4
Franklin	D-2
Fuquay-Varina	C-8
Gastonia	D-5
Goldsboro	D-9
Graham	C-7
Granite Falls	C-4
Greensboro	C-7
Greenville	C-10
Hamlet	E-7
Hatteras	D-12
Havelock	E-11
Hayesville	D-1
Henderson	B-8
Hendersonville	D-3
Hertford	B-11
Hickory	C-4
High Point	C-7
Hillsborough	C-8
Jackson	B-10
Jacksonville	E-10
Jefferson	B-5
Kannapolis	D-6
Kenansville	D-9
Kill Devil Hills	C-12
Kings Mountain	D-5
Kinston	D-10
Kitty Hawk	B-12
Lake Junaluska	C-3
Laurinburg	E-7
Lenoir	C-5
Lexington	C-6
Liberty	C-7
Lillington	D-8
Lincolnton	D-5
Longview	C-4
Louisburg	C-9
Lumberton	E-8
Madison	B-7
Maiden	C-5
Manteo	C-12
Marion	C-4
Mars Hill	C-3
Mayodan	B-6
Mocksville	C-6
Monroe	D-6
Mooresville	C-5
Morehead City	E-11
Morganton	C-4
Mount Airy	B-6
Mount Gilead	D-7
Mount Holly	D-5
Mount Olive	D-9
Murfreesboro	B-10
Murphy	D-1
Nags Head	C-12
Nashville	C-9
New Bern	D-10
North Wilkesboro	B-5
Ocracoke	D-12
Oxford	B-8
Pilot Mountain	B-6
Pinehurst	D-7
Pineville	C-8
Pittsboro	C-8
Plymouth	C-11
Raeford	D-8
Raleigh	C-8
Red Springs	E-8
Reidsville	B-7
Rich Square	B-10
Roanoke Rapids	B-9
Robbinsville	D-1
Rockingham	E-7
Rocky Mount	C-9
Roxboro	B-8
Rutherfordton	D-4
Salisbury	C-6
Sanford	D-7
Scotland Neck	C-10
Shelby	D-4
Siler City	C-7
Smithfield	D-9
Snow Hill	D-9
Southern Pines	D-7
Southport	F-9
Sparta	B-5
Spruce Pine	C-4
Statesville	C-5
Swanquarter	D-11
Swansboro	E-10
Sylva	D-2
Tabor City	F-8
Tarboro	C-10
Taylorsville	C-5
Thomasville	C-7
Troy	D-7
Valdese	C-4
Wadesboro	D-7
Wake Forest	C-8
Wallace	E-9
Warrenton	B-9
Warsaw	D-9
Washington	C-10
Waynesville	D-2
Whiteville	F-8
Wilkesboro	B-5
Williamston	C-10
Wilmington	F-9
Wilson	C-9
Windsor	C-10
Winston-Salem	C-6
Winton	B-10
Wrightsville Beach	F-9
Yadkinville	B-6
Yanceyville	B-7
Zebulon	C-9

South Carolina

Population: 3,505,707
(1990 Census)
Land Area: 30,207 sq. mi.
Capital: Columbia

Cities and Towns

Abbeville	F-3
Aiken	G-4
Allendale	H-5
Anderson	E-3
Andrews	G-7
Bamberg	G-5
Barnwell	G-5
Batesburg	F-4
Beaufort	I-6
Beech Island	G-4
Belton	E-3
Bennettsville	E-7
Bishopville	F-6
Blackville	G-5
Branchville	G-5
Calhoun Falls	F-3
Camden	F-6
Charleston	H-7
Cheraw	E-7
Chesnee	D-4
Chester	E-5
Chesterfield	E-7
Clemson	E-3
Clinton	E-4
Columbia	F-5
Conway	G-8
Cowpens	D-4
Darlington	F-7
Denmark	G-5
Dillon	E-7
Easley	E-3
Eastover	F-6
Edgefield	F-4
Enoree	E-4
Estill	H-5
Fairfax	H-5
Florence	F-7
Folly Beach	H-7
Fountain Inn	E-4
Gaffney	D-4
Garden City Beach	G-8
Georgetown	G-7
Goose Creek	H-7
Great Falls	E-5
Greenville	E-3
Greenwood	F-4
Greer	D-4
Hampton	H-5
Hardeeville	I-5
Hartsville	E-7
Hemingway	G-7
Hilton Head Island	I-6
Holly Hill	G-6
Honea Path	E-3
Isle of Palms	H-7
Jackson	G-4
Johnston	F-4
Jonesville	E-4
Kershaw	E-6
Kingstree	G-7
Lake City	F-7
Lancaster	E-6
Landrum	D-4
Latta	F-7
Laurens	E-4
Lexington	F-5
Liberty	E-3
Little River	F-8
Loris	F-8
Manning	G-6
Marion	F-7
McColl	E-7
McCormick	F-3
Moncks Corner	G-7
Mount Pleasant	H-7
Murrells Inlet	G-8
Myrtle Beach	G-8
Newberry	F-4
North Augusta	G-4
North Charleston	H-7
North Myrtle Beach	F-8
Orangeburg	G-5
Pageland	E-6
Pickens	E-3
Ridgeland	I-5
Rock Hill	E-5
St. George	H-6
St. Matthews	G-6
St. Stephen	G-7
Saluda	F-4
Santee	G-6
Seneca	E-2
Spartanburg	D-4
Summerton	G-6
Summerville	H-6
Sumter	F-6
Timmonsville	F-7
Travelers Rest	D-3
Union	E-4
Varnville	H-5
Walhalla	E-2
Walterboro	H-6
Ware Shoals	E-3
Wellford	D-4
Westminster	E-2
Whitmire	E-4
Williston	G-5
Winnsboro	E-5
Woodruff	E-4
Yemassee	H-5
York	D-5

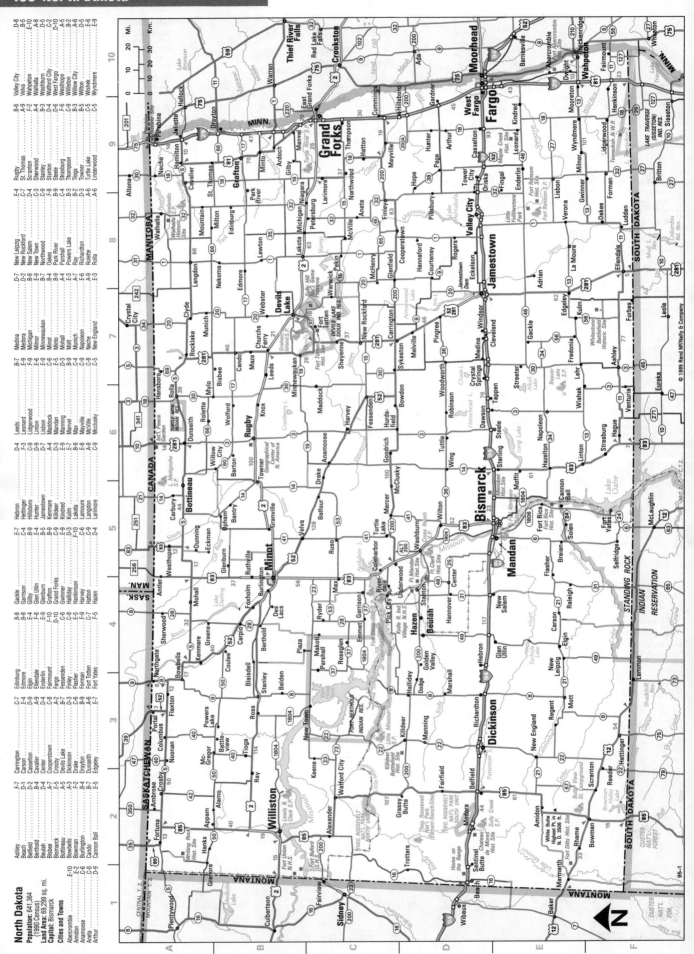

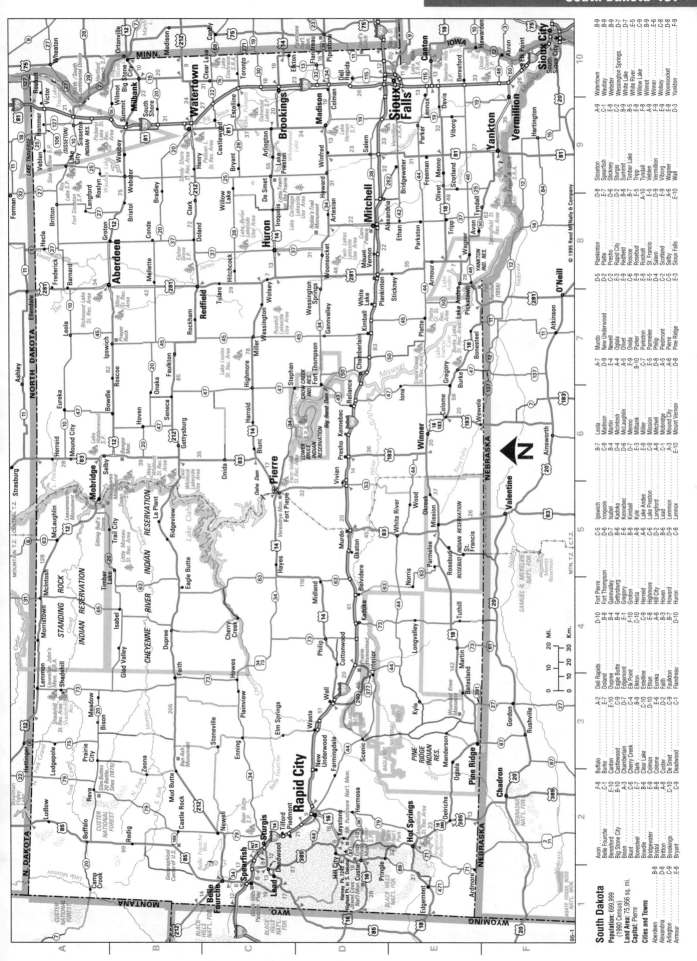

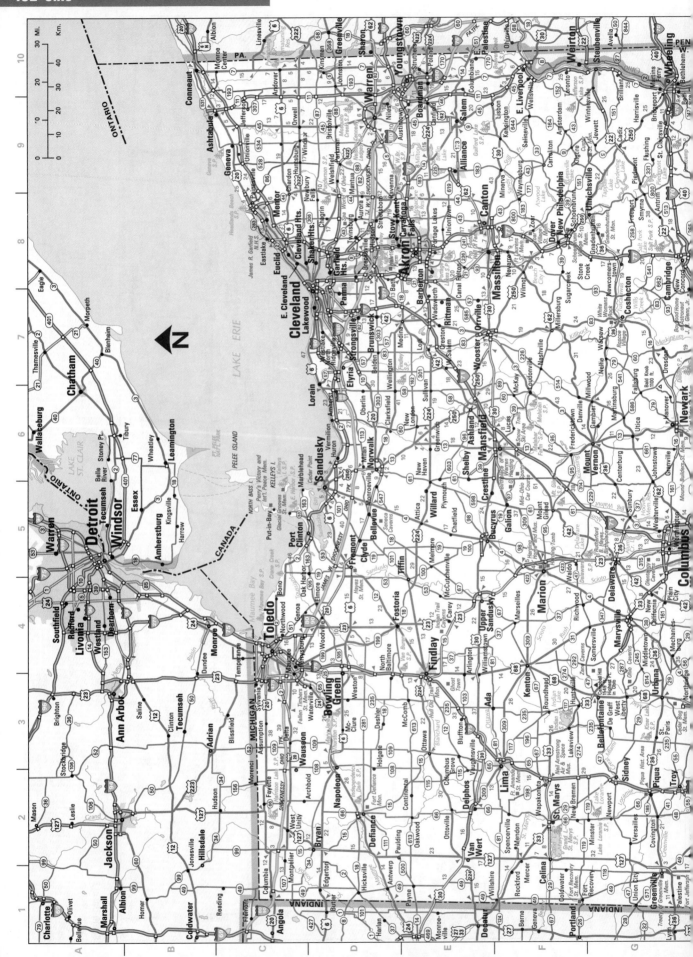

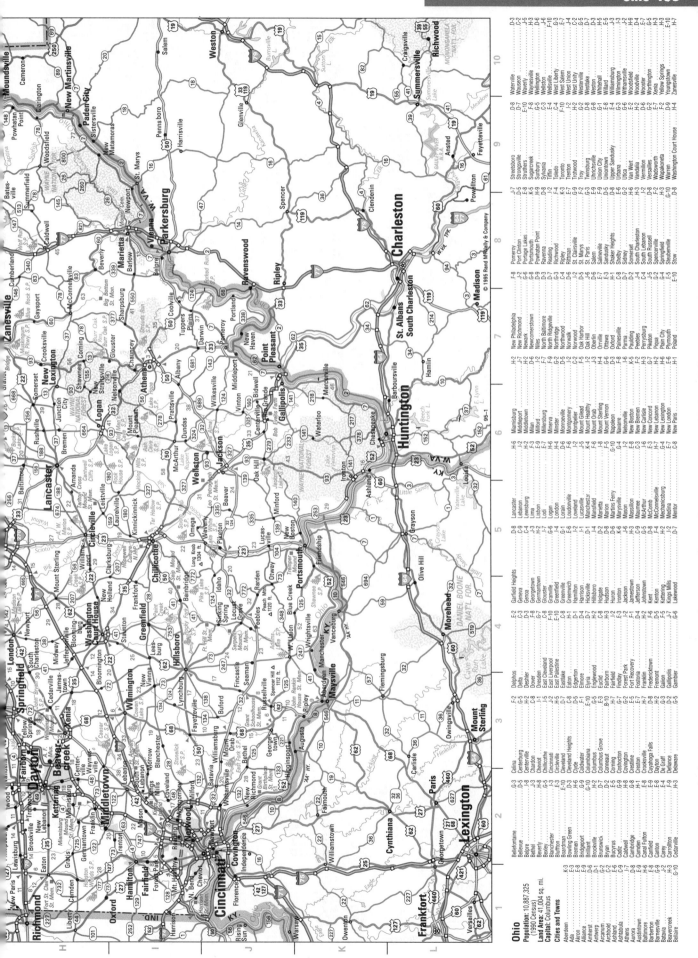

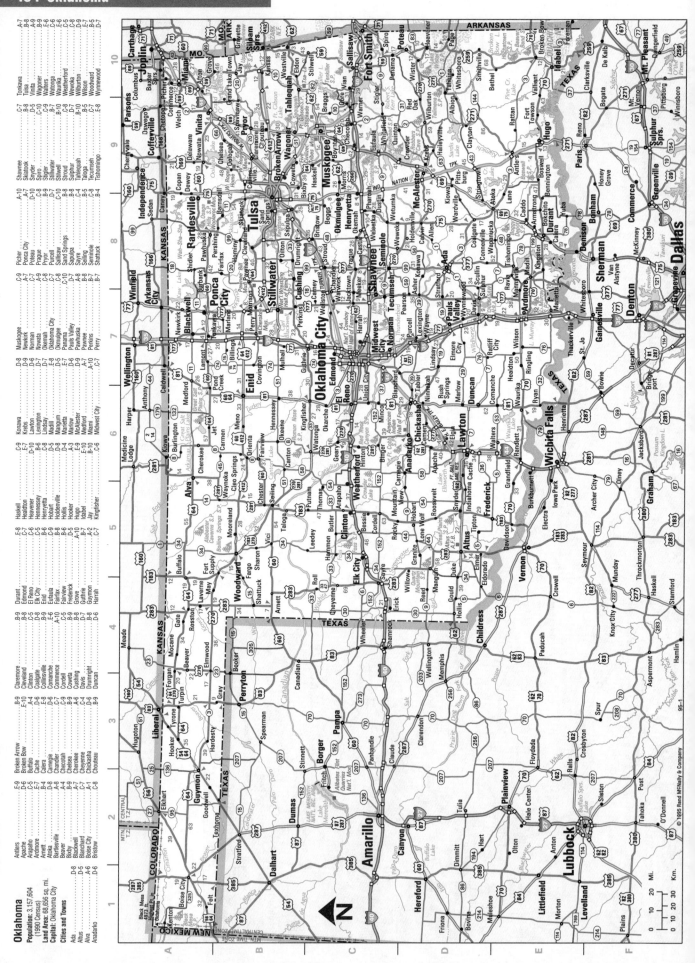

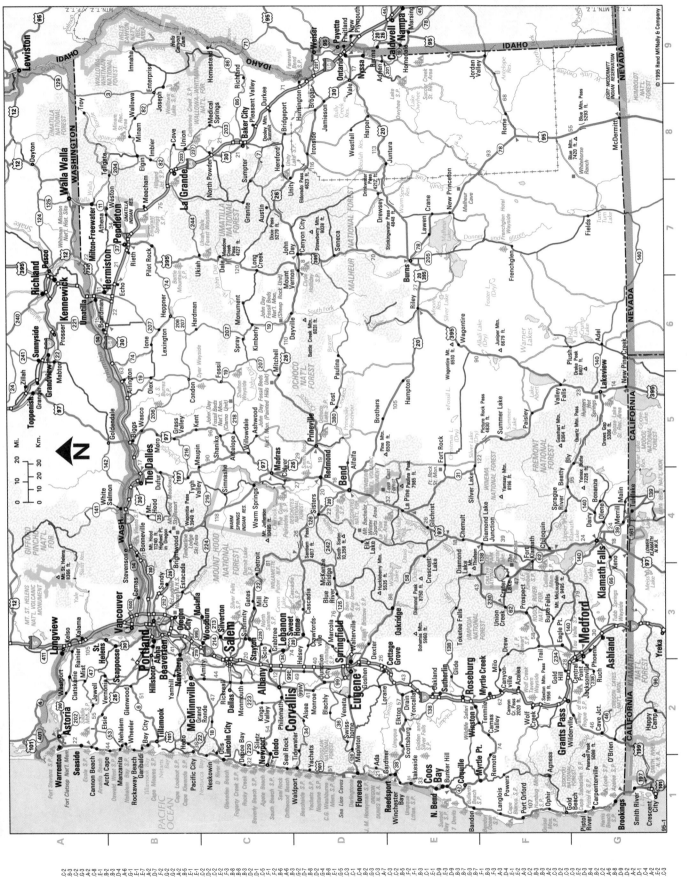

© 1995 Rand McNally & Company

Oregon

Population: 2,853,733
(1990 Census)
Land Area: 96,187 sq. mi.
Capital: Salem

96-1

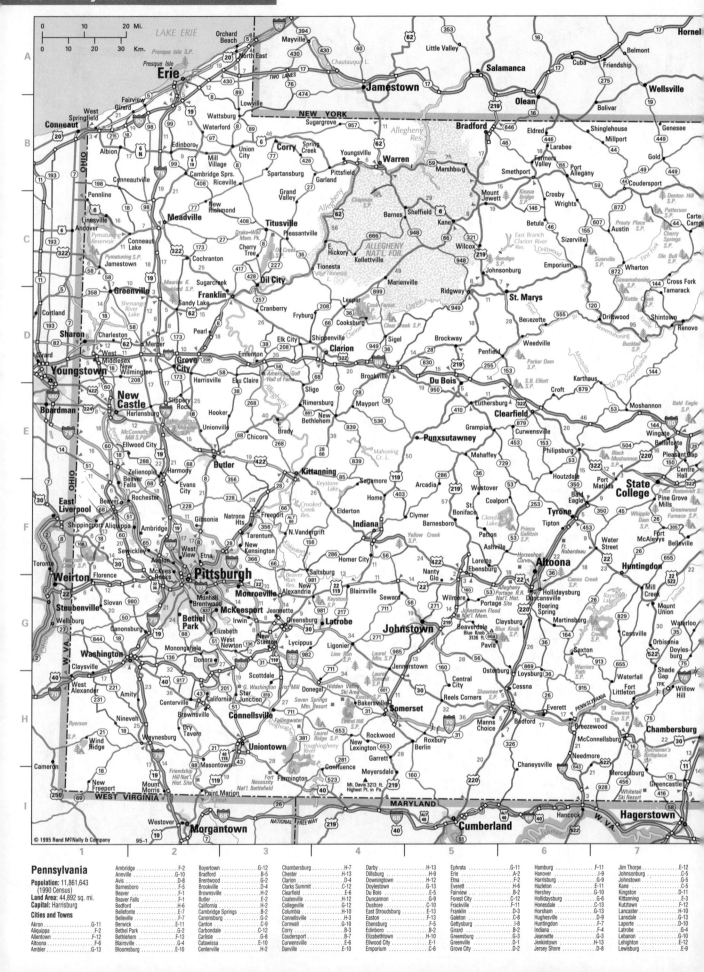

Pennsylvania

Population: 11,861,643
(1990 Census)
Land Area: 44,892 sq. mi.
Capital: Harrisburg

Cities and Towns

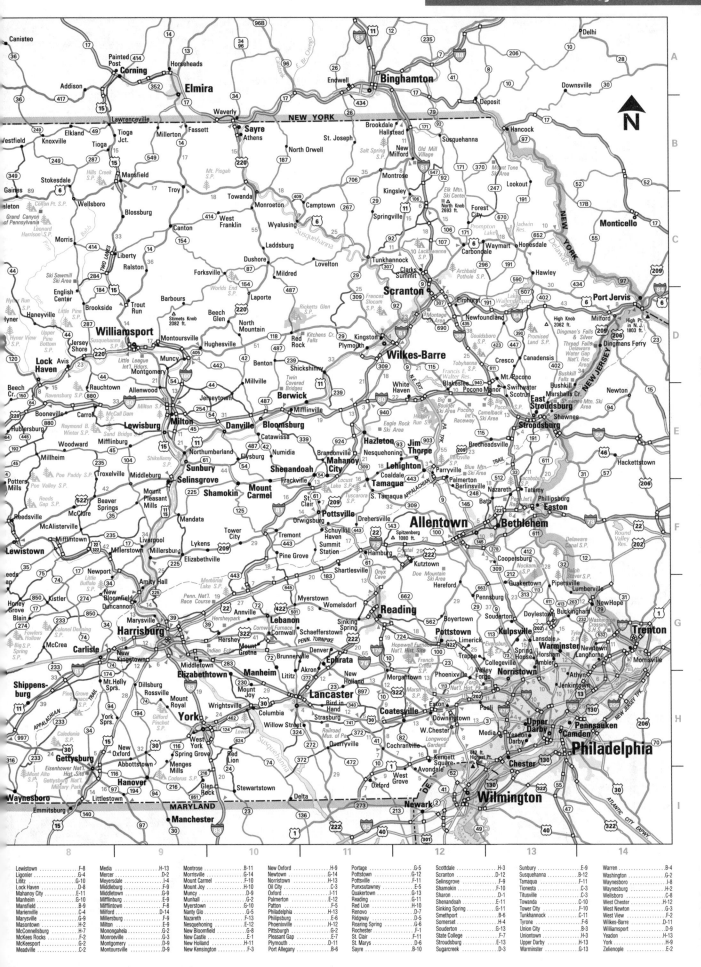

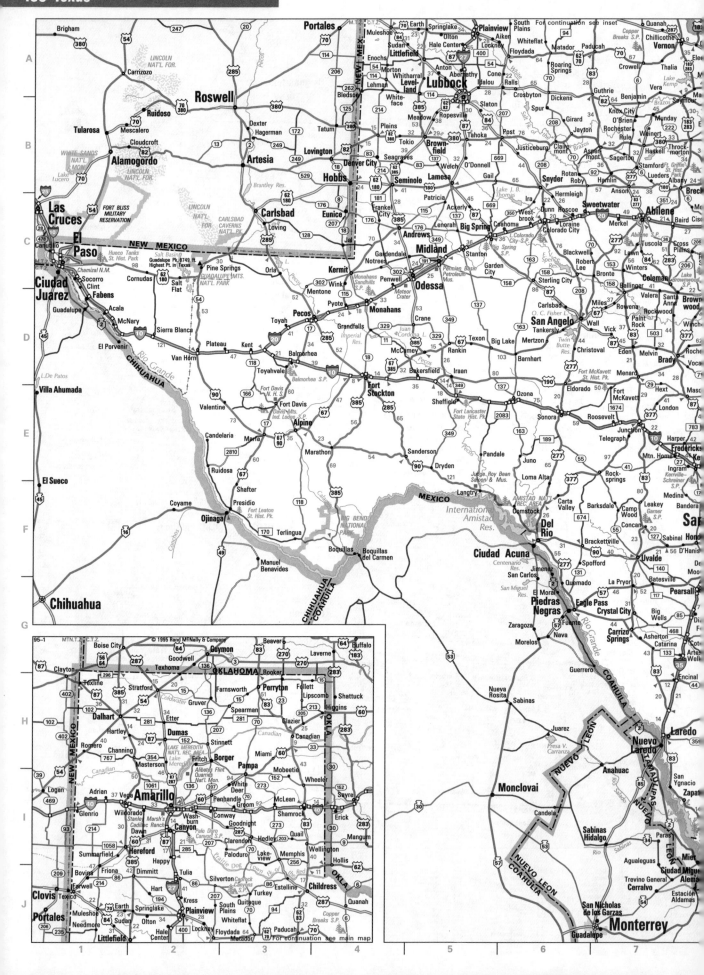

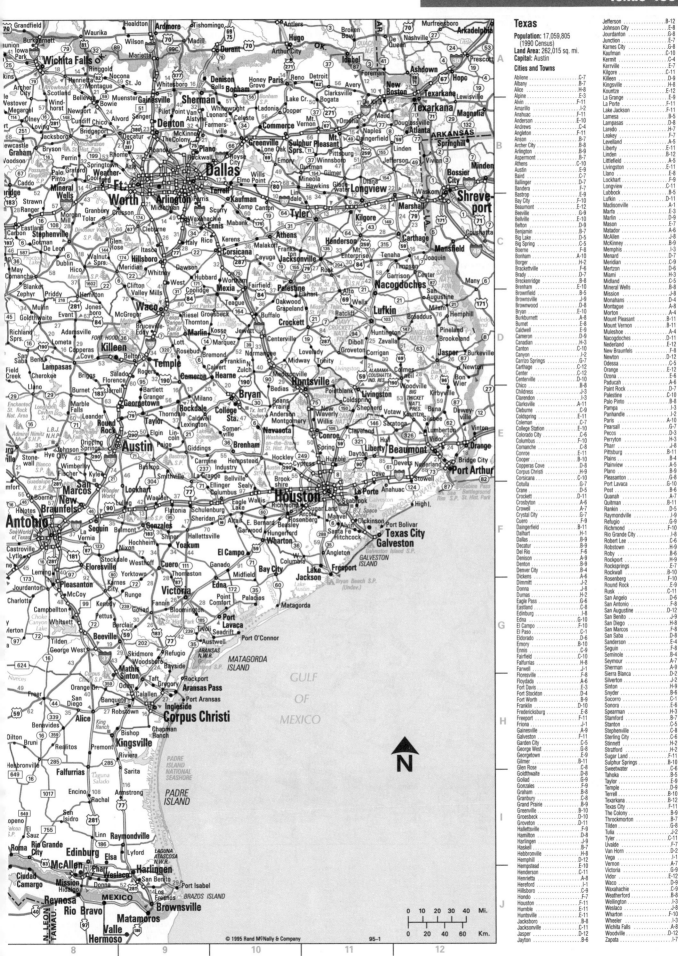

Texas

Population: 17,059,805 (1990 Census)
Land Area: 262,015 sq. mi.
Capital: Austin

Cities and Towns

0 10 20 30 40 Mi.
0 20 40 60 Km.

95–1

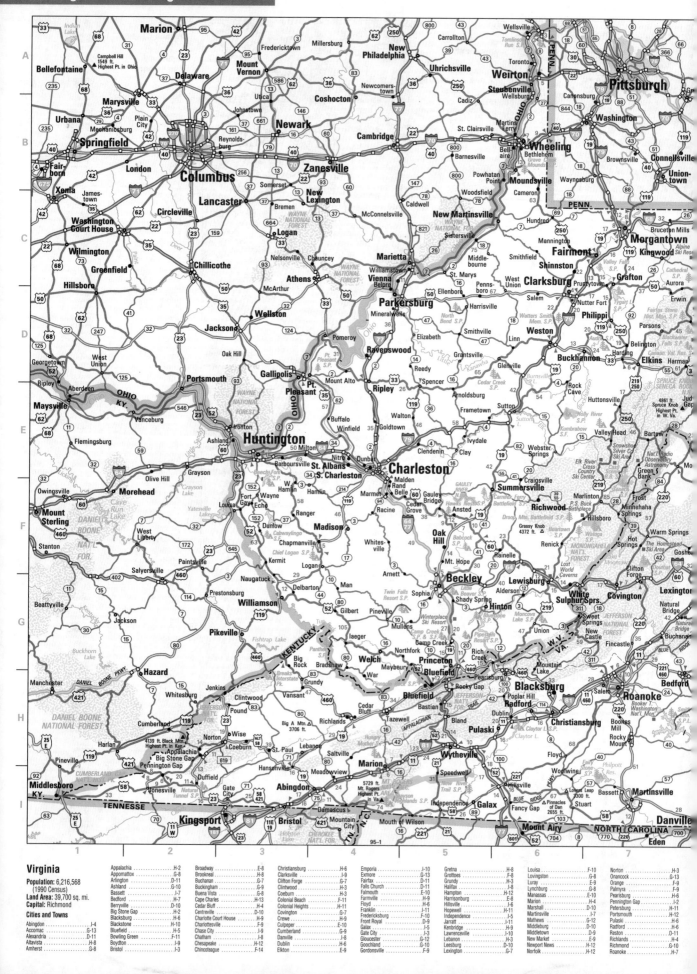

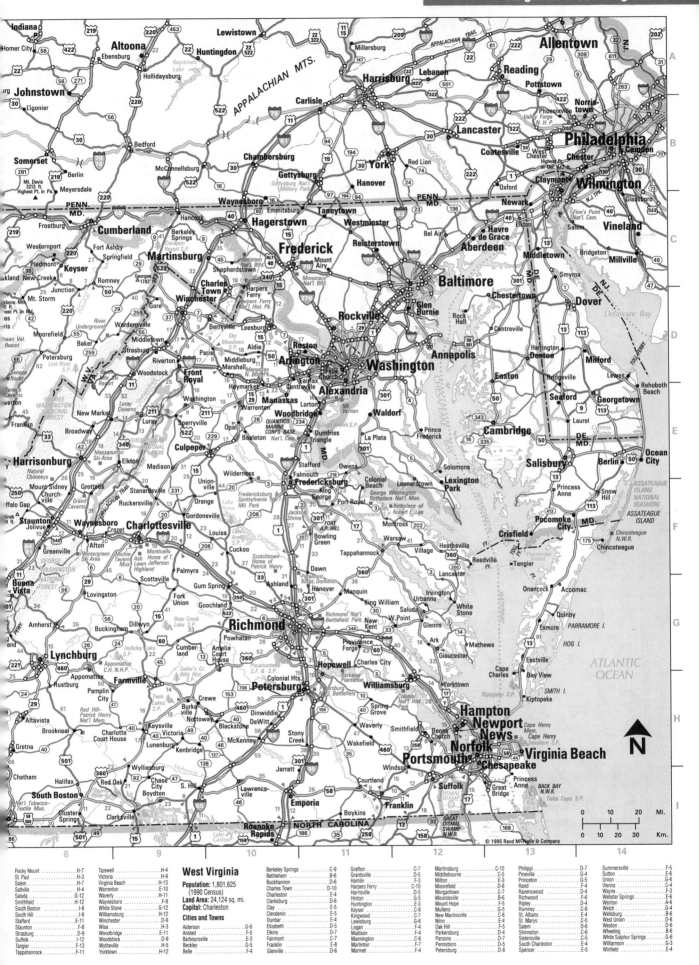

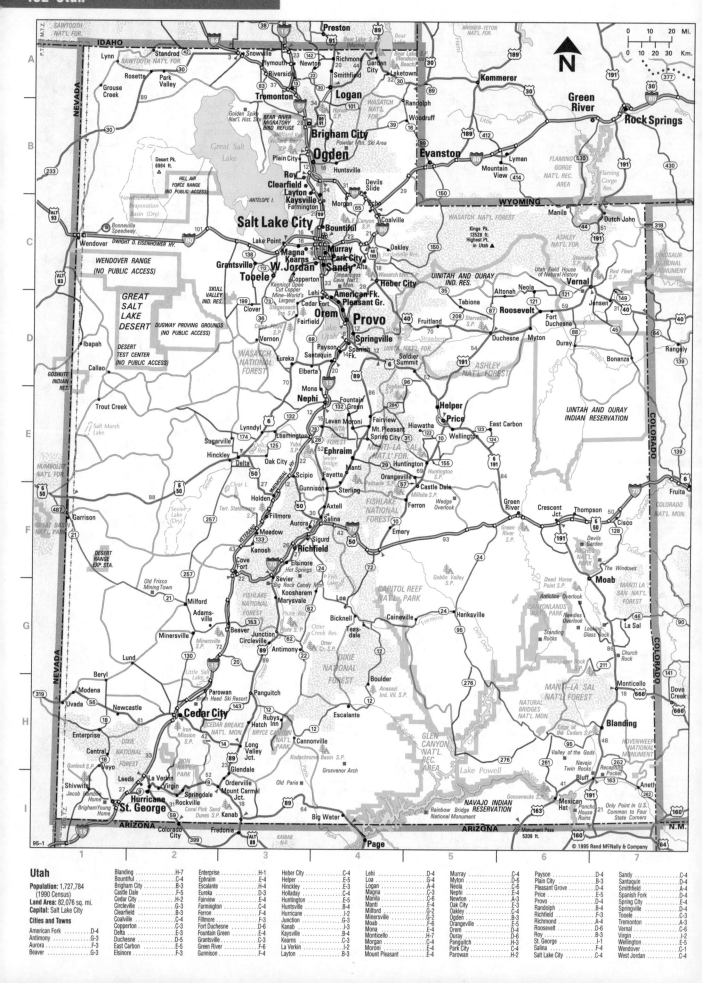

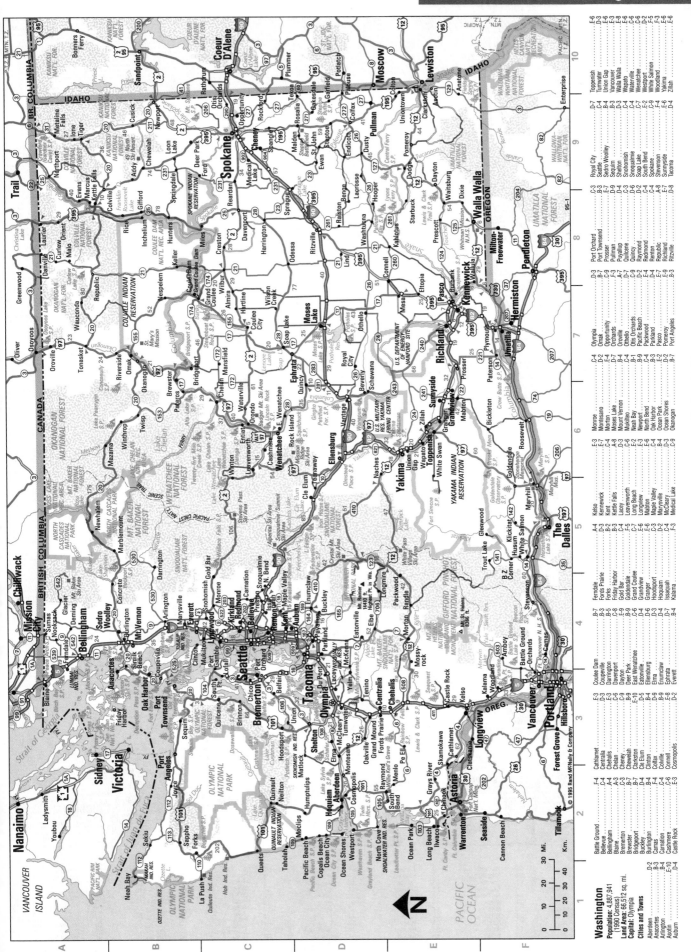

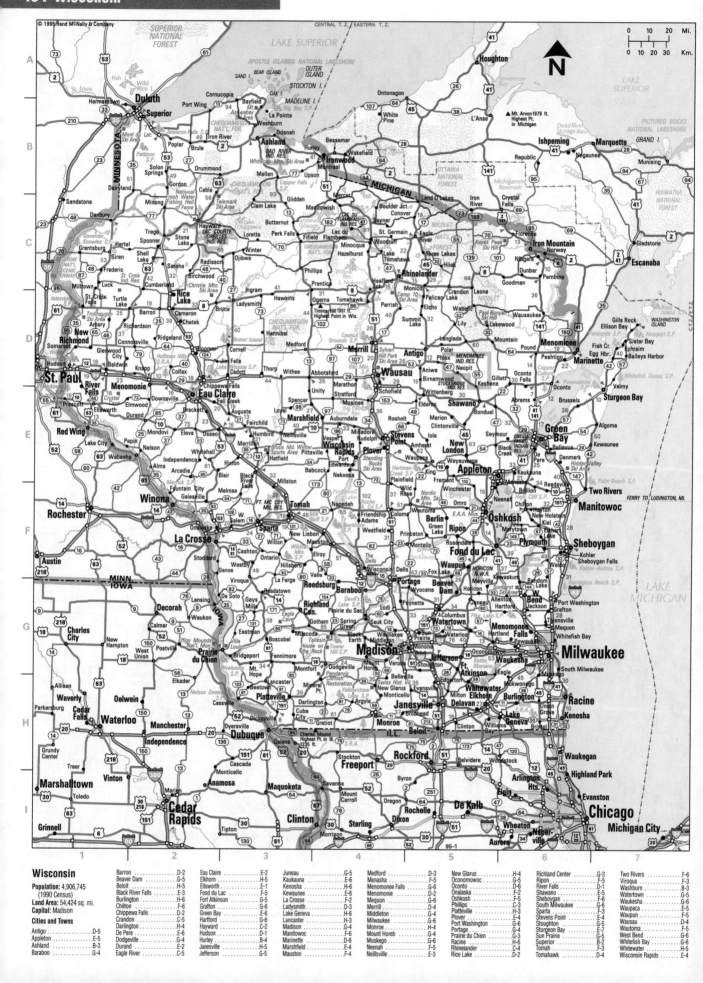

© 1995 Rand McNally & Company

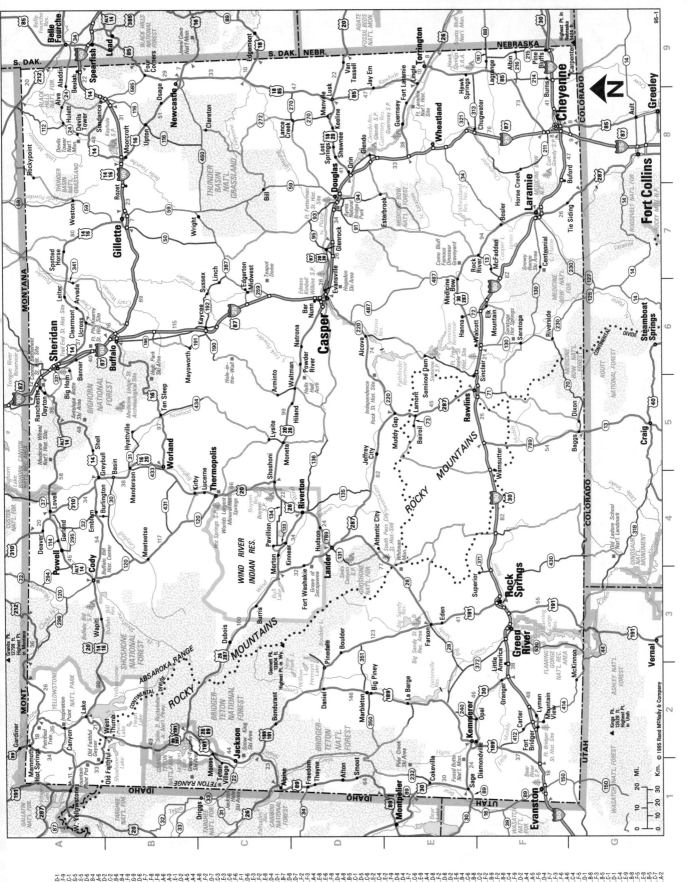

95–1

© 1995 Rand McNally & Company

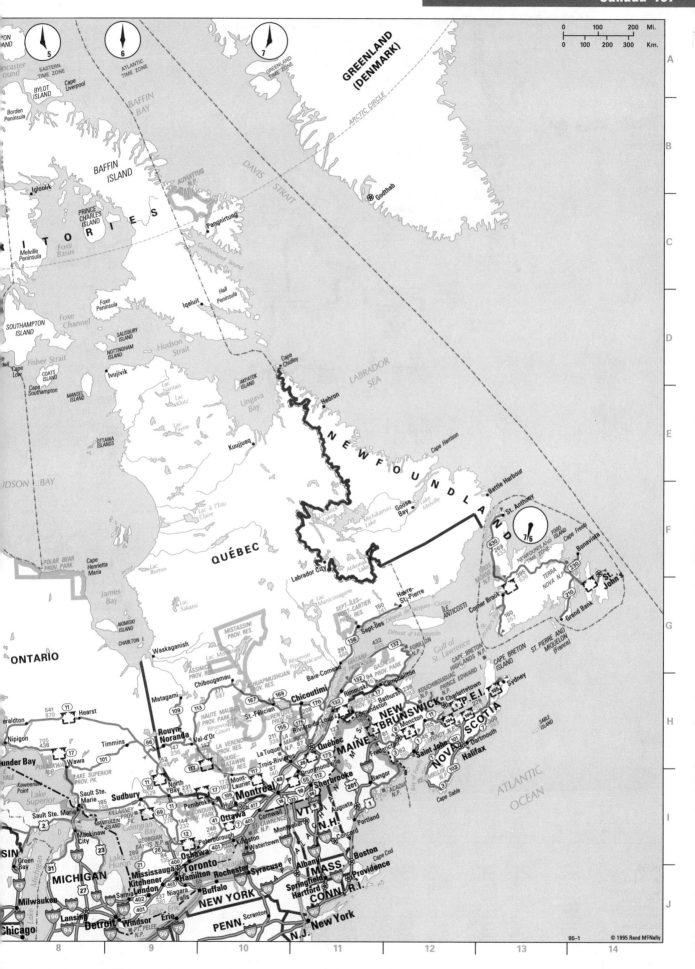

GREENLAND
(DENMARK)

ARCTIC CIRCLE

EASTERN
TIME ZONE

ATLANTIC
TIME ZONE

GREENLAND
TIME ZONE

BAFFIN
BAY

DAVIS
STRAIT

BAFFIN ISLAND

PRINCE
CHARLES
ISLAND

Igloolik

Borden
Peninsula

BYLOT
ISLAND

Cape
Liverpool

AUYUITTUQ
N.P.

Pangnirtung

Cumberland Sound

Godthab

Melville
Peninsula

Foxe
Basin

Hall
Peninsula

Foxe
Peninsula

Iqaluit

SOUTHAMPTON
ISLAND

SALISBURY
ISLAND

NOTTINGHAM
ISLAND

Foxe
Channel

Hudson
Strait

Ivujivik

AKPATOK
ISLAND

Cape
Chidley

Hebron

LABRADOR
SEA

Cape Harrison

Fisher Strait

Cape
Low

COATS
ISLAND

Cape
Southampton

MANSEL
ISLAND

Lac
Nantais

Lac
Klotz

Ungava
Bay

Kuujjuaq

N E W F O U N D L A N D

Battle Harbour

Lac
Payne

OTTAWA
ISLANDS

St. Anthony

HUDSON BAY

Lac à l'Eau
Claire

Cape
Henrietta
Maria

POLAR BEAR
PROV. PARK

Goose
Bay

Lake
Melville

FORD
ISLAND

Cape Freels

Bonavista

NEWFOUNDLAND
TIME ZONE

7/6

Smallwood
Res.

Michikamau
Lake

430

269   433

St.
John's

James
Bay

Lac
Burton

QUÉBEC

Labrador City

Atikonak
Lake

GROS
MORNE N.P.

395
636

Corner Brook

TERRA
NOVA N.P.

230

210

AKIMISKI
ISLAND

Lac
Sakami

Lac
Manicouagan

Havre-
St-Pierre

160
257

Grand Bank

ST. PIERRE AND
MIQUELON
(France)

CHARLTON I.

Waskaganish

MISTASSINI
PROV. RES.

SEPT-ÎLES-
PORT-CARTIER
PROV. RES.

150
241

ÎLE
ANTICOSTI

Détroit de Jacques-Cartier

SABLE
ISLAND

ONTARIO

ASSINICA
PROV. RES.

Réservoir
Pipmuacane

291
468

Sept-Îles

Détroit d'Honguedo

CAPE BRETON
HIGHLANDS N.P.

ATLANTIC
OCEAN

Chibougamau

371
597

FORILLON
N.P.

Gulf of
St. Lawrence

CAPE BRETON
ISLAND

Matagami

113

St-Félicien

CHIBOUGAMAU-
ASHUAPMUSHUAN
PROV. RES.

169

170

Chicoutimi

167

138

432

132

94

Baie-Comeau

132

MATANE
N.P.

GASPÉSIE
PROV. PARK

Campbellton

PRINCE EDWARD I.

Sydney

541
870

11

Hearst

109

HAUTE MAURICIE
PROV. PARK
Réservoir
Gouin

155

208

175

Rimouski

132

RIMOUSKI
PROV. RES.

238

Bathurst

KOUCHIBOUGUAC
N.P.

Charlottetown

P.E.I.

228

NOVA

Rouyn-
Noranda

66

147
236

Val-d'Or

142

LA VÉRENDRYE
PROV. RES.

211
339

LA
MAURICIE
N.P.

87

Rivière-du-Loup

123

Edmundston

Moncton

2

NEW
BRUNSWICK

145

104

SCOTIA

101

Dartmouth

705
438

Nipigon

Timmins

17

101

Wawa

152
244

117

ROUGE-
MATAWIN
PROV. RES.

La Tuque

140

8

MAINE

Fredericton

62

FUNDY N.P.

Saint John

KEJIMKUJIK
N.P.

103

Halifax

Thunder Bay

PUKASKWA
N.P.

185
298

YALE

LAKE SUPERIOR
PROV. PK.

Sault Ste.
Marie

KILLARNEY
PROV. PK.

80

North
Bay

231

129

172

107

Mont-
Laurier

105

40

Trois-Rivières

55

Québec

173

20

Drummondville

112

Sherbrooke

201

ACADIA
N.P.

Bangor

1

Cape Sable

Keweenaw
Point

Lake
Superior

Sault Ste. Marie

69

11

Sudbury

MANITOULIN
ISLAND

GEORGIAN
BAY IS. N.P.

Georgian
Bay

Pembroke

ALGONQUIN
PROV. PK.

278

232

Hull

Montréal

16

10

66

VT.

5

N.H.

Augusta

Portland

MICHIGAN

2

Mackinaw
City

269

21

26

135

248
399

Peterborough

401

ST. LAWRENCE
ISLANDS N.P.

Cornwall

7

Montpelier

ALB.

Concord

Green
Bay

31

23

Lake
Huron

12

Kingston

Watertown

MASS.

Boston

Cape Cod

Milwaukee

27

Oshawa

400

Toronto

Lake Ontario

Rochester

Syracuse

Albany

Providence

R.I.

43

Mississauga

Hamilton

403

London

NEW YORK

Niagara
Falls

Springfield

Hartford

CONN.

Chicago

96

Lansing

94

Detroit

75

Windsor

Sarnia

402

237
381

Lake Erie

Erie

80

79

Scranton

New York

N.J.

PT. PELEE
N.P.

PENN.

95-1

© 1995 Rand McNally

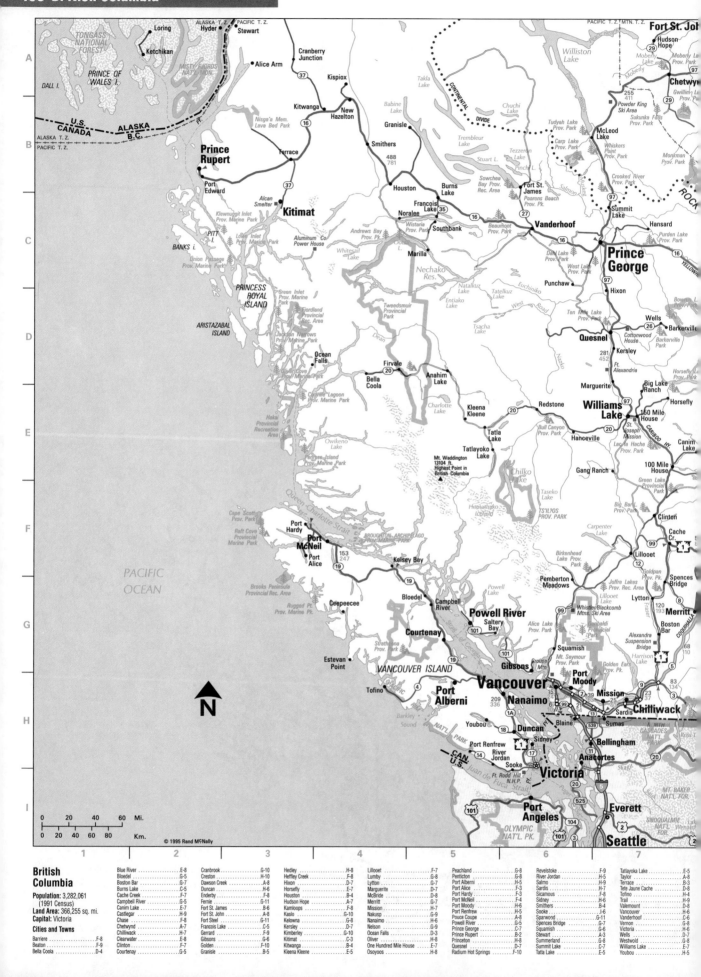

© 1995 Rand McNally

British Columbia

Population: 3,282,061 (1991 Census)
Land Area: 366,255 sq. mi.
Capital: Victoria

Cities and Towns

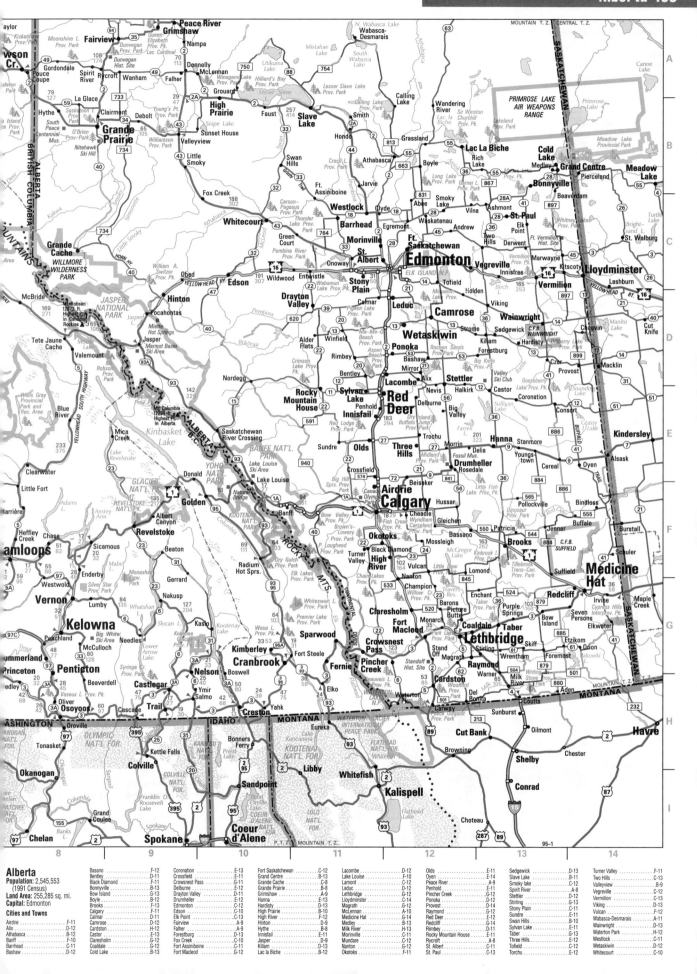

Manitoba

Population: 1,091,942 (1991 Census)
Land Area: 251,000 sq. mi.
Capital: Winnipeg

Cities and Towns

Amaranth	H-9
Angusville	H-7
Arborg	G-10
Ashern	G-10
Austin	I-9
Baldur	I-9
Beausejour	H-11
Belmont	I-9
Benito	F-7
Binscarth	F-7
Birch River	F-7
Birtle	H-7
Boissevain	I-8
Bowsman	F-7
Brandon	I-8
Camperville	F-8
Carberry	I-9
Carman	I-10
Cartwright	J-9
Clearwater	J-9
Cormorant	C-8
Cranberry Portage	C-7
Cross Lake	C-10
Crystal City	J-9
Darlingford	J-10
Dauphin	G-8
Deloraine	J-8
Douglas	I-8
Duck Bay	F-8
Easterville	E-8
Elkhorn	I-7
Elm Creek	I-10
Elphinstone	H-8
Emerson	J-11
Erickson	H-8
Eriksdale	G-10
Ethelbert	G-8
Fisher Branch	G-10
Flin Flon	C-7
Gilbert Plains	G-8
Gimli	H-11
Gladstone	H-9
Glenboro	I-9
Grand Rapids	E-9
Grandview	G-8
Gretna	J-10
Gypsumville	F-9
Hamiota	H-8
Hartney	I-8
Hodgson	G-10
Holland	I-9
Inglis	G-7
Kenville	F-7
Killarney	I-9
La Broquerie	I-11
Lac du Bonnet	H-11
Langruth	H-9
Letellier	J-10
Lockport	H-11
Lowe Farm	I-10
Lundar	H-10
MacGregor	I-9
Mafeking	E-7
Manigotagan	G-11
Manitou	J-9
McCreary	H-9
Melita	I-8
Miniota	H-7
Minitonas	F-7
Minnedosa	H-8
Minto	I-8
Moose Lake	D-8
Moosehorn	G-9
Morden	J-10
Morris	I-10
Neepawa	H-9
Newdale	H-8
Ninette	I-9
Niverville	I-11
Norway House	D-10
Oak River	H-8
Oakville	I-10
Ochre River	G-8
Petersfield	H-11
Pierson	J-7
Pilot Mound	I-9
Pine Falls	H-11
Pine River	G-8
Pipestone	I-8
Plum Coulee	J-10
Portage la Prairie	H-10
Rathwell	I-9
Rennie	I-12
Reston	I-7
Rivers	H-8
Riverton	G-11
Robin	G-7
Roland	I-10
Rorketon	G-9
Rosenfeld	J-10
Rossburn	H-7
Russell	H-7
St. Georges	H-11
St. Jean Baptiste	J-10
St. Laurent	H-10
St. Malo	I-11
St. Pierre	I-11
Ste. Ann	I-11
Ste. Rose du Lac	G-9
Sanford	I-10
Selkirk	H-11
Shoal Lake	H-8
Snow Lake	C-8
Somerset	I-9
Souris	I-8
Sperling	I-10
Sprague	J-12
Steinbach	I-11
Swan River	F-7
Teulon	H-10
The Pas	D-7
Thompson	A-10
Tolstoi	J-11
Treherne	I-9
Tyndall	H-11
Virden	I-7
Vita	J-11
Wabowden	B-9
Wawanesa	I-9
Whitemouth	H-11
Winkler	J-10
Winnipeg	H-10
Winnipeg Beach	H-11
Winnipegosis	F-8
Woodlands	H-10
Woodridge	I-11

Saskatchewan

Population: 988,928 (1991 Census)
Land Area: 251,700 sq. mi.
Capital: Regina

Cities and Towns

Alsask	F-1
Arborfield	E-6
Arcola	I-6
Asquith	F-3
Assiniboia	I-4
Avonlea	H-5
Balcarres	G-6
Battleford	E-2
Beauval	B-3
Bengough	I-5
Bienfait	I-6
Big River	D-3
Biggar	F-3
Blaine Lake	E-3
Bredenbury	G-7
Broadview	H-6
Buffalo Narrows	A-3
Burstall	G-1
Cabri	G-2
Canora	F-6
Carlyle	I-6
Carnduff	I-7
Carrot River	D-6
Central Butte	G-4
Choiceland	D-5
Coronach	J-4
Craik	G-4
Creighton	C-7
Cudworth	E-4
Cumberland House	D-7
Cupar	G-5
Cut Knife	E-2
Davidson	G-4
Delisle	F-3
Duck Lake	E-3
Dundurn	F-4
Eastend	I-2
Eatonia	G-1
Elrose	G-2
Esterhazy	H-7
Estevan	I-6
Eston	G-2
Foam Lake	F-6
Fort Qu'Appelle	G-5
Glaslyn	D-2
Gravelbourg	H-3
Green Lake	C-3
Grenfell	H-6
Gull Lake	H-2
Hafford	E-3
Hague	E-4
Herbert	H-3
Hudson Bay	D-7
Humboldt	F-5
Indian Head	H-6
Ituna	G-6
Kamsack	G-7
Kelvington	F-6
Kerrobert	F-2
Kindersley	F-2
Kinistino	E-5
Kyle	G-3
La Ronge	B-5
Laflèche	I-3
Langenburg	G-7
Lanigan	F-5
Lashburn	D-1
Leader	G-1
Lloydminster	D-1
Lumsden	H-5
Luseland	F-1
Macklin	F-1
Maidstone	D-2
Maple Creek	H-1
Martensville	F-4
Meadow Lake	C-2
Melfort	E-5
Melville	G-6
Midale	I-6
Milestone	H-5
Montmartre	H-6
Moose Jaw	H-4
Moosomin	H-7
Mossbank	H-4
North Battleford	E-2
Naicam	E-4
Nipawin	D-5
Nokomis	G-5
Norquay	F-7
Outlook	G-3
Oxbow	I-7
Pelican Narrows	B-6
Pense	H-5
Pierceland	C-2
Ponteix	I-3
Porcupine Plain	E-6
Preeceville	F-6
Prince Albert	D-4
Qu'Appelle	H-5
Quill Lake	F-5
Radisson	E-3
Raymore	G-5
Redvers	I-7
Regina	H-5
Regina Beach	G-5
Rocanville	H-7
Rockglen	I-4
Rose Valley	F-6
Rosetown	F-3
Rosthern	E-4
St. Louis	E-4
St. Walburg	D-2
Saskatoon	F-4
Shaunavon	H-2
Shellbrook	D-4
Southey	G-5
Spiritwood	D-3
Star City	E-5
Stoughton	I-6
Strasbourg	G-5
Sturgis	F-6
Swift Current	H-2
Theodore	G-6
Tisdale	E-5
Turtleford	D-2
Unity	E-2
Wadena	F-6
Wakaw	E-4
Waldheim	E-4
Watrous	F-5
Watson	F-5
Weyburn	I-5
Whitewood	H-6
Wilkie	E-2
Willow Bunch	I-4
Wolseley	H-6
Wynyard	F-5
Yellow Grass	I-5
Yorkton	G-6

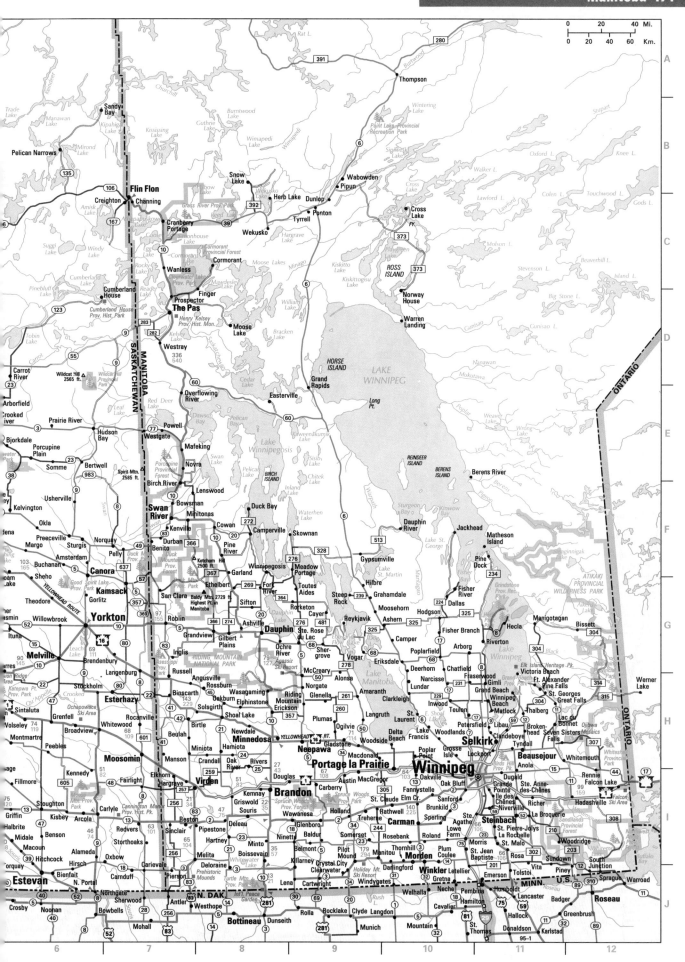

For continuation see map at lower right

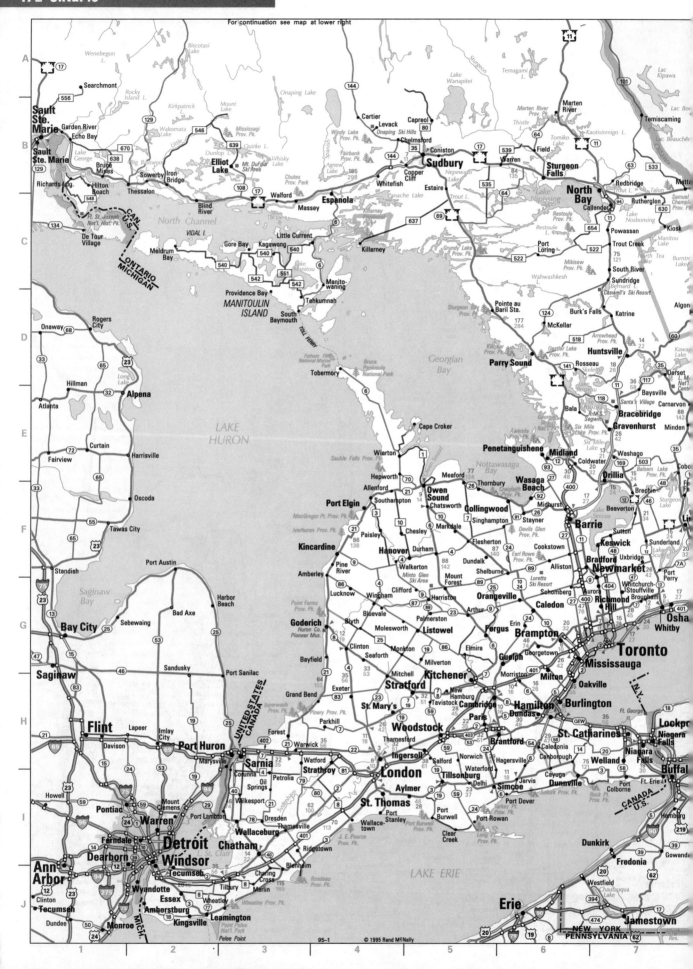

95-1   © 1995 Rand McNally

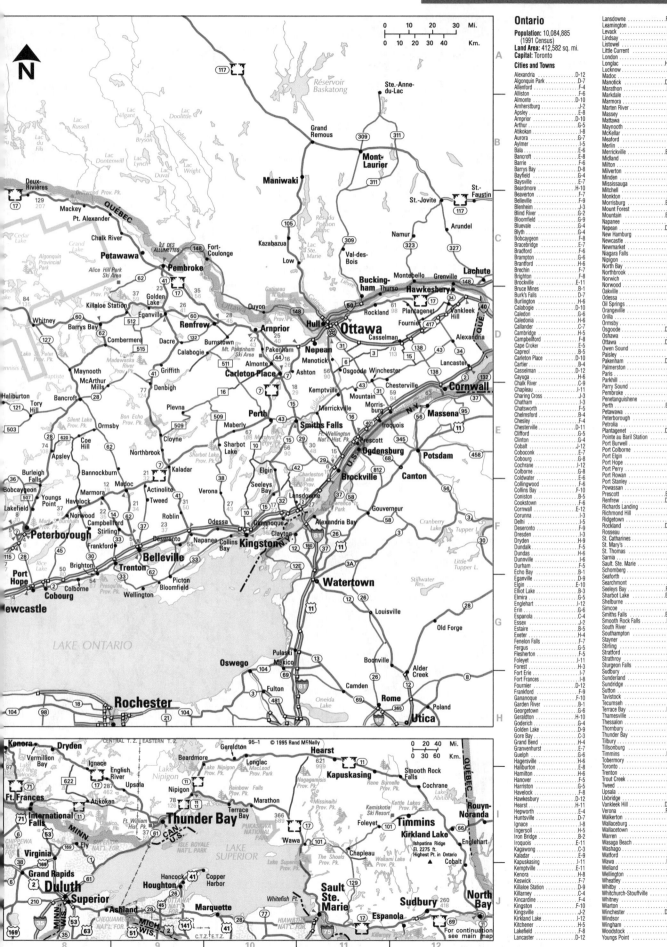

Ontario

Population: 10,084,885 (1991 Census)
Land Area: 412,582 sq. mi.
Capital: Toronto

Cities and Towns

For continuation
see main map

Québec

Population: 6,895,963
(1991 Census)
Land Area: 594,860 sq. mi.
Capital: Québec

Cities and Towns

© 1995 Rand McNally

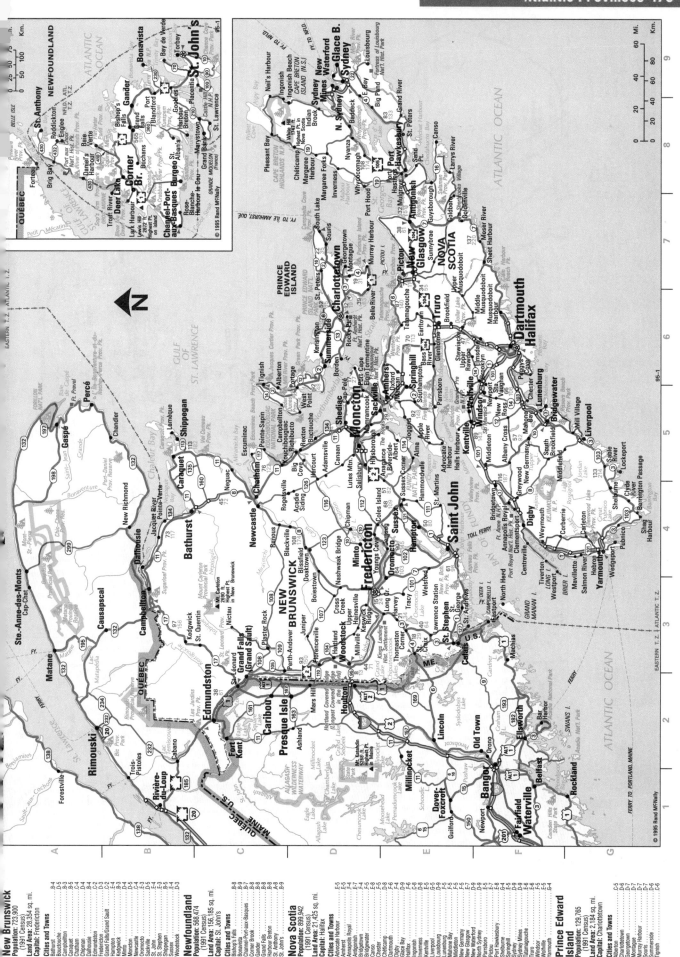

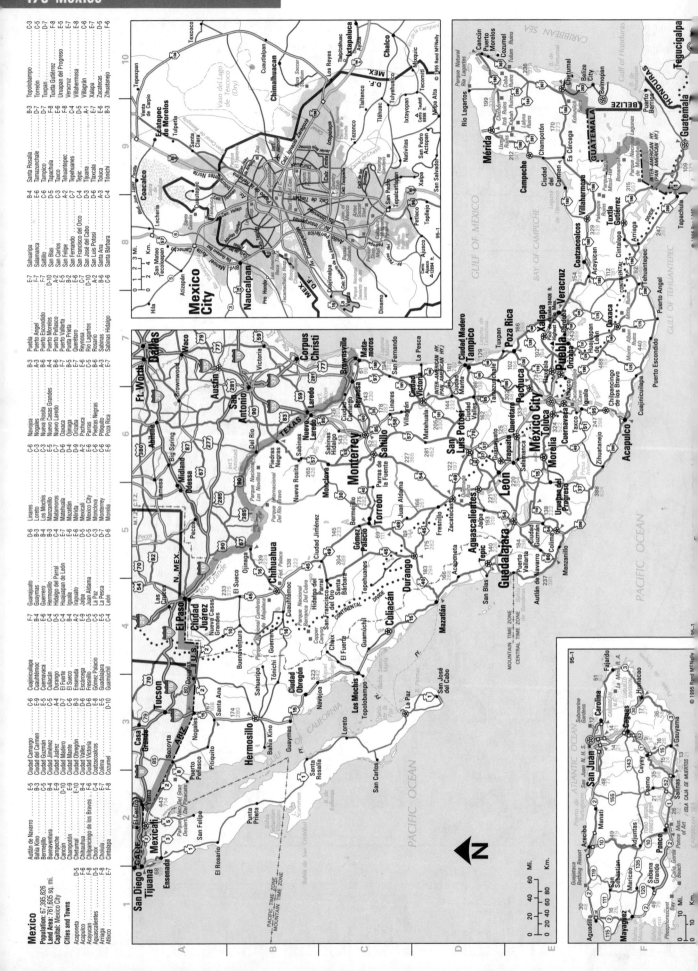